Research and Fieldwork in Development

Research and Fieldwork in Development explores both traditional and cutting edge research methods, from interviews and ethnography to spatial data and digital methods. Each chapter provides the reader with an understanding of the theoretical basis of research methods, reflects on their practice and outlines appropriate analysis techniques. The text also provides a cutting edge focus on the role of new media and technologies in conducting research. The final chapters return to a set of broader concerns in development research, providing a new and dynamic set of engagements with ethics and risk in fieldwork, integrating methods and engaging development research methods with knowledge exchange practices. Each chapter is supported by several case studies written by global experts within the field, documenting encounters and experiences and linking theory to practice. Each chapter is also complemented by an end-of-chapter summary, suggestions for further reading and websites, and questions for further reflection and practice. The text critically locates development research within the field of international development to give an accessible and comprehensive introduction to development research methods.

This book provides an invaluable overview to the practice of international development research and serves as an essential resource for undergraduate and postgraduate students embarking on development fieldwork.

Daniel Hammett is a Faculty Fellow at the University of Sheffield specializing in development and political geography, particularly in relation to citizenship and political protest, with a focus on sub-Saharan Africa.

Chasca Twyman is a Senior Lecturer in Geography at the University of Sheffield and Co-Director of the Sheffield Institute for International Development. Her research interests include the human dimensions of local and global environmental change, natural resource management, and governance and policy in Eastern and Southern Africa.

Mark Graham is a Senior Research Fellow at the Oxford Internet Institute, a Research Fellow at Green Templeton College and a Research Associate at the University of Oxford's School of Geography and the Environment.

One of the great paradoxes of contemporary social science is that researchers often head off into the field – whether the archive, the village, or the corporate headquarters – ill-equipped to conduct a rigorous and systematic field-based project. Research projects require both epistemological sophistication and an immersion in the multiple-methods appropriate to complex fieldwork settings. Research and Fieldwork in Development is an exemplary introduction to these questions and any scholar of development would do well to learn from this rich and compelling text.

Professor Michael Watts, Department of Geography,
University of California, Berkeley

Undertaking research on development themes is often daunting for the novice. Choosing from among the increasing number of guidebooks designed to help the researcher avoid becoming little more than a poverty tourist can be equally difficult. This very appealing new book by three experienced authors is distinctively coherent compared to edited volumes. It situates field research within a holistic context that problematises development (defined broadly as tackling poverty) and issues of positionality as well as the more familiar ethical concerns in accessible terms designed to assist rather than deter the aspirant researcher. The series of boxes containing invited 'reflections' on specific experiences by other authors enriches the main text. Hammett, Twyman and Graham have produced an invaluable guide for the uninitiated and those wanting to think more reflexively about their fieldwork practices.

Professor David Simon, Department of Geography, Royal Holloway
University of London, UK, and Mistra Urban Futures,
Chalmers University, Gothenburg, Sweden

This book is an excellent reference for those who are trying to decipher ways to conduct development research. Its value to students and practitioners lies in the ways in which it explains the differences in development research and development practice compared to research in other contexts.

Padmashree Gehl Sampath, United Nations Conference on Trade and
Development (UNCTAD) Headquarters, Geneva, Switzerland

Research and Fieldwork in Development

Daniel Hammett, Chasca Twyman
and Mark Graham

Routledge
Taylor & Francis Group

LONDON AND NEW YORK

First published 2015
by Routledge
2 Park Square, Milton Park, Abingdon, Oxon OX14 4RN

and by Routledge
711 Third Avenue, New York, NY 10017

Routledge is an imprint of the Taylor & Francis Group, an informa business

British Library Cataloguing in Publication Data
A catalogue record for this book is available from the British Library

Library of Congress Cataloging in Publication Data
Hammett, Daniel.
  Research and fieldwork in development/Daniel Hammett, Chasca
Twyman and Mark Graham.
  pages cm
  Includes bibliographical references and index.
  1. Social sciences – Research. 2. Social sciences – Fieldwork. 3. Social
sciences – Research – Methodology. I. Twyman, Chasca. II. Graham,
Mark, 1980– III. Title.
  H62.H23385 2015
  001.4 – dc23
  2014025144

ISBN: 978–0–415–82956–4 (hbk)
ISBN: 978–0–415–82957–1 (pbk)
ISBN: 978–0–203–64910–7 (ebk)

Typeset in Times New Roman and Franklin Gothic
by Florence Production Ltd, Stoodleigh, Devon, UK

Printed and bound by CPI Group (UK) Ltd, Croydon, CR0 4YY

CONTENTS

FIGURES

TABLES

CONTRIBUTORS

Daniel Hammett is a Faculty Fellow at the University of Sheffield specializing in development and political geography, particularly in relation to citizenship and political protest, with a focus on sub-Saharan Africa. He has conducted overseas fieldwork and organized fieldclasses in South Africa, Kenya, Morocco and Uganda and supervised students undertaking fieldwork in Zimbabwe, South Africa, Tanzania, Kenya, Ethiopia, Ghana, Jamaica, Mexico, Thailand and Bangladesh.

Chasca Twyman is a Senior Lecturer at the Department of Geography, University of Sheffield, and Co-Director of the Sheffield Institute for International Development. Her research interests include the human dimensions of local and global environmental change, natural resource management and governance and policy in Eastern and Southern Africa. She has conducted overseas fieldwork and supervised students working in Nigeria, Botswana, Swaziland, South Africa, Zimbabwe, Mozambique, Tanzania, Namibia, Kenya, Taiwan, Uganda, Madagascar, Vietnam, Democratic Republic of the Congo and Nicaragua.

Mark Graham is a Senior Research Fellow at the Oxford Internet Institute, a Research Fellow at Green Templeton College and a Research Associate at the University of Oxford's School of Geography and the Environment. His core research focuses on internet and information geographies and the intersections between ICTs and economic development. His current work looks at microwork, innovation hubs and other attempts to create 'knowledge economies' in the world's economic peripheries. He has conducted research and supervised students working in Rwanda, Kenya and Thailand.

Nicola Ansell is a Reader in Human Geography at Brunel University whose research has focused on youth experience of social and cultural change in Southern Africa.

George Barrett was a BA Geography student at the University of Sheffield and took part in a fieldclass to South Africa in 2013.

Charles Brigham is a specialist in technology and development who has worked in more than 30 countries, including projects with the United Nations and the World Bank.

Daniel Brockington is a Professor in Conservation and Development at the University of Manchester with field experience in Eastern and Southern Africa, India and Australasia.

Steve Cinderby is Deputy Director of the Stockholm Environment Institute at the University of York with participatory GIS research and training experience in various African and Asian countries.

Gordon Dabinett is Professor for Regional Studies and head of the Research Exchange for the Social Sciences at the University of Sheffield.

Lawrence Dritsas is a Lecturer in Science, Technology and Innovation Studies at the University of Edinburgh, specializing in the history and sociology of science in Africa.

Marc Fletcher is a Post-Doctoral Fellow in the Department of Sociology, University of Johannesburg, with experience of researching identity and football fandom in a number of African countries.

Jeff Garmany is a Lecturer at King's Brazil Institute, King's College London, with extensive field experience working on poverty and political governance in Brazil.

Katherine Gough is a Professor of Human Geography at Loughborough University with field experience in Latin America, West and East Africa, and the Asia-Pacific region.

Benjamin Hennig is a Senior Research Fellow in the School of Geography and Environment at the University of Oxford specializing in visualizations of geographical concerns.

Gijsbert Hoogendoorn is a Senior Lecturer in the School of Geography, Archaeology and Environmental Studies at the University of Witwatersrand specializing in urban and tourism geography in South Africa.

Simon Jones is an Associate Dean at Leeds Metropolitan University with extensive experience of designing and leading academic field teaching.

Nina Laurie is a Professor of Development and the Environment at Newcastle University with extensive research experience working on gender, development and social movements in Latin America.

Sian Lazar is a Lecturer in Social Anthropology at the University of Cambridge with extensive ethnographic research experience in Argentina and Bolivia.

Emmanuel Letouzé is a PhD candidate in Demography at UC Berkley whose work focuses on big data for development.

Harriet Lowes was a BA Geography student at the University of Sheffield and took part in a fieldclass to rural Kenya in 2014.

Fiona McConnell is an Associate Professor in Human Geography at the University of Oxford with extensive fieldwork experience in India.

Andrew Maclachlan was a BA Geography student at the University of Sheffield and took part in a fieldclass to South Africa in 2013.

Emma Mawdsley is a Senior Lecturer in Human Geography at the University of Cambridge with extensive research interests and experience in Africa and Asia, notably India.

Paula Meth is a Senior Lecturer in Town and Regional Planning at the University of Sheffield whose work on gender, violence and inequality has involved fieldwork in South Africa.

Thomas Molony is a Lecturer in African Studies at the University of Edinburgh specializing in technology and development in Africa.

Jude Murison is a Senior Research Fellow in Governance and Regulation at the University of Edinburgh with experience of human rights and transitional justice research in central Africa.

Anna Niestat is Associate Director for Program and the Emergencies division of Human Rights Watch with experience of working in Africa, Asia, the Middle East and the Caribbean.

Siân Parkinson was a BA Geography student at the University of Sheffield and took part in a fieldclass to South Africa in 2013.

Thomas Smith is a Post-Doctoral Fellow in the Centre for Study of Childhood and Youth, University of Sheffield, with experience of field research in Tanzania.

Mary Suffield was a BA Geography student at the University of Sheffield and took part in a fieldclass to rural Kenya in 2014.

Lorraine van Blerk is a Reader in the School of the Environment, University of Dundee. Her work on youth, street children and HIV/AIDS has included research across Southern Africa.

Adam Whitworth is a Lecturer in Human Geography at the University of Sheffield who has carried out secondary data analysis on projects for South African, Welsh and British government departments.

Andrea Wilkinson is a PhD student in the School of Geography, Politics and Sociology at Newcastle University currently undertaking ethnographic fieldwork in Peru.

Glyn Williams is a Senior Lecturer in Town and Regional Planning at the University of Sheffield and has experience of development research in India.

Shane Winser is Geography Outdoors Manager at the Royal Geographical Society-Institute for British Geographers, providing expert guidance and leadership on field research training and safety.

Photo credits

Figures 1.1, 3.1, 4.1, 4.3, 5.2, 8.2 used courtesy of Daniel Hammett
Figures 7.1, 8.1 used courtesy of Mark Graham
Figures 2.1, 4.2, 6.1, 9.1 used courtesy of Robert Bryant
Figure 9.2 used courtesy of Chasca Twyman
Figure 5.1 used courtesy of Katherine Adams
Figure 7.2 used courtesy of Siân Parkinson

ACKNOWLEDGEMENTS

As with many books, this work has benefitted from the contributions of many colleagues and friends who have supported the project in various ways. In particular, we would like to thank the editorial team at Routledge – most notably Andrew Mould, Faye Leerink and Sarah Gilkes – for all their support and assistance with bringing this book to fruition. A number of academic colleagues have commented on draft chapters and suggested topics or readings to include and we would like to thank Charles Pattie, Jess Dubow, Pat Noxolo, Greg Morgan, Monica Stephens, Stefano de Sabbata and Scott Hale for their inputs into this project. Our thanks also to the anonymous reviewers whose detailed comments helped us strengthen the manuscript.

ABBREVIATIONS

BREAD	Bureau for Research in Economic Analysis for Development
BRICS	Brazil, Russia, India, China and South Africa
DARG	Developing Areas Research Group (of the Royal Geographical Society)
DFID	Department for International Development
ESRC	Economic and Social Research Council
FBO	faith-based organization
GIS	geography information system
GPS	global positioning system
HEI	higher education institution
HEIF	Higher Education Innovation Fund
ICE	in case of emergency
IRB	Institutional Review Board
MDG	Millennium Development Goal
NEPAD	New Partnership for Africa's Development
NGO	non-governmental organization
OSM	OpenStreetMap
PGIS	participatory geographic information system
PLA	participatory learning and action
PRA	participatory rural appraisal/participatory reflection and action
REF	research excellence framework
RESS	Research Exchange for the Social Sciences
RRA	rapid rural appraisal
SAIMD	South African Index of Multiple Deprivation
SAP	structural adjustment policy
UN	United Nations
UPC	Uganda People's Congress
UYWS	Uganda Youth Welfare Service

INTRODUCTION

What does a development researcher – or, at least, someone researching development – do? Popular representation might suggest they are lone anthropologists 'discovering' the secrets of an Amazon basin community or public health specialists conducting a health needs analysis survey of a refugee camp. Such perceptions often draw on romanticized and stereotypical representations of poverty, development and the 'exotic other'. In reality, development research encompasses a vast array of methods, approaches, contexts and activities. This book reflects this variety of experiences and draws on the range of our own engagements with the ideas behind and practices of development research. Our hope is that the reflections and discussions presented here will provide you with a guide to navigating these processes and practices. While reading this book, it is important to remember that development research is not only about 'boots-on-the-ground' practices and experiences, but a much more holistic understanding of what development is, what research is and how development research can be – and is – conducted in many different ways and places. In short, this book will not provide you with *the* way to do development research but will provide you with an awareness and understanding of a range of debates, approaches and strategies that you can draw on as you develop your research topic, questions and methods.

Underpinning this book is our recognition that development studies – and associated disciplines including geography, anthropology, politics and economics – is grounded in research and fieldwork. From undergraduate dissertations and fieldclasses through postgraduate research degrees to academic books and journal articles, research is paramount. Such research findings not only provide the basis for the advancement of academic theory and knowledge but can – and should – inform the evolution of development planning, policy and practice. This book is intended to help you understand the processes of designing, conducting and presenting rigorous, reliable and credible research to a range of audiences wherever your 'field' may be. For some, the most obvious manifestation of their fieldsite may be a remote Tibetan village or a vibrant urban market in Kampala, for others, it may be a museum or library archive, the corridors of government and civil society offices, the pages of a census or a demographic and health survey or the online chatrooms and social media presence of community organizations and hometown associations.

Regardless of where this 'fieldsite' is located, the 'field' for development research begins at home, encompassing the literature that frames your engagement with research, the classroom teaching that inspires interest – and your own sense of curiosity. This departure point is the first component of the research process: research is a journey during which you will encounter logistical, ethical, linguistic and other challenges. This text is designed to assist along this journey, helping you plan and prepare for research, to understand the links between theory and methods, to develop an appropriate research design, to cope with and navigate the inevitable pitfalls and challenges of fieldwork and to engage with the processes of analysis, writing and dissemination. Our intention is to help you be well prepared – organized yet flexible, rigorous but sensitive – for the research experience.

Development and development research

In preparing to carry out development research, it is important to reflect on what is meant by 'development', 'international development' in particular, and the differences between doing development and doing development research. Chapter 2 (about contested terrain) provides an overview of evolving understandings and paradigms of development, highlighting various controversies in these shifting approaches and their implications for changing practices of development research. For the purposes of this introductory chapter, it is sufficient to observe that development, while understood in varied and ideologically framed ways, is concerned with combating poverty (broadly defined) and that understanding local, national and international contexts, experiences, priorities, challenges and opportunities is vital for efforts to reach developmental goals (for an introduction to this term and debates, see Black 2007).

From this starting point, we can say that development research is concerned with finding out about the causes, consequences and ways of tackling poverty. From this assumption, we can then begin to unpack the values, politics, histories and techniques that inform and are manifest in development research. We can also recognize the diverse drivers that motivate us to conduct development research and how these influence how we select and design research topics. Our own motivations for conducting research projects are multiple and, at times, contested. They include a desire to help others and an intellectual curiosity to understand more about the world around us. How about you? What are your motivations for doing development research? Is your interest based on a desire to help others? A means to alleviate guilt? An ideologically framed commitment to save the world? A means to develop a career? Intellectual curiosity? It may be you are driven by several interests and there is no 'right' or 'wrong' answer.

In this section, we reflect on the importance of these drivers and of understanding the differences between development practice and development research and the importance of development research for effective and sustainable development practice. This differentiation may seem self-explanatory – one is about 'doing' or implementation, the other is about 'understanding' or finding out. The division between the two is not always clear cut but understanding the differences between development practice and development research is crucial.

To illustrate this, we can reflect on Dan's experiences of convening and leading international development geography fieldclass trips to Kenya, South Africa and Morocco.

These courses involve a series of pre-trip lectures and then a ten-day fieldtrip during which students conduct small group research projects either with local guides (in Kenya or Morocco) or host non-profit organizations tackling urban development challenges (in South Africa). Lectures address contextual background materials, including historic and contemporary socio-political and socio-economic conditions and their importance to current development challenges, as well as specialized research design and methods training. Time is spent discussing students' understandings of development, of the nature and purpose of the fieldclass and what they expect from the trip. These conversations include discussion of the colonial gaze (see Clifford and Marcus 1986; Pratt 1992), of the danger of fieldtrips becoming 'safaris of the poor' and reiteration that the fieldclass is a development research training module.

This emphasis is intended to remind students that the fieldclass is not a practical, hands-on development project with immediate, concrete benefits and deliverables to the host community. Every year, however, some students comment that they do not feel that development research is contributing anything useful to the lives of the poor and would prefer to be doing a practical development project aimed at immediate poverty alleviation.

Box I.1 Reflecting on the purpose of development fieldclasses

Do you think a development fieldclass should only be concerned with developing research skills and knowledge?

What are some of the issues arising from 'wealthy Westerners' doing development research?

As a student on a development research fieldclass, you are faced with the dilemma outlined above – a number of your fellow students are complaining that they feel the research focus of the fieldclass is inappropriate and are asking to do something practical to aid local development. What are your views on this?

This moment of discomfort, of challenge to the pedagogic principles of a fieldclass, is a profoundly powerful moment for teaching and learning. It allows questions regarding the power of the colonial gaze and of the potentially extractive and exploitative nature of research to emerge. It also provides an opportunity to explore the importance of development research to good development practice.

The scenario in Box I.2, inspired by fieldclass conversations, underlines the role of development research in understanding the complexities of development practice. Implicit within the request (made in Box I.2) to move from research to practice is an awareness of the burdens (of time, energy, money) placed on participants and hosts by researchers and, therefore, the need to ensure research is well designed, effective and efficient (see Chapters 4 and 5). The course leader's questions raise another set of issues. First, that development research is vital to inform appropriate and sustainable development policies and practice. Without a solid understanding of the local context and local need, as well

Box I.2 Reflecting on the role of research in development

You are a student on a ten-day international development fieldclass to rural Kenya. The fieldsite is in a marginal, arid region, with very limited physical infrastructure, basic schooling and health care provision, low levels of literacy, limited provision of safe drinking water and sanitation services, high levels of absolute poverty and the majority of the community is reliant on the informal economy and subsistence livelihoods.

At the end of the fourth day you ask the course leader if the focus of the trip can be moved 'from just asking questions and looking at things' to 'actually doing something that will help the local community'. The course leader seems sympathetic, but asks you a question in return:

• What is it you want to do, what project?

They then proceed to ask you some follow-up questions:

• On what basis (why) have you decided to do this?
• How have you identified the needs and priorities?
• Who has helped determine these priorities? Who have you spoken to and who haven't you spoken to?
• Who designed the questions to ask about these priorities?
• What might be some of the (un)expected outcomes of the project?
• Who will benefit from this project and who might be excluded or suffer?
• How do you know this project is appropriate?
• When will the project be started and finished, how will it be financed and how will it be maintained?

What are your responses to these questions? Do these questions change your perspective?

as of broader (national, international, state, civil society) pressures and influences, development interventions often fail to deliver intended outcomes and can exacerbate existing challenges. Second, when conducting research and making recommendations you need to ensure reliability, fairness and credibility. Linked to this concern is a need to be aware of power dynamics: of your power as the researcher in framing a discussion and agenda – a key concern of post-colonial approaches to development – and of power relations and hierarchies among participants (who speaks and for whom and with what (personal) agenda) (see Chapters 2 (about contested terrain) and 3 ('Power, identity and the dynamics of research')).

Derived from these observations, you should see how research is integral to development policy and practice and good research is vital for best practice. Research is used throughout the development process, from baseline or feasibility studies and pilot projects, through ethnographic studies of communities or contexts to inform understanding and policy, to monitoring and evaluation strategies to ensure the effectiveness of development policies and practices. For postgraduate students development research may form the basis of a dissertation or thesis, involving a period of fieldwork of a few weeks through to a year or more. Undergraduate students are more likely to encounter development research on a smaller scale, either through group or individual research under the auspices of a fieldclass or a few weeks of research for a dissertation (cf. Fuller et al. 2006). Further development research opportunities may arise through working as a research assistant for an academic or development-related organization on a broader research project. However you experience development research, this book offers guidance to the designing, doing and writing up of development research and supports the development of transferable skills that can improve your employability.

The skills you develop will allow you to engage with development research across multiple scales, from cross-national comparative studies through national case studies to specific local challenges. These projects may entail the study of local, national or colonial archives, participatory methods with local community members, interviews with publics and elites, the completion of surveys to gather baseline data or to evaluate the outcomes of development projects and numerous other methods. At times, you may find yourself conducting such work as a lone and independent researcher, at others you may be part of – or even leading – a team of researchers. The heterogeneity of development research projects means there is no single approach or method to be adopted and the approach you adopt will be informed by -ologies (ontology, epistemology and methodology).

Discussion of -ologies (see Box I.3) can often be confusing, abstract and seemingly irrelevant to the practical aspects of research. They are, however, vital to determining how you understand and design research (a useful and accessible discussion of these concepts can be found in Chapter 4 of Grix (2004)). Understanding how we approach knowledge and meaning are integral to the ways in which we conceptualize research topics and questions, the types of data we want to collect and analyze and the methods we employ.

Box I.3 -ologies

Ontology – theory of being = what we assume exists

Epistemology – theory of knowledge = what we can know about what exists and how we know what we know. For instance, objectivism is the belief that meaning exists apart from consciousness. Subjectivism is the belief that meaning is created and imported from elsewhere and is produced free from ties with a specific object. Constructivism is the belief that all meanings are produced through specific interactions and that there is no objective truth.

Methodology – theory of method = suitability of methods used in empirical research

Method – how data are collected

Awareness and understanding of your personal view of the world, of what exists and what we can know about this (ontology and epistemology) are integral to the research design process. These understandings provide a frame for the questions asked, based on what we believe does/should exist and therefore what we can know about it and how to find this out. They dictate what kinds of question are asked, what types of data are sought and which methods are used to gather these data. Acknowledging your ontological and epistemological position will also help you to address concerns with how knowledge is produced.

The production of knowledge is a critical concern to development researchers. On a practical level, we are concerned that an inappropriate method or analytic tool or a poorly designed research project will produce a partial and inaccurate account of the matter being researched: in development work this can have long-lasting and costly consequences. At the same time, we are concerned with the positionalities implicated in knowledge production and how these affect our findings and recommendations.

In the field of development research, these concerns are bound up in historical and contemporary experiences of (neo)colonialism, imperialism, power and inequality. If we reflect on historical travelogues and other writings from explorers, missionaries and others involved (to greater or lesser extents) in efforts to 'develop', 'civilize' or otherwise 'uplift' peoples and cultures to European ideals, then specific representations and narratives of the 'savage and barbaric other' are readily apparent. These discourses (see Box I.4) firmly located the non-European as inferior and backwards, in need of civilization and upliftment that could only be realized with the support and assistance of Europeans. In this discourse, the development of non-Europeans was labelled as 'the white man's burden'. This

Box I.4 Discourse

Discourse refers to a way of communicating thoughts on a topic that influences and frames understanding. Discourses are powerful tools, shaping popular and policy approaches to the subject matter. For instance, think about this famous expression: 'One person's terrorist is another person's freedom fighter' and the ways in which the African National Congress and Nelson Mandela were viewed during the 1980s. To the South African government of the time, as well as other governments and many people around the world, the ANC and Mandela were terrorists. To others, they were an illegitimately banned political organization and an imprisoned political freedom fighter.

We continue to see the power of discourse in the production of popular and political understandings of topics including immigration, development and the global south. If you pick up a range of newspapers and read their coverage of the same topic or story you can quickly see the various discourses used to frame the same issue in very different ways. The discursive power of the press to shape opinion is tremendous and has significant influence on development policy and practice (see Smith 2006; Noxolo 2012; Hammett 2013).

historical pretext of development, rooted in paternalism, Eurocentrism and racism has remained influential, both in popular and policy approaches to international development in the West and as a legacy informing recipient's attitudes towards Western (development) interventions.

Development researchers are increasingly engaging with post-colonial theory, leading to the emergence of post-colonial approaches to development research. Integral to these practices have been critical reflections on the ways in which the global south has been imagined and constructed in Western thought and how associated racialized knowledges have perpetuated a sense of paternalism or trusteeship manifest in development approaches and interventions to 'help' those in the global south (see Mercer et al. 2003). These engagements have sought not only to recognize but to mitigate and invert the continued power relations within development, such as the continued construction and design of development priorities and research questions within the West.

Consequently, development researchers have begun to involve the 'researched' in the design and conduct of the research (see Chapter 2). These efforts attempt to shift, at least partially, control and power over the production of knowledge away from Westerners and into a more shared, co-production process led by those being researched. While such an approach is often unfeasible for student research projects, it is important to understand how our attitudes towards knowledge, towards those with whom we conduct our research and the power relations implicit in the design and conducting of research influence the processes and outcomes of our work.

Developing research and using this book

Research is a process, one that begins before the first data are collected and finishes long after the final data are logged. This process involves personal interests and beliefs (ontologies and epistemologies) and entails literature reviews, design of research questions and selection of appropriate methods, the completion of ethics and risk assessment procedures, the conducting of fieldwork, the analysis of data and presentation of findings. This book is designed to provide you with guidance throughout this process – from conceptualization to presentation. It is intended to act as a continual companion and guide throughout the research journey, but also as a resource that can be dipped in and out of as necessary at any point.

Each individual chapter links theoretical and conceptual concerns to the practicalities of research and fieldwork and draws on both the authors' and guest contributors' experiences of development fieldwork to provide examples of the links between theory and method as well as understanding why different methods are used in the field. While we cannot hope to cover every aspect of the fieldwork experience, the examples included here demonstrate the unexpected and unanticipated events that characterize development research. These encounters can be challenging and complex but often serve to remind us of the humanity of development research, of the often emotionally draining nature of this work but also the uplifting and invigorating moments.

The opening chapters of the book address a series of concerns that need to be addressed and considered while designing and developing a research project, broadly characterized as 'planning research'. The first chapter provides guidance and support to the initial stages of research design and associated skills, acknowledging the influence broader intellectual, political and economic factors have in framing the development of research projects. The second chapter, on the contested terrain of development fieldwork, critically locates development research within broader debates surrounding international development. These engagements should remind you that the field of international development, the priorities therein, and questions addressed are continually evolving. The third chapter proceeds to address a number of imperatives in the practical preparation of development fieldwork, discussing issues relating to the selection of fieldwork sites, working alone or in groups, and engaging with the practicalities of conducting research as part of a fieldclass. Chapter 4 then develops another set of concerns to be considered as integral to research and fieldwork, namely the importance of understanding power, identity and positionality within fieldwork.

Chapter 5 then addresses ethics in fieldwork. This chapter considers the core questions and practices relating to the ethical conduct of fieldwork and the legislative frameworks surrounding these and then reflects on wider ethical and moral components such as payment and reciprocity for participation and negotiating relations with field assistants. Chapter 6 then turns to address questions of risk and fieldwork, again engaging with the legislative concerns before addressing broader concerns relating to risk in everyday fieldwork practices, including mitigating risks to participants and minimizing risks of data loss or breaching of confidentiality.

The second part of the book, 'Collecting and analyzing data', addresses a series of methods and their use within development research. While this section cannot cover all possible methods that may be used in development research, we offer reflections on both 'traditional' social science methods and a number of emergent methodologies that are increasingly of interest to development researchers. The first chapter in this section, Chapter 7, explores verbal data, namely interviews and focus groups. In this chapter, consideration is given to different forms of interview, how to conceptualize and design interview and focus group questions and schedules, the different ways in which interviews can be completed and the importance of identifying and recording contextual and non-verbal information. Chapter 8 then focuses on ethnography and participant observation, considering the historical development of these approaches, a number of specific methods within this approach, how these can be utilized usefully in development research.

Chapter 9 considers the use of participatory methods in development research, linking these approaches back to the post-colonial turn in development and exploring why you might use such methods and when this would be an appropriate approach. We then identify a number of specific methods and explore how these can be utilized across a range of scales and topics. Chapter 10 addresses a range of methods for collecting and analyzing archival, documentary and visual data as a means to unpacking both historic narratives relating to development challenges and as a means of interrogating contemporary development challenges, policies and practices.

Chapter 11 shifts the focus to survey data and methods, considering how quantitative methods can be used within development research. In this chapter, we explore ways of using both existing, secondary datasets for development research, as well as how you might go about designing and conducting survey research to develop a new dataset, including questions of sampling and coding. Chapter 12 then explores how big data and social media research are increasingly important to development research, providing guidance on how these data can be collected and analyzed. Chapter 13 then explores how locational and spatial data can be used in development research and fieldwork to address a range of topics and concerns. The final chapter in this section, Chapter 14, addresses how you might think about integrating methods and analysis in the design of research and fieldwork can ensure robust and reliable data collection in the field. Key issues here include recognizing research as an iterative process with varying potential for triangulation of methods, data and analysis. We explore approaches to coding data as well as key tools for analyzing both qualitative and quantitative data.

The final part of the book, 'Presenting and writing up research', considers some of the challenges faced in the post-fieldwork phase of a research project. Chapter 15 begins by addressing some of the ways in which data – notably big data, social media and locational/spatial data – can be presented through visual means and the importance of such techniques for engaging both academic and non-academic audiences, while Chapter 16 engages with the complexities of writing up research for different audiences, from academic writing for dissertations and articles through to research reports, policy briefing notes and press releases. Here we also reflect on broader considerations of knowledge exchange and transferable skills developed through development research. A final chapter recaps the core messages presented throughout this book, reminding you that while the process of doing development research can be challenging it is also inspiring.

By ordering the text in this way, we do not suggest that the research process follows a regimented and linear path. While most research projects do begin with a design phase, leading into a data collection (or fieldwork) period and finish with a period of analysis and writing up, there are overlaps and circulation between phases in an iterative process. Indeed, we recognize that development research is complex and messy, involving unexpected developments, challenges and opportunities. In order to convey a sense of these complexities, we draw extensively on our own research experiences to provide examples, anecdotes and questions throughout the chapters. For the same reason, we have included short reflections from other researchers and students in which they reflect on their experiences of conducting research. The first of these reflections, from Nina Laurie and Andrea Wilkinson (Box I.5), gives you a sense of the multiple and dynamic concerns involved in planning and carrying out development research for both staff and students.

While the overall structure of the book replicates an 'ideal' research flow, each chapter is designed to stand alone so that readers can engage with and revisit them as appropriate during the research process. Cumulatively, this content should reassure you that development research can be challenging but immensely rewarding, emotionally draining but also uplifting, exhausting but stimulating. Overall, the book emphasizes that in the same way that efficient and effective development is founded on careful and considered research, effective research is rooted in strong research design and preparation. In emphasizing the importance of careful research design, we recognize that the best laid research plans are often thwarted by the realities of fieldwork and understand that – and have experience of how – research plans change rapidly in the field. However, a strong training and understanding of the research process will help you overcome such challenges and enhance your adaptability and innovation in conducting fieldwork.

Box I.5 Personal reflections

Supervisor Nina Laurie: Four days prior to departure to Peru for field-based supervision

I did my risk assessment form last night. I thought it would be a quick job: find the one I did last time and give it a swift update – 'staying with the same people going to the same places, not that difficult – a good task for an evening in front of the TV'. Not to be of course . . . no hard copy to hand and the electronic file 'risk assessment Peru 09' accidently overwritten by UK fieldwork interviewing volunteers returning from South America – well, at least the continent was sort of right. I turned the TV off: it was going to be a long night.

I used not to take risk assessment forms very seriously. For goodness sake, all those questions about my 'control measures to reduce the risk identified': it's hot, it's sunny so put on high-factor sun cream, how hard is that? I hate the whole colonial imaginary that these forms feed into: the re-inscribing of the global south as 'exotic' and 'dangerous' just boils my blood. The risk level coming into my office in Newcastle is probably higher – the so-called 'pedestrian crossing' in front of the university is lethal. Twenty years of experience in this job, seeing all sorts of situations, has taught me not to be glib. In case I needed reminding, Andrea's 'weekend news' email three weeks ago pushed the message home: 'Just a short email to let you know that Ben and I were in a car accident yesterday! It's been quite a week, two infections, an earthquake and a car crash!!! . . . I'll keep you updated, but nothing to worry about.' Nothing to worry about? Is she mad? . . . I did worry, I do worry, I will continue to worry. That is part of the relationship bit of the job.

For me, the student–supervisor relationship is always two way. It is not only about imparting field expertise: in fact, quite often the reverse, I learn more than I teach. The experiences of my students frequently challenge my thinking about and approach to fieldwork. In Andrea's case, her policy background, prior to the PhD, working with fair trade NGOs (non-governmental organizations) in different countries in Africa, has made me think more carefully about what constitutes participatory methods in different contexts and for what ends they are adopted. In her NGO days, she workshopped specific goals collectively in order to own an outcome that would improve coffee production and financing for the group. For her, participatory methods are never just forms of 'data collection'. Last night Andrea's influence on my learning was very practical – it focused my mind properly on my risk assessment form. I reflected on the possibility of taxi crashes, earthquakes and stomach infections and so took more time over the form's requirements regarding my 'emergency plan'.

Thinking in terms of an 'emergency plan' is also a good way to approach the shifting nature of 'the field' during fieldwork, which changes because of things that happen in specific locations and at particular moments. Through the changing relationship with our field and those with whom we share it, fieldwork has to become an iterative process.

Doctoral Researcher Andrea Wilkinson: Looking at the effects of climate change on coffee farmers in Peru – four months in the field (seven to go!)

I have been conducting fieldwork for over ten years in completely different circumstances to which I find myself currently. My background as a manager of an international development charity has taken me to a range of different countries from India to Costa Rica, Ethiopia to Swaziland looking at diverse issues. What has not changed during this time is the reason why I gathered these data and the scope I've had to do so. It was vital for me to collect high-quality baseline data in order to secure funds to run long-term projects, to involve local communities and coffee cooperatives in developing and owning project outcomes and thus determining the impact of the interventions.

What is vital for me is that my doctoral research should have value on a practical level. I want to work alongside NGOs and policy makers, cooperatives and coffee farmers to ensure that there is dialogue throughout the entire process, from development to dissemination. This has resulted in my current research taking on a whole new meaning for me. I am collaborating directly with a fair trade coffee company, Twin, based in the UK, which imports Peruvian coffee. Twin is a pioneer in the field and has the power to use my research to make a real and lasting difference to the lives of coffee farmers and, as importantly, their families and community.

Becoming a doctoral researcher has resulted in my questioning the validity of my past research. Previously, research was not my sole aim but only a small part of my job. I used to fly in somewhere to spend a few weeks in the field, conduct back-to-back interviews, focus groups and questionnaires and then fly back out again to analyze the mass of data I had just collected. This time I am taking a different approach. I have spent the last four months learning a new language in the country in which I will conduct my fieldwork. Instead of using translators, I am slowly immersing myself into my research fieldsite. My learning is organic and is growing naturally. I have learnt a huge amount from everyday conversations, which I have discovered give you completely different answers to those you hear when sitting with a pen, paper, Dictaphone and a tight time schedule.

I am about to embark on the pilot phase of my fieldwork, where I will live and work on a coffee farm, seeing first hand the impacts of climate change as opposed to hearing what people want me to hear. The farmers and cooperatives, as well as my collaborative partner, will help shape the direction of my research, which I am beginning to understand is fluid and not merely about baseline data from which I can justify funding for various projects. It is about providing a more holistic reflection of the reality of the issues facing coffee farmers, cooperatives, importers and businesses posed by climate change. It concerns what people are willing to do to support long-lasting beneficial change for coffee farmers, their families and of course the natural environment.

(Research is supported by the Economic and Social Research Council North East Doctoral Training Centre – Grant number: 300021026)

Part I

PLANNING DEVELOPMENT RESEARCH

1

DESIGNING RESEARCH

In order to give your research the best chance of being successful, it is important to ensure you are well prepared and have spent sufficient time developing a considered and coherent research design. Key research design concerns need to be addressed to ensure your research practice and findings are reliable, valid, trustworthy, credible and dependable and that your conduct is in keeping with ethical guidelines and expectations.

Alongside concerns with the quality of your research you need to address a further set of factors. For instance, your ontological and epistemological positions frame how you develop a research idea and the questions you ask. Practical considerations of when, where and for how long you can conduct fieldwork are also integral, not least for those with partners, families and mortgages to consider (Chapter 4). While much of the discussion below regarding research design focuses on theoretical and methodological concerns, we recognize that research does not occur in a vacuum and the realities of life will be pivotal in your decision making. These concerns derive not only from your 'home' life, but also due to the context of the research site where concerns over issues such as safety, security and surveillance may be important considerations (Chapter 6; Koch 2013).You should not shy away from these concerns as you develop and design your research, but rather accept these as integral factors to the way in which you operationalize your plans.

There are numerous ways in which you can approach the designing of your research project. Whichever path you take will involve a number of core building blocks and activities that fit together to provide the framework for your research activities. The sequence which you work through, conceive of and integrate these blocks will depend on a range of factors. These factors include the driving force behind your interest, the time available to plan the activities and the nature of the research itself. The process outlined below should not been seen as a one-size-fits-all process, but rather a guide to the different aspects to be considered in developing, and delivering on, your research proposal.

It is also important to acknowledge that research design is not a one-way street. It is an iterative and reflective process whereby new information, experience and evidence can be integrated into the research process. You should have a willingness to be flexible and adaptable in the questions asked and methods used in order to overcome challenges or to address emergent concerns. Siân Parkinson's reflections (Box 1.1) on the ways in which her fieldwork experience in Malawi challenged her understandings and expecta tions of gender empowerment practices should remind you that research is a learning process and you can – indeed should – develop your research project as you learn from the research experience.

Box 1.1 Siân Parkinson: Reflections on the importance of field research in increasing cultural awareness

Prior to conducting field research in Malawi I believed I had a critical and comprehensive understanding of the gendered dimensions of the development process built on extensive literature reviews and engagements with theories and reports of gender and development. Yet it became apparent to me while conducting research into opportunities for female empowerment provided by microfinance organizations in Malawi that this prior knowledge was limited by my lack of research experience.

The majority of the organizations I worked with in Malawi offered microloans exclusively to women. However, I was surprised to learn that in all of these cases the husband's permission had to be obtained as a prerequisite to receiving the loan. The organizations encouraged women to consult with their husbands, and were alike in stating that the loans would not be granted against the husband's wishes. My immediate feeling was that this was paradoxical: the microloans could help women gain economic independence, but this was dependent on the support of their husbands. I was initially frustrated to learn of this and so enquired as to the decisions behind why this was the case.

Respondents referred to the gender relations that operated within their culture, where the husband was considered head of the household and made the majority of household decisions. I realised that I had, mistakenly, not fully considered the cultural context in which the microfinance organizations were operating. Conducting empirical field research enabled me to think critically about and enhance my own cultural awareness. Additionally, I had not considered the implications on gender relations if the women were to receive loans without their husband's support. Responses from interviews highlighted the potential issues concerning this, where female economic independence might be viewed as a threat to masculinity. Even though my main focus was on female empowerment, these findings were strengthened by interviewing both men and women involved or implicated in these development interventions. Involving both genders in such research is advocated by theories of gender and development, but I had no experience of the practical significance of this prior to conducting this field research. It was clear from my findings that this approach provided me with a broader set of insights than if I had only interviewed women.

In order to begin the journey of developing and designing a research project it is important to identify the key building blocks required. Box 1.2 identifies these, providing a short comment on each and indicating where in this book further guidance on each component can be found.

Box 1.2 The building blocks of research design

Block A: Overarching research theme/topic – this is the 'big picture' or 'big question' you want to address in your research. Essentially, this is the heart of the project. This overarching theme will, in conjunction with your -ological approach, determine the kinds of method you can use in your project, as different types of question/theme will need different types of data (who, what, where, why, how, when . . .). See following sections in Chapter 1 and also Chapter 2.

Block B: Research questions – these are the more defined and focused questions your research will address in order to understand/address the overarching research theme. These are not the questions to be asked in your interviews, surveys, etc., they are the bones – the skeleton – that gives structure to your project (i.e. in order to address A, I need to find out about B1, B2, B3, and B4). See following section in Chapter 1.

Block C: Methods – different types of question, and different epistemological and methodological approaches, require different methods: there is no point in using a national-level census to find out how people in a specific village understand climate change and livelihood adaptation! See chapters in Part II for further guidance on different methods and when to use them.

Block D: Ethics, risk, and logistics – If we lived and worked in an ideal world, this block would be of minimal importance. In reality, it is always a key determinant of our research as it informs the scope of the research questions addressed, the methods used, sample size targeted, etc. Items to consider here include the length of time available for the project (and for data collection in particular), where these data must be collected and how (and the time and economic costs involved, including visas, travel and accommodation costs, other resource needs), risk and safety (for instance, if there are travel advisory notices against travel to the area in which you wish to conduct research this may prevent or severely restrict your research project as there are important implications for travel insurance, risk assessments, etc.), ethics, local context and sociocultural norms. See Chapters 4, 5 and 6.

Block E: Literature review – this may inform the development of your research topic and/or questions or may be framed by an existing interest and topic decision. Either way, a rigorous and sound knowledge of existing studies, theories and understanding is required in order to develop appropriate and adequate research questions and to locate your work within existing knowledge. See subsequent sections of Chapter 1.

Block F: Dissemination and outputs – whether this is simply your dissertation or thesis or includes academic publications, research reports, policy briefs and other materials for both academic and/or non-academic audiences, it is important to consider what you plan to do with the data you have collected. Implicit in these reflections are concerns with the impact your work may have on theory, policy and practice, but also in relation to what promises you may make and then need to deliver on for your hosts and supporters. See Part III.

To illustrate the ways in which these building blocks can be addressed and brought together, a number of Dan's experiences are included here (Box 1.3). As you will see, there is no single 'best' way to design a research project; research projects are products of numerous negotiations throughout the process.

Box 1.3 Examples of coming to the research

Dan: 1. I was approached to contribute a chapter, based on previous work on the role of satire in political expression, to an edited book addressing how resistance and governmentality were being experienced in sub-Saharan Africa. In previous work, I had focused on the production of satire and, in broad terms, popular and government responses to these products. In reviewing the literature for this earlier work, I was struck by Dittmer and Gray's (2010) comment that a continued focus on the role of elites in producing and interpreting satire meant little attention was paid to audiences' understandings of and reactions to such products.

My contribution to the book was therefore designed to offer an initial investigation into audience responses to a set of controversial cartoons by the South African cartoonist, Zapiro. Ideally, this research would have involved interviews with the cartoonist and key figures in political parties and civil society organizations who had been outspoken in their responses to the 'Rape of Lady Justice' cartoon series, as well as focus group discussions with demographically varied groups who were exposed to the cartoon. However, because of the time lag between publication of the cartoon and book chapter request (the cartoon was published in 2008, the chapter requested in 2012) and a lack of time and finances to provide for extensive fieldwork, an alternative approach was required.

Consequently, I drew on existing interview data from previous work with the cartoonist and an accessible repository of public responses through the online comment sections of the cartoonist's and newspapers' webpages. These postings were identified, archived and subject to discourse and content analysis in order to explore the range of audience responses in relation to debates around democracy, citizenship, race, gender and violence (cf. Hammett 2010). The constraints of time and money were key factors in determining how this research was conducted and imposed limitations on the data collected and conclusions made (not least regarding whose voices were being considered and who was able/chose to participate in online exchanges in particular forums). This work was, nonetheless, able to develop a first, albeit partial, engagement with popular sentiments towards the cartoons and the ideologies and political sentiments mobilized.

Dan: 2. With a senior colleague, I designed a 12-month research project addressing citizenship education in South Africa. This work represented a step-change both for my own research away from race and identities in South Africa, and my colleague's work on citizenship in the US. The process began with a series of (in)formal conversations about research interests and the need for me to secure a research grant as my existing fixed-term contract was coming to an end. Questions emerged from these discussions as to how current citizenship education policy and practice envisaged 'good' citizenship in the post-apartheid state and how school teachers and communities engaged with these ideals. From this starting point, we began a literature search to identify both key sources on citizenship education and relating to education and citizenship in South Africa. We compiled a bibliography of potential sources and constructed a library of journal articles and book chapters and began reading. Informed both by our previous research experience and the literature, we began to write the justification and context for the research as well as identifying the specific questions we wanted to address. Once these questions were known, we were then able to identify the kind of data to be collected and the most appropriate methods to do this (interviews, focus groups, ethnographic methods). We also identified the most appropriate funding option for the research, which provided a maximum budget for us to work to. With this constraint in mind, we began to develop the details of the method – the number of provinces to work in, the number of schools to work in within each province, how long to work in each school – while also thinking through specifics in relation to which urban area and then which schools to approach in each province, identifying contacts for ethics approval from each province's education department as well as at each school, the teachers we would want to speak to (specific subject areas of most relevance) and any other relevant parties we wanted to engage with.

Dan: 3. When I was contemplating the research focus for my MSc dissertation, my interest was piqued by an article in the popular press reporting on Cuban doctors and teachers working overseas as part of their government's agenda of 'international solidarity'. This report prompted me to begin searching for academic research on this topic. I found little of direct relevance, suggesting this research would be original. Broader engagements with the literature provided a frame to the project and guided the decisions regarding more specific research questions to be addressed. At this stage, a country-specific interest in South Africa conveniently coincided with the presence of Cuban doctors working in South Africa as part of the international solidarity agenda. This helped focus the questions to be asked (and thus the methods to be used) and to inform considerations of logistical considerations and access to/availability of potential research participants.

As these examples demonstrate, research projects are conceived of and designed in various ways. On occasion, the research design begins with an overarching theme which is broken down into a specific focus and then into research questions that determine the data required and the methods needed. Other projects are designed with an awareness of specific limitations to methods or data from the outset, meaning the questions to be addressed are – to an extent – dictated by these concerns. While there are many differences between each of the examples above, there are a number of commonalities: appropriate methods are used to provide the relevant data to address the research questions identified, the research logistics (of costs, time, travel, ethics, risk) are considered from the early stages of the research design, and existing literature plays a vital role in the design process.

When progressing through these decisions it is important to reflect on the material and emotional aspects of this process. A reflexive approach allows you to recognize research as an embedded practice in which your own interests and positionality, as well as broader socio-economic, political, cultural and other influences, contribute to the identification and dis-identification of what is deemed to be worthy of research (see Mauthner and Doucet 2003; Elmhirst 2012). It is also important to recognize that this process is framed and informed by the broader, and continually evolving, landscape of development research and practice.

The shifting terrain of development research

Approaches to, priorities in and understandings of development will inform your own research engagements. They are also strongly influenced by changes in the broad political landscape. Researchers have contributed to these debates and shifts and are also affected by the outcomes of these debates – perhaps most obviously through changing funding availability and priorities. Approaches to and ways of doing development research have also been influenced by changes to the academic landscape as particular paradigms have waxed and waned in their dominance over academic practice and discourse. These experiences are not universal: they are mediated and informed by national political and academic contexts that privilege different theories/theorists, types of knowledge, types of question asked and methods used. These differences are further complicated by the intricacies of specific (sub)disciplinary interests and priorities that further influence who does research and on what topics: Bourgois' (1990: 43) review of the complexities in difference between American and European anthropological work illustrates these issues succinctly. In his article, Bourgois demonstrates how the different political framings of the discipline and understandings of ethical practices influence the questions asked and the methods used. Whether you realize it or not, your own training in and engagement with development research will have continued these practices, developing within you a particular disciplinary/institutional/national set of approaches to development.

Evolving approaches to knowledge and the changing influence and dominance of specific paradigms over time are also powerful influences over development research. Recent shifts towards post-modern and post-colonial approaches to development research reflect and embody broader trends in social science research, not least in emphasizing the subjective nature of research. This awareness stands in clear contrast to early anthropological studies whose authors positioned themselves as objective observers of people and places, a standpoint that was common to much development research at the time. Current thinking in the social sciences rejects assumptions that 'out there' can be separated from 'in here' and that the researcher is an impassive, impartial observer who stands back from emotions, ideology and beliefs and makes detached, objective recordings of 'the field'. Instead, we are encouraged to reflect on the academic, ideological and other experiential knowledge and beliefs that inform our academic practice as these inform how we view and record the field.

Taking this awareness a step further, it is important that we not only think about how our background and experience frame our view of the world and of research, but also how our presence in the field – and other people's reactions to our presence – influences the construction of knowledge. In other words, it is incumbent on us – as researchers – to recognize that we are not simply going out and 'harvesting' data and information, but rather that the process of research is inherently collaborative and that knowledge is co-produced through the interactions between the researcher and participants (Kaufmann 2002). That we are encouraging you to adopt such reflective practices and recognize social science research as subjective is, itself, a result of the current dominant paradigm in the social sciences.

Dominant paradigms in social science research influence the types of epistemology and methodologies deemed suitable and appropriate for doing development research. In broad terms, we can identify three main research paradigms that frame the majority of development (and, more broadly, social science) research: positivism, post-positivism and interpretivism. The summary of these three approaches provided below is derived from the accessible discussion in Grix (2004: 78–88).

At one end of the spectrum, *positivist* approaches seek to explain social phenomena and to develop causal theories through the use of scientific methods to study the social world. The concern here is with 'facts' rather than 'values' and the empirical measurement of society from an objective position. At the other end of the paradigm spectrum, *interpretivist* approaches seek to develop understanding (rather than explanation) grounded in a view of the social world as constructed through interaction and that our understanding of the social world is produced through our interpretation of it. As a result, you are seen as integral to the production of knowledge about society and your analysis is subjective (informed by your opinions, experience and ideology).

In between these two positions, are the approaches of *post-positivism* and *critical realism*. Unsurprisingly, this middle ground incorporates aspects of both *positivist* and *interpretivist* approaches in seeking to understand and explain social reality. In seeking to develop such understandings, critical realists view structure and agency as mutually constitutive and seek to understand and explain how different causal mechanisms exist and work. (There are many more paradigms and approaches, should you wish to explore these in greater detail: an accessible starting point is Chapter 5 in Grix's (2004) *The Foundations of Research*.)

These concerns inform the topics and questions that are deemed important at different historical junctures and influence the availability of research funding. If certain topics are *de rigueur* at a particular moment in time then it is likely that funding will be channelled towards projects engaging these issues. While these funding imperatives will have a strong bearing on what development research is done, the questions asked will also be driven by what you see as interesting or important as well as broader engagements with topics that provoke political anxiety and are deemed of policy relevance (Elmhirst 2012). In addition to your own interests in determining your research focus and questions, you should also consider whether/how broader political agendas are implicated in your research. For instance, if you are working on someone else's project or are trying to/have secured external funding, you may want to reflect on who is funding/supporting the research and how your work will fit to or be used in their agenda. In turn, this raises the important issue of for whom is the research being done and whom will it benefit. Completing your dissertation or thesis or getting a paper published will help your career and may satisfy a personal curiosity, but are there any other benefits to other groups (not least those involved in the research)?

This last concern should encourage you to consider how research participants could be involved in the conception, design and conduct of the research and, where possible, in a mutual analysis of data and sharing of findings (for a lengthy discussion of this drive, see the 2011 special issue of *International Journal of Social Research Methodology* (volume 14, issue 6), including the papers by Gobo (2011), Ryen and Gobo (2011) and Uddin (2011)). Such engagements are important, not only to ensure a robust and sensitive research design and practice, but also to ensure that findings are relevant, useful and accessible to multiple stakeholders (see, for instance, the reflections of Simon et al. (2007) on how they would have changed an international health research project to make it more locally relevant to different stakeholders). This concern resonates with the current drive within British academia to increase the accessibility of academic research to and impact on the general public. The specifics of the debates around the 'impact agenda' lie beyond the focus of this book (for insights into these discussions you may wish to read Williams 2013 or see the debate between Rachel Pain et al. (2011, 2012) and Tom Slater (2012) in *Area*), but they are part of a longer history and set of debates that are important here, relating to the purpose of academia/academic research and the politics of knowledge production.

Such debates are not new: anthropologists have viewed engaged scholarship as part of their social responsibility for several decades (Bourgois 1990) and critical geographers in the 1990s argued that academics had a political responsibility to not only develop critical insights grounded in both theory and everyday life but that these should be aimed at progressive social change to the benefit of the marginalized and silenced (for contributions to this debate in the 1990s see Blomley 1994; Chouinard, 1994; Tickell, 1995; Bassett 1996). Such debates have continued and resonate strongly with the post-colonial turn in development research, as researchers have sought to navigate a delicate path between advocacy, academia and activism, as expressed by Speed (2006) and Petray's (2012) advocacy of critically engaged activist research. This approach, they contend, is an important method for engaging with issues of power and resistance, noting that approaching research as an activist can facilitate access and the depth of understanding developed while also contributing to campaigns for social justice and helping to strengthen activist movements through helping them gain a deeper understanding of the movement and its practices. Martin Hammersley's (2000) text (*Taking Sides in Social Research*) provides an informed and balanced discussion of ideas of partisanship and bias in social science research and is worth reading (see also discussion and sources in Box 1.4).

Questions relating to the nature and role of research are particularly pronounced for development researchers. Views on the topic range from advocates of activist research (as outlined above) to those who view their research as part of an abstract intellectual responsibility for the academy to be at the heart of knowledge production. Critics of this later position would argue that this approach is redolent of imperialism and is a practice of exploitative extraction (and (re)production) of knowledge from the global south for the benefit of researchers based in the global north. These concerns clearly resonate with

Box 1.4 Activist academia

Reflections on the merits, drawbacks and challenges of engaging in activist academia highlight a range of concerns that researchers need to be aware of in developing a research strategy engaging with issues of activism and protest or with an explicit social justice agenda. The articles noted here provide a range of perspectives and experiences that are informative for anyone contemplating such a research approach:

Chatterton, P. (2006) '"Give up activism" and change the world in unknown ways: or, learning to walk with others on uncommon ground', *Antipode*, 38: 259–81.

Davis, D. (2003) 'What did you do today? Notes from a politically engaged anthropologist', *Urban Anthropology and Studies of Cultural Systems and World Economic Development*, 32: 147–73.

Dyrness, A. (2008) 'Research for change versus research as change: lessons from a *mujerista* participatory research team', *Anthropology and Education Quarterly*, 39: 23–44.

Harvey, D. (1995) 'Militant particularism and global ambition', *Social Text*, 42: 69–98.

Hay, I. (2001) 'Critical geography and activism in higher education', *Journal of Geography in Higher Education*, 25: 141–46.

Lees, L. (1999) 'Critical geography and the opening up of the academy: lessons from "real life" attempts', *Area*, 31: 377–83.

Petray, T. (2012) 'A walk in the park: political emotions and ethnographic vacillation in activist research', *Qualitative Research*, 12: 554–64.

Routledge, P. (1996) 'The third space as critical engagement', *Antipode*, 28: 398–419.

Speed, S. (2006) 'At the crossroads of human rights and anthropology: toward a critically engaged activist research', *American Anthropologist*, 108: 66–76.

recent contributions promoting a post-colonial approach to development research and the importance of ensuring that development research is not only beneficial to the academic conducting the work, but that the findings are accessible to and useful for stakeholders outside the academy. While such debates are increasingly common across the social science, they are particularly pronounced for those working on development topics as it is impossible to disassociate fieldwork from imperial histories and projects (Abbott 2006). It is important, therefore for you to reflect on your position in the field and how the 'field' is a far from neutral space, as demonstrated in Fiona McConnell's reflections (Box 1.5) on negotiating her positionality in relation to academic and activist roles while working with Tibetan exile groups.

Bearing these concerns in mind, it is essential to reflect on how your fieldwork practice relates to these histories, to question how your positionality embodies and is influenced

Box 1.5 Fiona McConnell: Negotiating activist/academic roles in researching exile politics

It's a chilly March evening in the foothills of the Himalayas and I'm sitting in a café with a group of Tibetan activist friends. They're getting ready to put up posters across town advertising their list of favoured candidates for the 2006 elections to the Tibetan parliament in exile: candidates who take a *rangzen* or pro-independence stance on the future of Tibet. Flour and water paste in being mixed in large buckets, brushes are being assembled and bundles of A3 yellow posters – pungent with chemical ink – are ready to be attached to walls, notice boards and shop windows. I've frequently gone out 'postering' around Dharamsala with this group of friends to advertise film screenings, talks by former political prisoners and candle-lit vigils. But this evening I say I'll give it a miss. The reason for this trip to India is to research the exile Tibetan election system and, given the close-knit nature of exile politics, this evening's activities are just a bit too close to my academic work for comfort. It's a moment when the often fuzzy boundary between my activism and my research sharpens into focus and needs to be carefully negotiated.

The relationship between activism and academia is one that has received productive critical attention over the years in geography (Blomley 1994; Routledge 1996; Ruddick 2004). In my case, I was not researching the practices, discourses or even organizations associated with Tibetan political activism (at least not in the conventional sense of the term). Rather, my focus was – and continues to be – on the state-like functions of the exile Tibetan government, which was established in India in 1959. While certainly not a panacea for the often unequal power dynamics of ethnographic research, especially in the global south, there are a number of ways that engaging in activism alongside ethnographic research can be productive and my continued participation in Tibetan activism shaped my research experience in several aspects. Being white, non-Buddhist and Northern Irish, I am automatically an outsider in the Tibetan community. However, my long-term involvement with Tibetan campaigning NGOs in the UK and India was a key connection with many respondents in Tibetan settlements scattered across India. Having a Tibet activism 'track record' meant that I was an accepted part of an already known community, rather than a difficult-to-place-stranger.

Activism is about collaboration rather than dependency, on ideas sharing rather than instruction and, with activism continuing 'back home', it is a long-term commitment rather than an activity confined to the field. As such, activism has the potential to be an alternative form of 'giving back' to the communities in which we work. There are the beginnings of research fatigue within the exile Tibetan community and I found the likes of organizing information evenings for Western tourists, speaking to Tibetan radio about my involvement with the Tibet movement and collaborating with Tibetan colleagues to organize demonstrations a potentially progressive and mutually empowering alternative to conventional practices of financial recompense.

While I was open about my activist work to all my respondents, there were occasions, like the one noted above, when my role as activist and that as researcher became 'messy'. Although *injis* (Westerners) engaging in the wider Tibetan movement is actively encouraged, I had to tread carefully when it came to domestic exile politics. Being seen out election campaigning with what is sometimes labelled a radical group may have compromised my access at the Tibetan government headquarters. There was a phrase several exile government officials used to describe how they negotiated involvement in pro-independence activism with their job for an institution that was advocating for a different future for Tibet that resonated with my own experiences. They spoke of having 'two hats'. Deciding which to put on when and where was a revealing aspect of my fieldwork, sharpening my awareness of the nature of exile politics and my own positionality.

by history, to recognize that your fieldwork will have (often unintended) impacts on the host community and to consider how you can reduce the 'voyeuristic' potential of fieldwork (cf. Abbott 2006). These reflections bring to the fore questions of history, positionality, power, race, language and ethnicity that are important for you to reflect on in relation to your own fieldwork expectations and practice (and which are explored further in Chapter 3). These reflections and practices are important contributions towards the broader concern expressed by Linda Smith (1999) and picked up by other scholars since (for instance Rogers and Swadener 1999; Denzin and Lincoln 2005; Denzin et al. 2008; Ndimande 2012) for a decolonizing of methodologies.

The historical ties between exploration, empire and knowledge production continue to inform research questions, practice and ethics. Critical scholars have argued that social research has served to reinforce and replicate colonial relations and practices. In particular, scholars such as Smith (1999) have interrogated how the dominance of social research agendas and questions by researchers from the global north perpetuate particular practices of knowledge extraction and production based upon particular epistemologies. The challenge from anticolonial scholars is for researchers to question the underpinning beliefs and assumptions to their work and to recognize, value and utilize indigenous knowledge, methods and theories. How you position your work and engage with such debates will have an important bearing on your research practice and engagement with such debates is useful in reflecting on how you produce knowledge – from conceiving the research question through the design of the research to the methods used and your practice in the field (see Figure 1.1).

Developing your research idea

This leads us back to a starting point – of how to identify a research topic. If you are working as a research assistant, have a project-linked scholarship, are completing a placement-based dissertation with a host organization or undertaking a masters international course linked to a Peace Corps placement (Box 1.6), the topic of your

Figure 1.1 Negotiating questions of positionality and colonial history is a complex but integral part of development research, whether you are working as a lone researcher or part of a diverse fieldclass group

research may be imposed on you. In other scenarios, you will have much broader scope to identify your own research topic and specific project(s) to pursue. In these situations, there are three main categories of influence over research topic choice: personal, social and academic (see Blaikie 2000). For post-colonial development approaches we can also add the influence of the host community to this list.

In broad terms, your identification and selection of a research topic will be driven by a number of the following drivers:

- *Personal interest or experience* arising from previous travel or work experiences or resulting from a story or image found in the popular press or social media that has stimulated a personal interest.
- A desire to help *solve a problem* or achieve (developmental) goals – this may be a problem you have first-hand experience of or may be a challenge that others face and which you want to help resolve.
- As a *social obligation* – this may be because you are driven to address a social need and/or a desire to give back to society.
- *Academic interest* and gaps noted by or evident in existing literature and work on a topic – papers will often have a note to this effect in their conclusions (i.e. that 'further research is need on Q' or 'while this paper did not address X this is an avenue for further exploration').
- The requirements or focus of an *external* call for funding applications.
- As an *update to previously conducted research or questions* – whether your own previous work or that of others that you are seeking to update, expand or supplement through another case study.

Box 1.6 Linking 'service' with study

One avenue to doing development research is through combining a period of volunteering or 'service' with your studies. This may be realized in a number of ways, perhaps individually organized with a non-governmental organization (NGO) or faith-based organization (FBO), through a placement-based dissertation integral to a course or through a programme such as the Peace Corps' masters international. This last option, available only to American citizens, provides an opportunity to undertake a college (university) degree including a Peace Corps service period, undertaken once coursework is completed, that can be used as the foundation for any remaining thesis or paper requirements. Details of such opportunities can be found here: www.peacecorps.gov/volunteer/learn/whyvol/eduben/mastersint. Examples of MI dissertations are available here: www.peacecorps.gov/volunteer/learn/whyvol/eduben/mastersint/mitheses.

Alternatively, you may seek to organize your own affiliation with an NGO or FBO through which to carry out a research project. If you pursue this option, it is important to understand the nature and values of the organization you will be working with (and how these may influence both the design of your research and how you are viewed while in the field), what you can realistically offer to them, what demands they are making on you and what resources or other 'burden' you are placing on them to support your research (and if they are willing to provide this). You must also ensure that you have approval, including ethics clearance and completed risk assessments and support for such a project from both the host NGO or FBO as well as your academic institution.

While not exhaustive, this list of influences on research topic selection covers many of the common drivers to selecting and designing a research project. These categories are not exclusive and the interaction of multiple drivers often leads to the development of a research proposal. Whichever drivers are relevant to you, at the heart of the matter will be your sense of curiosity – of being interested in both the questions asked and the various answers you may get to these (White 2009).

Linked to your identification of a research topic is your research strategy. A research strategy is linked to your ontological and epistemological stance and the resultant relations between theory, methods and data. Depending on your conceptual approach, you will seek to answer certain types of question and have specific research aims, each of which will be suited to particular research strategy. There are four commonly recognized research strategies, each associated with a different way of viewing knowledge and understanding (see Table 1.1). The *inductive* approach aims to establish universal generalizations based on theories generated from accumulated data. The *deductive* approach begins from a set of findings to develop understandings and theories from this, while the *retroductive* approach uses hypothesis testing to identify the underlying causes of the phenomena.

Table 1.1 *Research strategies*

Typical type of question suited to	Aim	Process	School of thought
Inductive			
What	Develop universal generalizations	Generate theory from data	Modernism, positivism
Deductive			
Why	Test theories	Test theory against data	Critical rationalism
Retroductive			
How	Discover underlying mechanisms	Work from data to test hypotheses	Scientific rationalism
Abductive	Describe and understand from actors' perspective, development of grounded theory	Discover and explore everyday life	Hermeneutics, social phenomenology

Source: Derived from Blaikie 2000: 101

An *abductive* approach seeks to develop an understanding of the social world and actors relations to and understandings of this.

In turn, your research strategy informs your approach to the research itself. For inductive and deductive approaches, the research questions are the foundation of the research and all else flows from this, while retroductive and abductive approaches allow for a more iterative process whereby the research question evolves and changes during the research process. Depending on the strategy used and type of question being explored, different approaches to doing the research will follow. If you are using an experimental approach, this is suited to addressing 'how' and 'why' questions. A survey can be useful in approaching 'who', 'what', 'where', 'how many' and 'how much' questions, while archival approaches are useful for 'who', 'what', 'where', 'how many' and 'how much' questions. Case studies are generally useful for 'how' and 'why' questions.

As part of this decision-making process, you need to identify your research questions as these determine your research approach and methods. Different types of question provide for varying forms of engagement and understanding. 'What', 'why' and 'how' questions are suited for description, explanation and understanding change. 'What' questions are generally aimed at the discovery of information. 'Why' questions seek to explore the causes, reasons and relationships between factors and to explain these. 'How' questions are concerned with understanding and bringing about change. Categories of questions such as 'who', 'where, 'how much/many/far' serve to provide additional layers of understanding and detail to the 'what', 'why' and 'how' questions. Ultimately, however, your research questions define the scope, focus and nature of research: the specific wording of your questions can have significant effect on research activity.

In constructing a research question, you must think through a number of key concerns. Your epistemological position (how you know what you know) will influence how you conceptualize your research, the questions asked and kinds of data sought and with which methods. Therefore, as you think through your overall research aim you need to identify the (sub)questions that you need to address in order to meet the overarching aim. You must also address the logistical, resource, time and other factors that affect how, where and when you can conduct the research. You will then need to consider what kinds of data are available within these constraints: a lack of suitable data will fundamentally undermine a research project if the wrong approach is chosen.

With these concerns in mind there are a number of ways in which specific research questions can be developed and honed. A common process to this end is now summarized:

1   Brainstorm ideas and areas of interest and the kinds of question needed to explore and unpack these.
2   Review ideas to identify main topic and discard inappropriate/unanswerable questions.
3   Consider how to phrase remaining questions.
4   Expose assumptions and respond to these.
5   Place a boundary on the scope of the work.
6   Separate major and subsidiary questions.
7   Review literature and make sure all questions add to your project (theoretical, empirical, methodological).

In developing your strategy, approach and questions, you will engage with and negotiate a series of concerns and influences. Central to this process is a need to understand the practicalities of doing research – which is where this book is of help – as well as the theoretical, conceptual and contextual framing of your research topic and field location. In order to develop this understanding, it is imperative to engage with existing literature on the topic and the context being studied.

Literature reviews and horizon scanning

Awareness of and engagement with existing literature and studies is vital to development research and practice; as White (2009: 7) notes, 'Research does not, *should* not take place in a vacuum.' History has shown how failures to engage with previous studies, knowledge and understanding can fatally undermine development interventions. The same is true for development research. In development research, a literature review can serve a number of purposes, including positioning your research in a broader, relevant body of knowledge (including theoretical frames and previous empirical findings) and informing your research questions.

Research questions can emerge from your literature review in direct response to a stated acknowledgement of scope for further research, because you want to challenge orthodoxy and existing understandings or because there is a specific gap in the literature that your research is intended to address. In other instances, your engagement with literature may be driven by your choice of research topic or question. Regardless of how you come to the literature review, this is a vital part of the research design process. Your literature review plays an important role in honing specific questions and provides a theoretical framework as well as conceptual and contextual background and understanding (for a longer discussion see White 2009). In short, a literature review:

* presents a summary of existing knowledge in the field
* critically interrogates this knowledge
* frames and informs your own research design
* contributes to the development of your research questions.

While the emphasis in a literature review for a student project will be on academic literature, many reviews do – and often should – include other sources including government documents, reports and policies, research reports and policy documents from national and international civil society organizations, media materials and other 'grey literature'. Given the breadth of materials to locate, constructing a literature review can be a daunting task. While there is no single, magical solution to this, here are some useful tips on locating relevant literature:

* Determine combinations of keywords relevant to your project area to use in your searches (often a lack of relevant materials being discovered can be overcome by slight changes in the search keywords).
* Use academic search engines to identify relevant academic literature (do not just use GoogleScholar! If you do use GoogleScholar, remember there is a 'cited by' link under each article – you can use this to trace 'forward' other research using articles you find relevant to your own work).
* As you locate relevant articles, download and save these to a library and construct a bibliography for these.
* If you see specific authors emerging as key people in the field, search their publication histories for other relevant materials.
* If certain journals emerge as common outlets for relevant materials, browse their archives for further materials.
* Use the reference lists in the articles to identify other sources and for ideas for other key authors or journals.

Once you have identified and accessed this literature, you can use this to give the theoretical and conceptual frame for your research. In order to assist in this process, it is important to clearly and logically name and file your readings, either through a bibliographic program such as Endnote or through your own filing system. While this does take time initially, it will save a lot of frustration and time later on in the research process when you are trying to revisit your readings and consolidating your reference list.

When you are reviewing the literature relevant to your topic, it is important to utilize good academic practice both in note taking and filing/archiving. Therefore, in making notes be sure to not only accurately record bibliographic information but also to clearly differentiate between summary notes, quotations and your reflections on the source content. You are not seeking to simply summarize the main argument of the book/article, but to draw out key findings and locate these in relation to other works, both empirical and theoretical. Remember, your literature review is not simply a summary of existing works – a good literature review will critically examine and reflect on the literature, identifying key trends, omissions, contradictions and disagreements. Your aim is to build a narrative that draws on these, critically, to both frame and inform your research findings and to underpin the argument you make from your research.

Other sources of guidance

There are many other texts dealing with research methods, some focused on specific methods or realms while others opt for a broad sweep and introduction to the field of research. For students working in the development field, this text provides expert views and guidance on the research process, from project conceptualization through to dissemination practices, covering a number of methods and tools along the way. A number of other texts provide further engagements with development research that you may find useful to consult alongside this text. Although lacking an explicit focus on the challenges and encounters associated with doing overseas fieldwork and generally focused on 'traditional' social science research methods, texts including Phillips and Johns' (2012) *Fieldwork for Human Geography* and Perecman and Curran's (2006) *A Handbook for Social Science Research* may be useful companion texts. To help you think through how to develop a research topic and research questions, you may find White's (2009) *Developing Research Questions* of assistance, as well as Grix's (2004) *The Foundations of Research*.

For readers with primary concerns relating to fieldwork practices and more applied forms of development research, you are likely to find that Laws et al. (2013) *Research for Development: A Practical Guide* offers further insights into particular considerations and practices of more applied forms of research. Many readers will find the range of short yet detailed chapters in Desai and Potter's (2006) *Doing Development Research* useful starting points for further engagements with specific considerations in the planning and practices of research. Another useful collection that engages with the practicalities of doing development research is Scheyvens' (2014) edited book *Development Fieldwork: A Practical Guide* (2nd edition). For those wishing to develop more detailed understanding of ethnographic methods, Atkinson et al. (2001) *Handbook of Ethnography* provides an extensive and detailed engagement with ethnographic methods and practices with relevance to development research.

You may also find various online resources of use to you, including those provided by the Developing Areas Research Group (http://www.devgeorg.org.uk/) and the Centre of Global Development at the University of Leeds (http://www.polis.leeds.ac.uk/centre-global-development/about-centre/researchers-development-network/resources/).

What to take from this chapter

From these introductory comments, we hope that you realize the importance of carefully conceptualizing and designing your research while recognizing that you will also need to be flexible and adaptable in your research practice in order to respond to emergent issues, challenges and opportunities. This chapter has guided you through some initial debates around development research and highlighted the importance of good research to good development practice. In so doing, we have emphasized that there is no single 'correct' way of doing development research, but rather demonstrated that there are multiple drivers and approaches to how we conceive of and do our research and that negotiations of these influences – and other concerns such as time and money – are an integral part of the research process.

We have then linked these debates to the process by which research is designed, reiterating the need to think through the various approaches to and building blocks of research design through a series of real-life examples. The skills needed in this process – of priority setting, negotiation of competing needs, communication skills, summarizing and reviewing literature, constructing arguments, logistics, time management – are all important transferable skills. Throughout the process of designing and doing development research you will develop a wide range of skills that enhance your employability. There are the more obviously transferable skills such as interviewing and surveys that resonate with the demands of market research companies, as well as the broad suite of skills and understandings developed through constructing contextually informed research designs through to the range of presentational skills involved in the variety of dissemination methods that provide a set of skills transferable to many employers.

If you want to explore the ideas covered in this chapter in more detail, you may find these further readings useful:

Blaikie, N. (2000) *Designing Social Research*. Cambridge: Polity Press.

Desai, V. and Potter, R. (2014) *The Companion to Development Studies* (3rd ed.). Abingdon: Routledge.

Scheyvens, R. (2014) *Development Fieldwork: A Practical Guide* (2nd ed.). London: Sage.

Watts, M. (2006) 'In search of the Holy Grail: projects, proposals and research design, but mostly about why writing a dissertation proposal is so difficult', in Perecman, E. and Curran, S. (eds.) *A Handbook for Social Science Field Research: Essays and Bibliographic Sources on Research Design and Methods*. London: Sage.

2

THE CONTESTED TERRAIN
OF DEVELOPMENT

As outlined in the introduction, development research is a vital component of development policy and practice. Just as development research informs, monitors and evaluates development challenges and interventions, so broader trends and framings of development as a concept and set of practices influence development research. This chapter engages with these changes and considers how they have affected the shaping of development funding streams and priorities as well as research agendas, with clear implications for the possibilities of doing – how we do and on what we do – development research.

Before we go any further, we should make clear that we assume that you have previously engaged with debates and concepts relating to development. If not, you will find some suggested readings at the end of this chapter that will help you develop an understanding of development debates and concerns. In this chapter, we will provide a quick summary overview of the history of development as a refresher. But, before we do that, there are two activities (Boxes 2.1 and 2.2) for you to engage with as quick reminders of the complexities of development and development-related debates and practices. By inviting you to reflect on how you define and how you measure development, our aim is to ensure you are thinking critically about key concepts and questions and that you transfer these critical engagements to your research design and practice. It is also important that you recognize that how you understand development will have a direct bearing on the kinds of question you want to explore, the types of knowledge you seek to produce and the outcomes you would like to achieve through development research.

A brief history of development

Contemporary development concerns, at the international level at least, are rooted in political and economic responses to the global devastation wrought during World War Two. Harry Truman's 1949 inauguration speech as President of the United States of America was a seminal intervention at the time, defining what international development was focused on and the importance of this concern in international politics and social obligations. Before considering the implications of Truman's speech for international development since 1949 it is important to acknowledge a much longer history that is implicated in contemporary development challenges and your engagement with development research (see Figure 2.1).

Box 2.1 The complexity of defining development

How would you define 'development'? If you were asked to explain what development is and how it can be achieved in one minute, what would you say?

In thinking about development, how would you quantify or measure development? At what scale would you prioritize development strategies? If you were promoting development, what would you prioritize – an overall improvement in development indicators due to dramatic improvements for a few or through more modest improvements for many? Would you pursue policies aimed at benefiting all equally? To what extent would you advocate an approach to development that involves a radical redistribution of wealth and resources? To what extent would you adopt a free-market approach to development?

Are there certain rights and needs that should be prioritized in development? What different aspects of life can be considered in relation to ideas of 'development'? Can you have political (rights/development) without economic (rights/development)? What is the value of increased economic development without social and political rights?

Would you agree with Black's (2007) argument that development is simply a fictional concept given prominence in academic and theoretical spheres with little relevance to the daily lives of those it is supposed to be helping? Or with post-/anti-development approaches that view development as an (invalid) Western discursive construct?

Box 2.2 The challenge of measuring development

Select five countries from the following list and rank them in order of development:

Albania, Poland, Laos, Ghana, Angola, Peru, Argentina, Botswana, Vietnam, Morocco, Lebanon, Barbados, Haiti, Oman, Nepal.

How did you rank them? What criteria did you use? (Popular image? Economic indicators? Human Development Index score? Measures of inequality?) Why did you use this approach and set of indicators/measures?

How else could you have ranked them? What other measures of development could you use? How do the rankings differ when you use these alternative indicators? What does this tell you about what development 'is'?

Figure 2.1 How you engage with development research, from the questions you address to the dynamics of fieldwork, is informed by broader historical processes and contemporary relations

Development emerged as a political concern in the 18th century. Advocates of Enlightenment thinking championed ideals of progress and freedom that were rooted in the European experience. The resultant Eurocentric view that the standards of living and values of the great European powers represented the highpoint of development informed popular and political world views. The emergence of empiricism and a belief in the universality of science and knowledge provided an intellectual basis for proponents of utilitarian ideals and views of development. Utilitarianism promoted an ideological position that decreed that 'good' was whatever brought the greatest happiness to the greatest number of people.

As a consequence, early development thinking was modelled on the European experience and fostered a belief that Westerners had a duty of obligation to 'uplift' other parts of the world to European standards of living. This belief not only informed philanthropic endeavours but was used as a justification for colonialism. Popular and political discourses of the time spoke of the 'white man's burden', a moral obligation placed on the 'developed' and 'civilized' nations to uplift the 'barbaric' and 'backwards' countries in the global south. Colonialism was frequently justified as a 'civilizing mission', a rhetoric that was deployed as a key rationale for the expansion and maintenance of empires and colonialism. Colonial powers argued that the subjugation of populations to colonial rule was necessary to promote development and maintain security. Such attitudes were inherently paternalistic and often overtly racist and served to distract from other economic and political concerns of empire (see Craggs 2014).

Philanthropic bodies and missionaries frequently mobilized similar rhetoric to explain and justify their own interventions and practices. The prominent role of religious foundations in early development endeavours primarily focused on a 'civilizing mission' through religious conversion. Over time these engagements grew to encompass the provision of health and education facilities tied to religious orders and missionary's efforts to religious conversion. While the exact focus and manifestation of these engagements – both by states and philanthropic/missionary societies – evolved over time, the ideology of Enlightenment and political and economic concerns of colonial powers continued to inform development projects and interventions into the mid-20th century.

Development interventions, both by states and non-state entities, were thus inherently ideological. Political and economic self-interest have been consistent influences over much development funding and practice. The 'development' of the colonies by European powers was never altruistic. Development interventions promoted a skewed and partial process that sought to maintain power relations and to use the colonies to support the growth and power of the empire. Infrastructural development was skewed towards export of raw materials from the colonies, with transport and communication networks developed to this end rather than promoting internal development. These colonial-era development strategies have left serious challenges for post-colonial states that inherited unbalanced economies that were dependent on primary community exports and infrastructural networks ill suited to internal and regional communication and development.

Although development aid had flowed from major powers to colonies and poorer nations since the 19th century to support these infrastructural and, later, social development projects it was not until the 1940s that development funding and policies gained prominence. The Bretton Woods conference in 1944 marked the establishment of the International Monetary Fund, World Bank and International Trade Organization as core institutions for supporting development policies and funding in the post-World War Two period (Moyo 2009; Craggs 2014). In the aftermath of World War Two, the signing of the US-funded European Recovery Program (commonly known as the Marshall Plan) in 1946 is regarded as the first significant large-scale international development aid programme. The success of this programme in Europe meant this approach became a blueprint for the subsequent expansion of aid funding to support economic development and poverty alleviation across other continents (Moyo 2009).

The focus and ideology of the Marshall Plan informed the tenor of subsequent international development policies. The Marshall Plan was inherently neoliberal, focusing on economic development and infrastructural and industrial modernization, emphasizing the need to remove trade barriers and promote free trade and the market. It was also a clearly political venture: the Marshall Plan was not only concerned with Europe's economic recovery, as a key market for US products and exports, but also with preventing the spread of Communist influence in Europe. As Western governments began to place emphasis on their duty to take responsibility for promoting the economic (and, to a much lesser extent, social) development of poorer nations, these emergent policies were also framed by neoliberal ideology as well as the accelerating demise of colonialism.

Emergent international development policies in Europe and the US therefore continued to embody neoliberal attitudes to development. They were also used to promote geopolitical strategic agendas hidden behind Enlightenment period rhetoric of progress and upliftment. These geopolitical concerns are evident in Truman's 1949 speech, which clearly connected American values with Enlightenment thinking and positioned the emergent development agenda as an essential antidote to Communist thinking and expansion while emphasizing the responsibility of the West to help poorer nations develop:

> [We must] embark on a bold new program for making the benefits of our scientific advances and industrial progress available for the improvement and growth of underdeveloped areas. More than half the people of the world are living in conditions approaching misery. Their food is inadequate. They are victims of disease. Their economic life is primitive and stagnant. Their poverty is a handicap and a threat both to them and to more prosperous areas. For the first time in history, humanity possesses the knowledge and skill to relieve the suffering of these people.
>
> (Truman 1949)

The United Nations' (UN) approach to development at the time was solely focused on economic development: their first major report on economic development was strongly Keynesian and was preoccupied with achieving full employment in post-colonial societies. This stance was rapidly adapted to economic realities and embraced issues of taxation, land reform, governance and service delivery while remaining focused on the acceleration of economic growth as *the* driver for development (Jolly 2005: 52).

The values emphasized in Truman's speech and UN Development Reports were reflected in the aid and development policies of Western governments during the 1950s (cf. Black 2007; Moyo 2009). Development thinking and policies in this period were dominated by classical economic theory and efforts to apply the experiences of the West to other contexts (Moyo 2009; Potter et al. 2012). Allied to this focus was a Western-/Eurocentrism that discursively positioned the West as embodying 'modernity' and a standard of development that other parts of the world should both aspire to and be encouraged to reach from their existing positions of un-/underdevelopment. Consequently, Western development policies focused on economic development based on the opening up of markets and stimulation of economic production and consumption, with development measured through economic growth.

Among the influential theories framing development thinking at this time was Rostow's 'take-off model' through which he argued that all countries either had or could progress from a state of underdevelopment through the (pre)conditions for 'take off' (i.e. economic development) before a drive to maturity and period of mass consumption. Implicit in this developmental approach was a belief that Western development processes were the only and best ways to develop and that all countries could follow the same pathway (see Rostow 1956, 1959). Critics of this approach highlight not only the flawed Western-centrism of this approach but also that Western economic development was largely predicated on the exploitation of labour and natural resources in the global south. However, in the 1950s Rostow's thinking – allied to other Keynesian economic approaches – were pivotal in setting development and aid agendas.

At this time Western governments approaches to development were also informed by geopolitical concerns associated with the Cold War and decolonization. Development funding and interventions were therefore used as political tools to foster the spread of capitalism (with the market being positioned as the key driver of economic growth within development policies) and to cultivate and maintain 'friendly' governments as allies in the Cold War. The historical context was conducive to such practices. The new generation of political leaders in the former colonies were eager to drive the development of their newly independent countries through accessing the capital and skills required for industrialization. The easiest way to access such resources was through alliances with the wealthy nations of either the capitalist West or communist East. To this end, political leaders would – at least superficially – adopt the political ideology of their benefactors. As a result, international development policy and practice developed as an inherently political realm (cf. Black 2007: 16).

The emphasis on market-led economic growth was manifest in a focus on top-down approaches to international development in the 1960s – the UN's 'Decade of Development'. Launched in January 1961 by US President Kennedy the UN hoped that a concerted effort to tackle basic development challenges would alleviate poverty within a decade (Jolly 2005: 52). Economic growth was the core focus of the UN's policies at this time, alongside an emphasis on infrastructural development as an enabler of market access and logistical efficiency. As a result, development research and investment were directed towards large-scale infrastructural and highly technical interventions. While these technocratic, top-down approaches continued to dominate Western government's development policies and practices, a number of challenges began to emerge to these practices. First, a broader and more holistic view of development as encompassing social and political factors began to emerge through engagements with human rights and initial efforts towards realizing a more equitable and redistributive society. Around this time a more holistic approach to development studies as a cross-disciplinary field of research concerned with tackling poverty and inequality also began to emerge (Potter et al. 2012).

Challenges to conventional development thinking and approaches based on modernization theory were promoted by more radical scholars working in and on Latin American and Caribbean contexts. These scholars argued that existing development agendas served to maintain the status quo and reinforced the West's privilege by fuelling Western development through the exploitation of former colonies, a process that kept these nations poor and dependent (Potter et al. 2012). As Chilcote (1974: 4) summarizes, these radical scholars argued that 'the economy of certain nations is believed to be conditioned by the relationship to another economy which is dominant and capable of expanding and developing', with the 'dependent' economy remaining underdeveloped as it feeds the progress of the dominant economy. This school of thought, subsequently known as 'dependency theory' (see Box 2.3 for further readings on this topic), gained purchase in radical and critical approaches to development and development research, but remained marginal to development policy and practice at the time.

The dependency theory approach challenged both the fundamental assumptions of key models of modernization theory, such as Rostow's (1959) take-off model, and the overriding concern with economic indicators of development. The emergence of a cross-disciplinary field of development studies and growing influence of dependency theory

Box 2.3 Dependency theory

Strongly rooted in Latin American studies, dependency theory remains a powerful discourse within development studies. The core idea to this approach is that the pace and type of growth and accumulation enjoyed by countries in the global south are determined by external (i.e. Western political and economic) interests and forces. These external interests are thus seen as maintaining these relations and power asymmetries, which distort or hinder growth in the global south, in order to maintain power and support their own growth trajectories:

Chilcote, R. (1974) 'Dependency: a critical synthesis of the literature', *Latin American Perspectives*, 1: 4–29.

Frank, A. (1973) 'The development of underdevelopment', *Monthly Review*, 18: 4–17.

Fernandez, R. and Ocampo, J. (1974) 'The Latin American revolution: a theory of imperialism, not dependence', *Latin American Perspectives*, 1: 30–61.

Santos, T. (1970) 'The structure of dependence', *American Economic Review*, 60: 231–36.

contributed to understandings of development as a more holistic concern than simply economic growth. This more holistic understanding of development in turn required a more diverse and sensitive set of indicators, encompassing not only economic factors but also markers of social development such as health and education outcomes. Among the range of indicators that were generated to meet this demand the Human Development Index has emerged as one of the most widely used, providing a composite statistical measure of development incorporating life expectancy, education and income.

This impetus towards a more holistic understanding of development, coupled with growing criticism of top-down development approaches and the privileging of economic growth as the *raison d'être* of development policies and funding, contributed to a shift in development thinking in the 1970s. At this time, governments and philanthropists began to promote interventions targeted at the poor that aimed to reduce inequalities through social redistribution. This change in emphasis reflected a broader concern with social justice and humanistic approaches to development (Potter et al. 2012). As a result, international development projects were increasingly designed to tackle multiple areas of concern including education, health, sanitation, water, food and various forms of human security. While these interventions were intended to reduce both poverty and inequality, they remained top down in implementation, were overly complicated in terms of logistics and implementation and administratively cumbersome. These challenges, as well as shortcomings in training and capacity building, undermined many development interventions in the 1970s.

The election of conservative leaders in three of the major Western powers – the UK (Thatcher), USA (Reagan) and Germany (Kohl) – in the early 1980s reflected a strong neoliberal, conservative political shift among major international development donors. This political shift had significant ramifications for development policy, practice and research. The move towards social justice and wealth redistribution as a tool for promoting development and overcoming inequality in the 1970s was cast aside. The growing power of neoliberal orthodoxy among key political leaders as well as in the practices of multilateral development institutions including the World Bank and International Monetary Fund led to a refocusing of development priorities. Development spending and policies declined in importance for many donors as conservative policies emphasized the role of the market and sought to reduce state power and welfare spending. As a consequence, bilateral development funding and interventions were scaled back and increasingly tied to strategic (geo)political agendas linked to security concerns and the promotion of free-market, neoliberal ideology.

Similar imperatives were evident in the rhetoric and practices of the major multilateral development agencies. Informed by neoliberal ideology, these organizations advocated the free market as the key driver of development and located this at the heart of their development policies – most notably in the structural adjustment policy (SAP) programme. SAPs became the predominant form of multilateral international development support and funding during the 1980s, requiring those states receiving development funding to commit to a market-oriented approach to development and undertake significant changes to internal structures and spending. SAPs focused on three objectives of reducing inflation, addressing deficits and promoting economic growth (Jolly 2005: 54–55). To this end, they required recipient governments to significantly streamline and scale back their state apparatus, reduce taxation and state expenditure, notably on social service delivery, and embrace privatization and free market capitalism. The outcomes of SAPs were poor. While inflation was generally reduced, economic growth was rarely stimulated and service provision was deeply compromised, contributing to the labelling of the 1980s as 'a lost decade for development' in Latin America and sub-Saharan Africa (Jolly 2005: 55).

Just as significant shifts in the political landscape in the early 1980s resulted in a dramatic change to international development policy and practice among major donors, further dramatic upheavals in the geopolitical context of the 1990s contributed to a reorientation of the international development agenda. The ending of the Cold War in 1991 signalled a major shift in global geopolitics, contributing to a decoupling of international development aid from foreign policy agendas. The privileging of ideological and security concerns during the 1980s contributed to a rather short-termist view of international development and a reactive mode of development aid focused on emergency relief. The 1990s witnessed a move towards longer term, planned development assistance.

A waning of conservative dominance of Western politics and a global economic upturn led to a reinvigoration of political and popular interest in development as a holistic concern. In both popular and political understanding, poverty resulted from a combination of economic, social and political factors and development was not solely driven by economic growth. Instead, the focus of development policy and research moved towards broader concerns with improving living conditions and the expansion in provision of and access to essential services such as health care, education and housing (Sen 1988). Development policy thus shifted towards integrated development solutions in which recipients were treated as partners holding agency, knowledge and power. At the same

Box 2.4 Amartya Sen and *Development as Freedom*

Amartya Sen is viewed as a key public intellectual, contributing to discussions around political economy and social philosophy. At the core of his argument in *Development as Freedom* (Sen 1999) is the contention that it is a lack of entitlements – or freedoms – to particular 'goods' that are at the heart of development challenges. To explain this claim, Sen argues that famines are not caused by an absolute lack of food but are due to the fact that some people (due to poverty, political marginalization, etc.) do not have an entitlement to access food. In other words, economic growth or social modernization can act as a *means* for expanding individual freedoms and the potential for development but the realization of this possibility is influenced by a range of determinants including socio-economic arrangement and political and civil rights. Based on this understanding of development, Sen's approach argues that *unfreedoms* – the lack of freedom and entitlement to meet needs in the realms of political freedoms, economic facilities, social opportunities, guarantees of transparency and protective security – undermine the possibilities of development. Instead, there is a need to realize each of these sets of freedoms in order to strengthen development outcomes.

Gasper, D. (2000) 'Development as freedom: taking economics beyond commodities – the cautious boldness of Amartya Sen', *Journal of International Development*, 12: 989–1001.

Sen, A. (1999) *Development as Freedom*. Oxford: Oxford University Press.

Sen, A. (2013) 'The ends and means of sustainability', *Journal of Human Development and Capabilities*, 14: 6–20

time, the UN began to produce human development reports, informed by Amartya Sen's work (Box 2.4) which conceptualized human development as the strengthening of capabilities and freedoms (Jolly 2005: 55).

This period also saw the emergence of alternative aid modalities to traditional north–south donor–recipient relations. The growing profile of the G77, NEPAD (New Partnership for Africa's Development), the spread of south–south development linkages and the emergence of the BRICS (Brazil, Russia, India, China, South Africa) all contributed to changing power dynamics within and control over international development policy and practice. The increasing power and influence of these non-traditional development donors and investors is dramatically altering the development landscape and has triggered new fields of research focused on these donors and their activities. At the same time, these developments are changing the ways in which development research is done and the response received by development researchers in the 'field', as Emma Mawdsley reflects (Box 2.5).

These organizations embodied a reinvigoration of many of the post-colonial ideals of the 1960s, representing a collective desire to refashion development agendas and policies

Box 2.5 Emma Mawdsley: Researching the 'rising powers' in a rapidly changing world

I have been involved in various forms of research in and around the 'global south' since 1992, particularly in India. A lot of this research was explicitly focused on uneven power relations, voice and agency, including as part of joint projects into 'whose ideas count' within transnational NGO networks and the World Bank. It was also conducted – to the best of my ability, if certainly imperfectly – in ways that reflected feminist and post-colonial sensitivities to the various relational and structural inequalities that I embodied and represented. Here I reflected the wider community of critical geographers, and the intense debates that were continuing and newly opening up about positionality and power. As this debate revealed, and I certainly experienced, this is not to say that I, as a Western researcher, was always in a straightforward position of 'power'. But there was no doubt that by and large my status as a UK/Western academic opened doors and helped persuade people to talk to me: from generous but largely perplexed small farmers in the valleys of the Himalaya, to intrigued senior government figures in powerful Delhi ministries (somewhat to my shame, in this context, the 'Cambridge' label was an advantage that I wasn't above using). Indian friends and colleagues and I would talk about the uncomfortable differences we faced in access, especially to elite and official respondents.

In the last few years my work has turned increasingly to the question of how the (so-called) 'rising powers' are driving new meanings and cultures of international development, opening up new opportunities and challenges for poor people and poor countries and increasingly reshaping the ideas and institutions of global development governance. The backdrop is, of course, a tectonic shift in global power and there is much to celebrate and welcome in an increasingly multilateral world.

It is hard to put my finger on how these changes are altering the research landscape and I do so cautiously. My recent research has been oriented around interviews and discussions with senior government officials and other business and NGO elites. New and very exciting research sites are opening up in the many and various south–south forums, both state and non-state, and across and between different partners. While I have continued to meet with generosity and openness from many, I can feel the ground shifting. My identity as a white, and particularly white *British* citizen, is by and large not enabling of trust, empathy or identification in the context of an increasingly vocal and explicit rejection of the 'traditional donors'. The association (however chagrined I am by it) with arrogant, neo-colonial and now 'submerging' development powers, as well as a degree of suspicion about my intentions, does seem to close down some spaces and dialogues or colour other interactions. I've talked about this with two of my PhD students also working on the new development landscape, Sungmi Kim from Korea and Danilo Marcondes from Brazil, and they agree that their 'Southern'/Asian identity allows them to move

> more easily and openly in researching the 're-emerging' development actors. Of course, there are other continuities and changes and the situation is more complex than this short outline suggests. While I might have felt misunderstood or different or uncomfortably privileged earlier, I am increasingly likely to find myself more uncomfortably excluded or eyed askance now. However plaintive I feel about it, I also wryly appreciate the salutary experience of this occasional and partial feeling of being marginalized: it's about time.

free from Western dominance and to design 'home-grown solutions to home-grown problems'. Broadly characterized as anti- or post-development approaches (see Box 2.6), these engagements were driven, in part, by authors concerned with the centrality of local knowledge and culture as core to representational practices and understandings, of the need to critically engage established and dominant knowledge and orthodoxies (as driven/determined by Western thought) and to promote local movements and approaches to development (Escobar 1992). In particular, these engagements challenged the discursive power of development as a representational practice that produced knowledge and understanding about the 'global south' as un-/underdeveloped and the political, social and cultural power this holds. This understanding was used to challenge orthodoxies of understanding, policy and practice (i.e. to resist the imposition of political, economic and infrastructural development policies/projects from the West) and instead to promote locally developed knowledges, understandings and practices of development (Escobar 1992).

Further challenges to Western dominance of international development agendas emerged from a post-modern turn in development studies. A questioning of the meta-narratives of development thinking and unpacking of these discourses sought to provide scope for multiple, divergent and contradictory understandings to be developed. These developments contributed to a shift towards more contextually appropriate and specific development interventions, rooted in local needs and requirements rather than on donor desires.

The turn of the millennium provided a seminal moment around which many governments and development activists mobilized on issues relating to international development. Popular campaigns such as the Jubilee Campaign utilized this date change as a focal point for activism and popular pressure for political action on international development and to push for a change in attitudes and systemic inequalities. This date change also acted as a focal point for the international community to reflect on the relative successes and failures of international development. This process culminated in a number of declarations and pronouncements that resonated with Truman's 1949 speech as well as the idealism of the UN's Decade of Development campaign in the 1960s. The most prominent expression of these ideals, and the concept which currently dominates popular and political engagements with international development, are the Millennium Development Goals. Comprising eight overarching goals (see Box 2.7) the MDGs are intended to emphasize partnership, cooperation, moral and humanist obligations and a meta-narrative of development for global security.

Box 2.6 Arturo Escobar and post-/anti-development approaches

The post-development approach views development as 'a Eurocentric Western construction in which the economic, social and political parameters of development are set by the West and are imposed on other countries in a neocolonial mission to normalize and develop them in the image of the West' (Potter et al. 2012: 24). Layered on to this critique of development are concerns that the external definition of development means local voices, needs and ideas are excluded or silenced. Therefore, development remains a Western agenda and one that positions the global south as 'abnormal', 'inferior' and 'dysfunctional'. At the heart of this approach is a concern not only with developing local solutions or 'solutions from below' but also with pursuing non-capitalist approaches to development and social justice.

A range of academics and theorists are associated with anti-/post-development approaches, including André Frank and Jan Nederveen Pieterse. Prominent among these scholars is Arturo Escobar, an anthropologist whose work focuses on Latin America, development and political ecology. One of Escobar's key contributions to post-development thinking has been to understand development as a culturally constructed discourse (see his 1995 book *Encountering Development*) and to push for a post-development agenda. His work has advanced post-development thinking by drawing in a range of new concerns to the critique of development and emphasizing the importance of local organizations and social movements in advancing these agendas.

Escobar, A. (1995) *Encountering Development*. Princeton: Princeton University Press.

Rahnema, M. and Bawtree, V. (eds.) (1997) *The Post-Development Reader*. London: Zed Books.

The success of the MDGs has been extensively questioned and there is growing recognition that very few of the MDG targets will be met by the target date of 2015. However, advocates have noted the success of the MDGs in providing an accessible and understandable frame through which to quantifiably measure development and to talk about international development in popular discourse. Sceptics have questioned the ways in which these goals have skewed development interventions in particular ways and have introduced some relatively arbitrary goals without thinking through the implications of these. Many critics also point to the failure of development policies to make progress towards meeting many of the MDGs.

Depending on your research focus and questions, you may find the MDGs useful in providing background information and data to your work or you may wish to develop questions that relate to the MDGs. These questions may relate to specific goals and

Box 2.7 Millennium Development Goals

The MDGs continue to dominate donor approaches to development funding in the second decade of the 21st century. They also exert a powerful influence on development research agendas through the availability and prioritizing of research funding. The UN's website provides some useful background to the MDGs, the specific goals (listed below) and progress towards these (see http://www.un.org/millenniumgoals/).

The eight MDGs are:

MDG1   Eradicate extreme poverty and hunger
MDG2   Achieve universal primary education
MDG3   Promote gender equality and empower women
MDG4   Reduce child mortality
MDG5   Improve maternal health
MDG6   Combat HIV/AIDS, malaria and other diseases
MDG7   Ensure environmental sustainability
MDG8   Develop a global partnership for development

outcomes, perhaps through a case study of the outcomes of a particular intervention or through analysis of survey data relating to one of the goals. Some of Dan and Chasca's postgraduate students have, in recent years, drawn on the MDGs to develop research projects. These students have engaged with a range of topics, including work to understand the outcomes of efforts to improve maternal health outcomes in Tanzania (MDG5) and critically evaluating the success of advocacy and policy interventions to promote gender equality and empowerment through land rights (MDG3).

The ideals underpinning the MDGs were reiterated in the 2005 Paris Declaration on Aid Effectiveness. This declaration further stressed the importance of recipients owning and leading development policies, a need to understand the heterogeneity of recipient needs (both between and within states), and a focus on improving accountability and governmentality. These values appear likely to be at the core of the post-2015 development agenda. As we move towards a post-MDG development framework, the UN, national governments, multilateral institutions and civil society organizations are participating in conversations and discussions to identify how to frame the next development agenda. Just as previous changes in broader political frames and concerns have been vital in informing development policies, practice and research, so too will this new agenda.

What to take from this chapter

This chapter has provided a brief engagement with the complex and contested frame within which development research is conducted. The brief history of development should indicate to you how the idea of 'international development' is a discursively powerful but contested and value-laden idea. Discourses of development, often linked to historically contingent notions of 'modernity' and 'civilization', have created powerful narratives and representations of people and places. Over time these discourses have worked to locate *who* needs to be developed, *how* they need to develop, and *who can help* secure development, rendering certain ways of doing and thinking about development (in)valid.

You should also recognize that approaches to development, and the indicators used to measure development needs and poverty, have changed over time, framed by shifting economic and political imperatives. These changes, in turn, influence how development research is conceived, conducted and communicated – often in multiple and contradictory ways depending on stakeholder's interests and priorities. Thus, it is important to be conscious of the often unintended outcomes of development projects and research, outcomes that may be controversial and have lasting negative impacts. Thinking back to the differing understandings of development and associated priorities, it is easy to see how a development project that stimulates economic growth in one region, through, say, intensification of irrigated farming of export crops or development of a hydroelectric dam, may realize certain developmental benefits for some groups according to some development indicators but may have tremendous negative impacts on other groups or regions. As Black (2007: 10) notes, 'the numbers of poor people in whose name development is justified are greater than they were when it was invented [taking Truman's 1949 speech as the invention of this term], and in many cases their poverty stems directly from the havoc it has wreaked on their lives.' In order to understand how we come to this situation, it is useful to reflect on the historical narrative that has brought us here.

What we hope you take from this chapter is recognition that your research is not conducted in isolation. Your research – and thinking on development in general – is informed by a longer and contested history of development and development research, encompassing both the shifting intellectual agenda of the West but also the legacies of (neo)colonial experiences in the global south.

If you want to explore the ideas covered in this chapter in more detail you may find these further readings useful:

Desai, V. and Potter, R. (2014) *The Companion to Development Studies* (3rd ed.). Abingdon: Routledge.

Escobar, A. (1995) *Encountering Development: The Making and Unmaking of the Third World*. Princeton: Princeton University Press.

Potter, R., Binns, T., Elliott, J. and Smith, D. (2008) *Geographies of Development: An Introduction to Development Studies*. Harlow: Pearson.

Williams, G., Meth, P. and Willis, K. (2014) *Geographies of Developing Areas: The Global South in a Changing World*. Abingdon: Routledge.

3

POWER, IDENTITY AND THE DYNAMICS OF RESEARCH

Having recognized the contested terrain of development fieldwork, this chapter explores the politics of development research. Beginning with a discussion of power dynamics in development research attention is paid to the importance of reflexivity and positionality in research, including the imperative to understand and negotiate the expectations and perceptions pertaining to sources of funding, political contexts and ascribed identities in the field. These discussions are grounded in concerns around the design, conduct and representation of research, including arguments surrounding community participation in the research design and presentation processes, as well as power relations surrounding gender, sexuality and language.

What is power and why does it matter?

Power is a critical dimension of all research (Brydon 2006). It pervades all that we do and every interaction that we have in the field. Power is the ability or capacity to do something or act in a particular way and the capacity or ability to direct or influence the behaviour of others or the course of events (Gregory et al. 2009). Power is inherently complex and often difficult to disentangle, it is always relational and it can be enabling as well as obstructive. Power is both personal as well as institutional and it is political, ideological and economic too.

Power in development has a complex history. Crush's (1995) seminal text provides a useful overview of the power of development, including a comprehensive review of how power has permeated development interactions from the days of the Enlightenment, through the post-Truman era up to the present day. He draws on Escobar's (1995) critiques of the ways in which development discourses have been constructed to legitimize the voices of western experts while marginalizing those of local people. This accords with Said's (1978) notion of 'orientalism' and 'othering' and England's (1994) questioning of how we can incorporate voices of 'others' in research without colonizing them in a way that reinforces domination. Chapters 1 and 2 outline this complex and unconformable history of academic tradition (Pieterse 2010; Halperin 2013) rooted in exploration (Power 2003) and colonial administration (Hodge 2007). Throughout the book, debates around power cross-cut all our discussions about research design and methodology. For example, in Chapter 9, you will see how even present-day approaches such as participation need to be critically reflected on to ensure they do not reinforce hidden or invisible power structures.

Later in this book we engage more deeply with debates about whether we should be doing research in the global south (Scheyvens 2014) and reflect on arguments about 'researching the distant other', a term first coined by Corbridge (1993). Throughout these discussions our premise is that these are important debates, but ones that should make our research better, more transparent and more relevant to the world, rather than cutting ourselves off from real-world complexities.

One of the starting points for conducting transparent and reflexive research is to think carefully about power. Gaventa (2006) provides a useful description of how power can be visible, hidden and invisible. Visible power is observable through formal rules and structures. It can also be hidden in terms of who is able to set the agenda and who is included or excluded from decision-making processes. Power can also be invisible, shaping how individuals think about the world: Box 3.1 gives further details.

Another useful framework for thinking about power is the framework from Eyben et al. (2006) of consenting to power, contesting power, expanding power and transforming power. They draw on a range of work to expand their discussion, as summarized in Box 3.2.

These conceptualizations of power resonate from the international arena down to an individual's personal politics. When doing research, who you are, how you behave in the field and how you interpret this, all affect your research and its outcomes. Power pervades project design (were your research participants involved in setting the agenda?), as well as how you conduct your research and interpret your results (what hidden and invisible power relations have influenced your data and its interpretation?; were your research participants involved in the analysis process?). As Brydon (2006: 27) reminds us, research contexts are:

> [R]iddled and cross-cut by relationships of power, from those between the sponsors of the research and the researchers, and between the researcher and the researched, to power relationships within the culture of the research setting, relationships between classes and clans, landholders and landless, educated and illiterate, elders and juniors, women and man, rich and poor.

Research is never value neutral or unbiased.

So, during fieldwork you are likely to embody, in certain ways, a position of power and privilege, but also a position of powerlessness in relation to others. Furthermore, when thinking about power, you must not think of research participants as powerless but as people (and institutions) that have agency and power (in various ways – hidden, visible) over your research (see Figure 3.1). This relates back to debates around empowerment and the tensions between giving 'power to' and having 'power over' – both real and perceived – others. These are manifest in economic, political and ideological ways and hierarchies and can be complex to untangle (McEwan and Bek 2006; Hammett 2008).

In your research, you need to acknowledge the realities of power that you may come across in your fieldwork and help you ask the right questions to identify what sort of

Box 3.1 Forms of power 1

(Gaventa 2006: 29, adapted from VeneKlasen and Miller 2002)

Visible power: observable decision making

This level includes the visible and definable aspects of political power – the formal rules, structures, authorities, institutions and procedures of decision making . . . Strategies that target this level are usually trying to change the 'who, how and what' of policy making so that the policy process is more democratic and accountable, and serves the needs and rights of people and the survival of the planet.

Hidden power: setting the political agenda

Certain powerful people and institutions maintain their influence by controlling who gets to the decision-making table and what gets on the agenda. These dynamics operate on many levels to exclude and devalue the concerns and representations of other less powerful groups . . . Empowering advocacy strategies that focus on strengthening organizations and movements of the poor can build the collective power of numbers and new leadership to influence the ways the political agenda is shaped and increase visibility and legitimacy of their issues, voice and demands.

Invisible power: shaping meaning and what is acceptable

Probably the most insidious of the three dimensions of power, invisible power shapes the psychological and ideological boundaries of participation. Significant problems and issues are not only kept from the decision-making table, but also from the minds and consciousness of the different players involved, even those directly affected by the problem. By influencing how individuals think about their place in the world, this level of power shapes people's beliefs, sense of self and acceptance of the status quo – even their own superiority and inferiority. Processes of socialization, culture and ideology perpetuate exclusion and inequality by defining what is normal, acceptable and safe. Challenging this is complex but can help transform the way people perceive themselves and those around them and how they envisage future possibilities and alternatives.

power dynamics you might expect to find in the field. Mosse (1994: 499) believes that information does not just exist 'out there' waiting to be 'collected' or 'gathered', but is constructed, or created, in specific social contexts for particular purposes. If we can understand this, our role as effective researchers will be more easily achieved.

Box 3.2 Forms of power 2

(after Eyben et al. 2006)

Consenting to power

Do people consent to power because they are aware of their situation and feel they have no choice? Or are they socialized not to challenge power? This links to Bourdieu's use of the term 'misrecognition' to describe the process of mystification by which the powerful use their symbolic capital to prevent individuals from recognizing that their subordination is culturally constructed rather than 'natural'. However, people may choose to consent to power, rather than contest it, for rational reasons, so the complexity of disentanglement remains.

Contesting power

Understanding power as constructed through patterns for social relations that are reproduced through processes of socialization provides an agenda for contesting power that concerns changing the way people relate to each other – including changing the meaning they give to these relationships.

Expanding power

There is no finite amount of power and positive change can be achieved through identifying win–win situations. Power can be expanded by making alliances (with the powerful as well as others who have less power) and striving for cooperation rather than competition. Strategies for expanding power among those who are powerless might involve building horizontal alliances that link up different entry spatial points for change.

Transforming power

Many people in a subordinate position may question the way the world is ordered but do not organize themselves for strategic resistance because of the fear of the consequences should they fail. Those with power over people can use it as an empowering opportunity for both themselves and those seeking help. To be transforming, power with certain people or groups depends on their releasing their power within through a processes of self-interrogation or conscious raising. The difference between empowerment, and power over people, is highly subjective and shaped by the beliefs and values of all parties.

Figure 3.1 Throughout the research process, you will find yourself experiencing multiple power dynamics and relationships, bundled up in histories of colonialism, racism and development. Learning to navigate these is an integral part of being a development researcher

Reflexivity and positionality

Reflecting on the condition through which research is produced, disseminated and received is termed reflexivity. Positionality is the researcher's personal, social and cultural position and how these affect the entire research processes (Gregory et al. 2009). They are closely related. Reflexivity is the process through which we reflect on our positionality.

Taking positionality seriously means reflecting on your race, gender, nationality, age, economic status and sexuality and recognizing that it may influence the data gathered and the whole research process. You positionality will always bias your study in some way and therefore you need to acknowledge that it will and be transparent about the process (Marshall and Rossman 2006).

However, this area is fraught with complexities. For example, you may need to work out whether your gender affects who you can speak with during your research. Is it appropriate for a young man to interview older married women about personal issues in your research context? How you answer this question is complicated. Should you conduct the interview and acknowledge the potential limitations to the discussions you have or do you revert to suggesting that only women can interview women, only young the young and so on. This latter argument is deeply problematic, suggesting that only like can interview like. Nevertheless, you still need to be sensitive to realities of who you are working with and how who you are affects these interactions. These considerations may also influence decisions about whether or not you work with a research assistant and, if so, the ways in which their positionality will influence (positively and negatively) your research (Molony and Hammett 2007).

You may also need to make some difficult decisions. In some countries and societies, homosexuality is not culturally accepted and gay and lesbian researchers will need to consider how they present themselves within these contexts. Where legal issues come in (e.g. in Uganda), there may be implications for your own safety too. There is no prescribed answer to these dilemmas. You should seek advice and think through carefully the implications of the decisions you make.

There are other markers of identity that are inherently inscribed: gender, race. These identities, and the contextually specific ascribed meanings associated with these, will position you in multiple and often contested ways within the field. At times these positionings can hinder access and rapport in your research, at others they can be beneficial. The reflection in Box 3.3 by Marc Fletcher provides an account of how he examined his own positionality and the impact it had on his research in Southern Africa. As Marc notes, the outcomes can often be unanticipated – not least when your presence disrupts expected social relations and presences/absences in particular spaces.

Just as Marshall and Rossman (2006: 5) ask us to scrutinize the 'complex interplay of our own personal biography, power and status, interactions with participants and (the) written word', you need to do the same for any research assistants, translators, partners or family members with whom you are working. If you are working with a translator, you will need to acknowledge their positionality too. When Chasca was conducting her doctoral research in Botswana, she found an excellent translator who spoke five of the six languages needed within her research area. However, Chasca was concerned that a political link through her translator's older brother might cause unease within the communities with whom they were working. Chasca addressed this by ensuring she could discuss positionality issues with participants when the translator was not there. It was clear very quickly that participants did not align the translator with her brother (perhaps as she was young, female, a single mother and clearly less powerful) and as such she was not ascribed with his identity.

Another case is that of bringing family members on fieldwork with you. Scheyvens and Nowak (2003) provide a detailed account of taking partners and children on fieldwork. The extract below from Twyman et al. (1999: 316) reflects on how a postdoctoral researcher (Jean Morrison) found her positionality had changed when she returned to the field first pregnant and then with her daughter:

> During the second and third fieldwork periods, those respondents who remembered me from the previous year were more 'open' with me, that is, they were more willing to disclose information about their personal situations and relationships, and in return were more curious about my own circumstances. Often they would retell the story of when I had first arrived the previous year, where I had pitched my tent, how I had given someone a red hat, how I had

Box 3.3 Marc Fletcher: Race, power and the beautiful game

Before embarking on my research examining racial divisions in South African football fandom, I was aware that, despite the nuanced history of domestic football in South Africa, it is often constructed as a 'black game'. In preparing for my fieldwork, I read numerous sports ethnographies, specifically looking at the reflexivity of the researcher. Despite this preparation, I was unaware of the extent to which, as a white, middle-class Englishman, I would impact with my two Johannesburg-based case studies (a predominantly black, working-class supporters' club of a domestic football team and a largely white, middle-class supporters' club of an English club side).

With the domestic supporters, it soon transpired that I was the first white man to travel with them. Supporters at various matches, whom I had never met before, would grab my arm and ask to have a photo taken with me. My whiteness became increasingly significant as I became aware that the supporters' club was sometimes being referred to as the 'branch with the white guy'. I did not have the same impact with the supporters' club of the English team. Being in a mostly white group combined with my knowledge of the team as a childhood fan of the team meant that it was much easier to integrate with the supporters.

Through explaining my research to both case studies, supporters in both cases became increasingly aware of my work with the 'other'. After a few months, some of the domestic supporters asked me if I would bring other white football fans along. I reacted to this request uneasily as I was becoming increasingly aware of the impact that my presence was having in this group and did not want to impact the group any more than I had to. At the same time, a small number of the white football supporters began asking questions about my experiences at the local game as most had rarely been to domestic football (perceptions of local football as dangerous, disorganized and a space in which white people did not belong often emerged) and enquired into if I could take them to some games. I regularly had to make snap decisions about my positionality during the course of a single conversation, usually managing to get away with non-committal remarks. After these instances, I would further question my positionality in my research journal.

My research, especially some of the interview questions, was actively challenging the perceptions of some of the respondents to other groups. Although I initially resisted creating links between the two cases studies that would otherwise unlikely have been there, it also became increasingly difficult to deflect such requests. Eventually, after further requests from both cases, I decided to invite the white fans to a domestic game in May 2009; a small number accepted. This created an exciting opportunity to explore how participants' views of the other were in the process of changing. However, returning to the research sites in 2010 revealed that little had changed.

taken a sick child to the clinic and so forth, re-establishing and reaffirming relationships between us . . . Returning with a child, I was perceived to have attained adult status. Women in particular were more likely to question me, particularly about my pregnancy, and also about my daughter. In Botswana, children are viewed as a necessary component of a satisfactory life, therefore for someone to reach their late twenties and be childless is seen as unusual. The overwhelming majority of women I encountered during my third field trip were eager to pick up my baby and play with her, and she became a focus for discussions on issues of gender, fertility and childcare. The transient relationships that developed between myself and female respondents were characterized by a persistent tension between feelings of commonality regarding our gender identities (as mothers), and of difference regarding our cultural identities.

Who we are, clearly influences the whole research process and navigating the complexities of reflexive research and your positionality, without naval gazing, is challenging.

Ascribed, hidden and desired identities

The research stories and reflections above highlight issues about race, gender and nationality, but also how the researchers presented themselves to their participants and the different identities that they took on in their different roles, be it football supporter, researcher, or mother. These identities and how we understand them are an important part of deciphering power relations, and helping us reflect on our positionality. When conducting research we have multiple identities, whether deliberate or not, and understanding these can help you untangle further the complex power relations at play, allowing you to reflect more deeply on your research.

Identities of the researcher and the participants are multiple, shifting and sometimes contested. They may be:

- *Ascribed* – identities that are given to you, which you may or may not be comfortable with, for example the 'white guy' in Marc's reflection above.
- *Desired* – identities that you are trying to project, perhaps an objective, humble researcher in one context, or a professional knowledgeable efficient research in another context.
- *Hidden* – identities that you may not be aware of, or conversely ones that you may be trying to conceal.

These identities also help us to engage and disrupt the insider/outsider dichotomy. It is often assumed that the researcher is an 'outsider', whereas if they have a research assistant or translator they are 'insiders'. Researchers investigating their own cultures are similarly seen as 'insiders' and thus privileged in the research process (Corbin Dwyer and Buckle 2009). These assumptions are misleading and this outsider/insider duality is not clear cut; rather, such positions allow a 'betweenness' that implies that no one is an 'outsider' or 'insider' in any absolute sense (Twyman et al. 1999; Corbin Dwyer and

Buckle 2009*)*. When Chasca was conducting her doctoral research, she felt she was an outsider as she had a history that was unfamiliar to people. Her translator, however, had a background that people could relate to and understand more easily, but she was clearly very different from the people with whom they were working: she was not an insider. Therefore, when considering positionality, in this case, Chasca had to consider both their positionalities. Similarly, Chasca and Dan have found that working with students who go back to their home country to conduct fieldwork, even if this is in their 'home' area, is complex and not clear cut. Issues of class, ethnicity, gender and education often come into play, such that these negotiations of ascribed, desired and hidden identities can be just as complex if not more so, than a more obvious and different outsider (see Mandiyanike 2009).

All this reflection does matter. As Marshall and Rossman (2006: 86) warn, 'it is easy for an inadequately self-reflexive research to be appropriated by, and be complicit in, the process by which one group may marginalize another.'

Language

When you conduct your fieldwork you may be working in English or you may be working in another language, either through a translator, or because you speak another language. Whether you are multilingual or not, language plays an important role in how your data are produced (Crane et al. 2009). According to Sturge (1997), translation is a practice of intercultural communication, a task in which we 'understand other cultures as far as possible in their own terms but in our language'. Sturge (1997) further argues that transposing these terms into Western categories of understanding is a process that should subject to the same self-reflexive approaches as interpretations of other texts, however they are produced.

Transferring concepts and meaning through language and translation can be problematic but it can give rise to opportunities for 'in-between forms of understanding' (Smith 1996: 165). This can enrich our understanding of concepts and help us dig deeper in our enquiry. This makes the earlier discussion of research experience and the notion of positionality and reflexivity of both researcher and translator directly relevant. In such cases, the researcher is further 'displaced' and 'decentred' (and even distanced) from the source language, raising problems of 'authority' in translation, but perhaps allowing the potential, through the agency of a translator, for a more obvious view of the 'hybrid spaces' (Twyman et al. 1999).

According to Lefevere and Bassnett (1990: 1), 'translation as an activity is always doubly contextualized, since the text has a place in two cultures.' It is clear therefore that the value-laden processes of translation and transcription need to be examined fully in the contexts in which they were constructed: an integral part of these contexts is the autobiography of researcher and translator and the implications of power relations on the production of texts (Twyman et al. 1999). Twyman et al. (1999) and Crane et al. (2009) provide detailed accounts of how decisions about translation, and the texts that are produced, can be navigated and Watson (2004) provides some advice on learning vernacular languages in the field.

Managing expectations

When conducting research it is important to think about how to manage your own expectations, as well as those participating in your research, be they interviewee, key informant or host organization. You should always be realistic about what can be achieved and what cannot and this relates centrally to issues of power. It is usually better to be humble and to exceed expectations than to raise hopes and fail to meet them. You should also be aware that expectations may be heaped on you, unfoundedly, and you should do you best to address this sensitively.

When working with participants it is important to create a good impression. This is indispensable to the successful completion of the research. You should respect hierarchies of power, but not be restricted by them. In Botswana, official permits are required for research, which can be time consuming to obtain, but in South Africa, an open policy on research means these are not needed. Elsewhere, letters of permission to conduct research may need to be sought in person from federal government, state government, district government, regional chiefs and village chiefs, in the correct order. Having permission to conduct your research, and knowing what permission you need, makes a good impression. Power relations are also inherent in the spaces of your research and on a practical level you need to think about how and where you conduct your interviews. How your interviewees feel in relation to power hierarchies will influence their openness in interviews and place is important to consider (Elwood and Martin 2000).

You should always be professional but not arrogant. Dress appropriately for the context, respecting gender codes of dress as much as you are able and if you do not, be very aware of what this means to how you will be perceived. Poor people take pride in their appearance and will expect you to, even if you are living in basic conditions. Be polite and behave well; be prepared to listen and learn. Good impressions require flexibility and knowing when small talk and greetings are essential or sometimes necessarily brief with busy officials. When working in Nigeria, Chasca and her co-researchers would live in the villages they were working in and would attend church services on Sundays as this was an integral part of community life. This also gave them a chance to introduce their reason for being in the community. These church services were three to four hours long, but were essential for building up respect within the community.

Keeping promises is also an important part of managing expectations among participants. Chasca and another team of colleagues conducted research in an area of South West Botswana investigating livelihoods and policy changes in the region. They promised to return to feed back the analyses of the research and duly returned to present the findings for discussion with the community several months later. Some sheep farmers had requested that when they returned they would like to learn a bit more about farming from the researchers' home country, the UK. Chasca and the team collated some posters showing the local landscape in the Peak District and the key breeds of British sheep. The research participants found this exchange fascinating and quizzed the researchers in depth about grazing habits, land rights, herding customs, ownership, stocking rates, rainfall and so on. The researchers' own in-depth questions were turned back on them with the ensuing discussion revealing valuable triangulation about their initial research findings.

When working with hosts, it is important to make sure expectations from both sides are fully discussed prior to full commitment being made to the collaboration. As well as research topics and logistics, a key discussion point should be expectations from the partnership and the research. Between you, you need to specify who retains ownership of the data, who has access to the data and in what format, and what the data may be used for and when. You need to think through anonymity of the data relating to sensitive information and how this may be used by others. These issues are discussed in depth in Chapters 5 and 6 but warrant inclusion here in terms of managing expectations when working with host organizations.

It is also important to manage your own expectations. Scheyvens and Nowak (2003) discuss how to handle your own anxieties of the unknown (and known) and warn that you may be lonely or feel isolated at times. The majority of people conducting fieldwork on development-related topics will come into contact with people who are much poorer than themselves and who have difficulty in sustaining even the most basic livelihoods. Scheyvens et al. (2003) report that little is written of the challenges or the shock of these situations and we how prepare ourselves is important. Some of these issues are discussed further in Chapter 4.

You may also have to deal with the challenges of fieldwork with research colleagues. Chasca worked with a colleague in Nigeria who was an experienced researcher. One evening, a friendly comment from their host about how we burnt candles every night to do our evening work, and how that must be so expensive for us, suddenly got to my colleague. He found this simple, everyday comment about an inch of candle brought up the internal conflicts of inequality that he had been supressing during the fieldwork. He found the poverty around us deeply upsetting and struggled for a few days to come to terms with, and reconcile, this inequality.

In many situations, you will be in, and from, positions of extreme relative privilege and you will need to reflect on inequalities of economic power and wealth and how this affects your positionality and ascribed identities. This warrants reflection on the appropriateness of attire and behaviour but also includes 'not flashing the cash' in ostentatious ways that may offend, while being aware that being miserly could have the same outcomes.

What to take from this chapter

Power, as we have seen, is pervasive throughout the research process. It affects every part of the research planning process, every interaction in the field and the analysis and presentation of our findings. Our writing is authored by our positionality as a raced, gendered, classed and politically oriented individual and we are trying to represent our participants, the 'Other' (Said 1978) in our work. We can only do this by being vigilant about ethics, power, positionality and the politics of our work and by reflecting carefully about who we are and what influence that has on how we conduct research.

If you want to explore the ideas covered in this chapter in more detail you may find these further readings useful:

Crane, L., Lombard, M. and Tenz, E. (2009) 'More than just translation: challenges and opportunities in translingual research', *Social Geography*, 4: 39–46.

Eyben, R., Harris, C. and Pettit, J. (2006) 'Introduction: exploring power for change', *Institute for Development Studies Bulletin*, 37: 1–10.

Mandiyanike, D. (2009) 'The dilemmas of conducting research back in your own country as a returning student – reflections of research fieldwork in Zimbabwe', *Area*, 41: 64–71.

Moser, S. (2009) 'Personality: a new positionality?', *Area*, 40: 383–92.

Scheyvens, R. (2014) *Development Fieldwork: A Practical Guide* (2nd ed.). London: Sage.

4

THE LONE WOLF AND THE PACK: ENTERING THE FIELD ALONE AND IN GROUPS

Although research and fieldwork are common components to development-related under- and postgraduate courses, there is no single pathway into fieldwork or one ubiquitous experience to be gained. Despite the complexity and heterogeneity of fieldwork experiences, many postgraduate students are provided with generic research training courses often implicitly focused on doing research within the same country as the university they are studying at. In recent years, area studies and international development courses have increasingly provided more tailored methods training, engaging with the additional complex challenges faced in conducting research overseas. It is these concerns that this chapter seeks to explore. In so doing, we recognize that fieldwork – and the preparation for it – can be exciting, inspiring, frightening, messy and complex. This chapter should provide some guidance and assistance in negotiating these processes, both for those setting out to conduct research on their own and for those participating in fieldclass-based fieldwork.

To set the scene a little, consider the following narratives, each of which briefly details some of the considerations in approaching fieldwork:

Dan, reflecting on the morning of the departure day for a group fieldclass to rural Kenya: I am up early, working through my checklist even though I know everything is organized – visas secured, travel insurance in place, accommodation details and emergency contact numbers for Kenya and the UK, information packs with itinerary plus contact details, copies of passports, visas and insurance documents etc. printed and left with designated contacts in case of emergency, e-tickets printed for flights, arrangements in place for airport transfers, currency and credit card packed . . . and the list goes on. I check and recheck the list, drinking too much coffee and starting to worry about things beyond my control – will all of the students remember their passports, have those needing letters for the airline regarding nut allergies secured these (and has the travel agent informed the airlines of these allergies in advance), how many will fail to wear sun cream and a sunhat and suffer from sunstroke and heat exhaustion – have I got enough rehydration sachets for this eventuality and so on and so on. I am normally a calm traveller: having spent time in parts of Asia, Africa, Latin America and North America, I barely pause to think while packing and organizing my own research trips, but this time I am in charge of a fieldclass –

I am (alongside two other staff) responsible for 30 students, and cannot relax, instead thinking of ever more outrageous scenarios that may be encountered and how to deal with them.

Dan, reflecting on the morning of a day of fieldwork in South Africa in 2010: I sit outside my bedsit, drinking coffee in the sunshine and reading over the week's football paper. No stress, I tell myself, I need to be in Pretoria at 10am and there is no point in leaving from my accommodation in Melville, Johannesburg until about 9am – the journey should take about an hour, but trying to set off before 9am will simply get me stuck in the mania of rush-hour, so rather wait and have a more relaxed drive. I have the directions written out and the map pages marked, and having been before am not expecting to get lost. The car is fuelled, I have the notes and questions I need for the interview packed, as well as my laptop, camera and other bits and pieces needed for the archival work for the day. Now, who is it that the Kaiser Chiefs are rumoured to be signing this week?

These vignettes are drawn from Dan's research experiences in recent years. In each of them, you can note an awareness of various logistical considerations as well as more personal and emotional dimensions. Drawing from such experiences, we use this chapter to address various aspects of the process of setting up and organizing fieldwork. In so doing, we hope to impress on you the importance of being well prepared for fieldwork but also the need to be flexible and responsive to changes and challenges during your research.

This chapter should also remind you that 'The myth of the chameleon fieldworker, perfectly tuned to his exotic surroundings, a walking miracle of empathy, tact, patience and cosmopolitanism' that Geertz (1983: 56) identifies is precisely that – a myth. In reality, research is a complex, messy, exciting, challenging, nerve-wracking and exhausting experience during which you negotiate issues of positionality, ideology, safety, access, bureaucracy and logistics on a continual basis. We do not pretend that we are 'chameleon fieldworkers', neither do we suggest that our guidance here will turn you into these mythical beasts. Rather, we hope that our guidance and suggestions will help you to be well prepared for fieldwork, which, in turn, helps improve your fieldwork skills and research findings.

Approaching fieldwork

Fieldwork experiences are often recalled as some of the most inspiring and rewarding parts of development-related courses. For some students, their memories of fieldwork are drawn from group fieldclass visits to overseas destinations, while for others these experiences are the result of undertaking individual fieldwork for a dissertation or thesis. While fieldclasses provide many students with experience of group fieldwork, most students first encounter lone fieldwork through their dissertation research (Punch 1998: 158). Whether conducted alone or as part of an organized fieldclass, the experience of fieldwork can be exhilarating and fascinating, but it can also be frustrating, tedious, challenging and potentially dangerous. Good planning and preparation can enhance the positive experience of fieldwork while mitigating the potential for negative experiences. The planning decisions you make are integral to your entry into the research process: practical, logistical, ideological and emotional considerations inform how you actually carry out your research.

Definitions of fieldwork encompass learning activities conducted outside the classroom lasting for anything from an hour or two through day-long site visits to multiday residential fieldtrips through to longer term research placements or projects spanning many months or years (see Fuller et al. 2006). The focus here is primarily on longer duration fieldtrips and research projects, although the guidance presented will have relevance to shorter forms of fieldwork. Levels of supervision also vary tremendously between fieldwork experiences, in some cases, students will work under close staff supervision while in others they will work with greater independence or (in the case of dissertation research) be in the field alone, relying on remote supervision and contact with their supervisor in another country.

Fieldclasses often involve a range of teaching and learning styles, each requiring a different level of supervision and interaction. For instance, during a fieldclass to South Africa in 2013, Dan's students participated in dependent observation (a 'Cook's tour' of Bloemfontein led by a local academic and the course teaching team, in which staff acted as experts and interpreters to introduce and explain the local context and issues to students and respond to student questions), autonomous observation (self-led tours of the Apartheid Museum in Johannesburg, with students set a number of questions to engage with and reflect on as they moved through the museum at their own pace and engaging with the exhibition independently) and autonomous participation (students defined and undertook interactive data gathering using various qualitative methods in completing small research projects with a host non-profit organization in Bloemfontein with minimal oversight from course staff) (cf. Kent et al. 1997; Panelli and Welch 2005; Herrick 2010). Given the range of possible styles and types of fieldwork, we cannot provide a universal checklist of things to consider in preparing to conduct field research. However, we hope the discussions in this chapter provide useful guidance for those involved in both group fieldclass work and individual field research (see Figure 4.1).

Fieldclasses and group fieldwork

Within many development-related disciplines, fieldtrips are viewed as important pedagogical tools, providing experiences through which students gain a set of research skills – from observational to interactional – as well as critical thinking and logistical decision-making skills (Herrick 2010). While they do provide a tremendous teaching and learning opportunity (for discussions around this, see Kent et al. 1997; Fuller et al. 2006), they also present a range of challenges and dilemmas – ethical, practical and conceptual – relating both to relations and interactions with host communities as well as internal group dynamics. Behind the scenes, the logistics and organization of a fieldtrip including balancing cost, transport options, time, range and depth of experiences and engagements, quality of accommodation and value for money with the pedagogical outcomes and academic aims – are significant (see Figure 4.2).

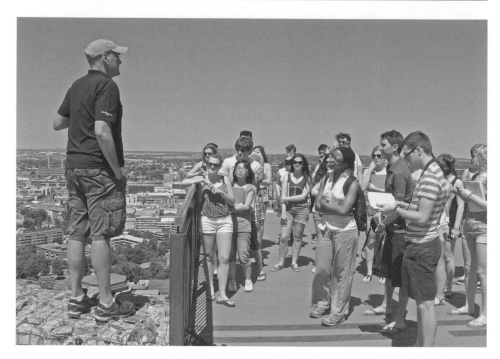

Figure 4.1 Learning and teaching experiences during fieldclasses can take many forms. In this situation, a local expert is providing a group of fieldclass students with an overview of local history and implications for contemporary development challenges

Figure 4.2 Fieldclasses can be among the most stimulating and rewarding learning and teaching experiences. With large groups, such as the group in this photo, students often build friendships and memories that outlast their degree programmes

Box 4.1 Decision making as a student on a fieldtrip

For each of the scenarios below reflect on you initial response to the challenge outlined. Once you have made an initial decision, think about what you based this on and the rationale for this choice. Then look through the 'factors to consider' listed after each challenge – are these things you considered? How do you respond to these additional considerations and do they change your decision?

Decision 1: You are going on an undergraduate development geography fieldtrip to a study centre in a rural village in Kenya. This destination has been used for this trip (and staffed by the same faculty) for several years. The field centre is also used for a postgraduate fieldtrip from the same university, as well as by several other UK, European and Kenyan universities. As a student on the course, what do you think the benefits to you might be? What might some of the problems be?

Factors to consider: Might the community be suffering from researcher fatigue?

What are the implications of multiple trips running to the same place, asking similar questions and using up the community's time?

Decision 2: In preparing for this fieldtrip, you have been asked to think about what you want to do with regards compensating the local community and research participants for participating in your research. How do you go about making this decision? What do you decide to do?

Factors to consider: What has been previous practice, both by your trip and others? What are the views of other students on their preferred practice? Is it possible to ascertain what the preference of the local community is and, if so, what is it? Are there ways in which you can ensure the fieldtrip's research is made accessible, relevant and useful to the local communities/hosts? At what scale do you decide to provide 'compensation' – to individuals, to a community development fund? What 'is' the compensation – do you give money, sugar, soap, gifts? How do you decide on the appropriate amount and does this stay the same for everyone you work with?

Decision 3: You are going on an undergraduate fieldtrip to South Africa. The focus of the fieldtrip is on understanding the dynamics behind urban poverty and inequality, with the expectation that you will gain greater sensitivity to the drivers of and outcomes of poverty. You are aware that a number of your fellow students want to stay on after the fieldtrip for a holiday. What are your thoughts on this? Would you want to stay for a holiday? What might be some of the unintended (negative and positive) outcomes of this?

Factors to consider: How would you feel if you or other friends on the trip were unable to afford the extra cost of staying on? Who takes responsibility for the logistics of making alternative insurance and logistical arrangements? Is there a danger that by staying on for a holiday, those students may focus on the holiday rather than the fieldtrip – and what might the implications of this be for group-based project work?

In planning and organizing a fieldtrip, decisions and negotiations are made that determine the nature and outcomes of the trip. For fieldclass teaching, these are made by academic staff, rather than students. The types of decision made and factors considered in reaching these decisions can provide you with useful insights into the kinds of ethical and logistical dilemma you may face in organizing your own fieldwork, whether as a lone researcher or participating in a group fieldclass.

These are just a few of the kinds of question involved in planning a student fieldtrip. By reflecting on and responding to these, you should see how similar decisions are involved in planning and designing your own fieldwork. As a student on a fieldtrip, these decisions are often made for you by staff members who balance economic costs and logistical considerations with issues pertaining to risk assessment, research ethics and the implementation of health and safety-related procedures and practices. In your own research, whether you are carrying this out alone or as part of a non-fieldclass group, you will need to negotiate these competing demands yourself and take responsibility for these.

Responsibility and conduct

The issue of responsibility is at the heart of many of the practical decisions made relating to fieldwork. During a fieldclass the lead staff member, supported by other staff, is responsible for the health and safety of participants. As a result, you are likely to find materials relating to risk assessment, code of conduct, ethics and health and safety concerns within a course handbook as well as in supplementary content from lectures, handouts and emergency contact cards being provided to you. (You may find these materials useful as prompts and guides for your own risk assessment, ethics and other paperwork you may be required to complete before conducting your own fieldwork. See also Box 4.2 for further guidance.) The reiteration of this often apparently basic health and safety information by the course leader may seem dull and mundane, but is important to meet relevant safety legislation and to ensure you know the basic context and ground rules.

At the end of the day while the fieldclass leader has a duty of care towards you as a fieldclass participant, you are responsible for your own actions and decisions. This means that once you have been provided with guidance and instructions, it is down to you to

---

Box 4.2 Further guidance on research conduct

We would also recommend you read the Developing Areas Research Group (of the Royal Geographical Society-Institute of British Geographers) *Ethical Guidelines*:

http://www.devgeorg.org.uk/?page_id=799; guidance on *Doing Development/ Global South Dissertations*: http://www.gg.rhul.ac.uk/DARG/DARG%20 dissertation%20booklet.pdf ; and publication on *Postgraduate Fieldwork in Developing Areas*: http://www.gg.rhul.ac.uk/ DARG/FieldWrkGuide.pdf.

ensure that you do not do things that endanger yourself or your fellow students. This principle – and the paperwork in which it is embodied through risk assessments etc. – must also underpin your own planning and conduct of personal, individual fieldwork for a dissertation or other project.

At a very basic level, this includes taking simple precautions and following basic safety guidelines provided by the fieldtrip leader (or, when carrying out your own research, following the same basic principles). For instance, a common concern when working in tropical areas is to reduce exposure to extreme heat. One of the consistent challenges faced on fieldtrips to East Africa is the need to wear a sunhat, sun cream and to cover up (to borrow from the famous Australian skin cancer advert of the 1980s – to 'slip, slop, slap' (you can see the advert here: http://www.youtube.com/watch?v=gAu5wCTEBt0)). Despite repeated reiteration by fieldclass staff on the trips Dan leads to Kenya, some students predictably suffer from sunburn or heat exhaustion by failing to follow these instructions. By failing to follow basic safety guidance, these students not only endanger themselves but also the other staff and students on the fieldclass. This may seem like a minor issue, but heat exhaustion and sunstroke can rapidly escalate into a serious first aid situation, particularly if you are conducting lone working in the field – one that can be easily avoided. Even without such a severe escalation, the lethargy and disorientation experienced by those suffering from heat exhaustion reduces their ability to carry out fieldwork and placing a greater burden on other students to complete any group-based work.

In preparing for a fieldtrip or to conduct your own fieldwork, it is important to find professional information about health and travel issues to help you plan for any vaccinations or other medication required before travelling and to take appropriate supplies with you for the field visit. Travel guidebooks usually offer some accessible and general health and wellbeing guidance and can be an easy place to start. If you are travelling to more remote regions or areas with extremes of climate, you would be well advised to consult more specialist publications for kit lists and guidance. You can also find advice from national health services or travel advice services, including the National Travel Health Network and Centre: http://www.nathnac.org/travel/; the UK's NHS FitforTravel service: http://www.fitfortravel.nhs.uk/advice.aspx; and the US government's travel advice pages: http://travel.state.gov/content/passports/english/country.html. The key advice is that you should seek professional medical advice from a doctor or travel nurse in good time before you travel.

As well as ensuring you are adequately prepared in terms of medical and health concerns, required kit, etc. you should also spend a few moments thinking about and preparing yourself for the stresses and pressures of working in a new and different environment (commonly referred to as 'culture shock') and how you might cope with these (see Box 4.3 as well as further content below and in Chapter 6). Culture shock is a common experience for those conducting research overseas, and in the development field in particular. Culture shock arises from intercultural contact that challenges your expectations, practices and coping mechanisms resulting in anxiety and a sense of discomfort and/or dislocation (Adler 1975). Your experience of culture shock will be influenced by factors including timespan of engagement, purpose and type of involvement and the degree of difference in cultural experience and context (see Ward et al. 2001). You may well find the experience of visiting different countries and working in areas of

> ## Box 4.3 Coping with fieldwork
>
> The stress of and emotional energy used during development research and fieldwork should not be underestimated. The intensity of fieldwork – whether as part of a fieldtrip or an independent research trip – can be exhausting due to levels of prolonged concentration, the challenges of negotiating new cultures and contexts, of constantly reflecting on methods, security and logistics, the stress of being away from home and accessing essentials such as food and water, as well as potentially coping with differences in climate. You may find the reflections of Heidi Postlewait, Kenneth Cain and Andrew Thomson, a group of development/humanitarian practitioners (2006), *Emergency Sex (and Other Desperate Measures): True Stories from a War Zone* (Ebury Press: London) an engaging and off-the-wall account of their experiences of negotiating the stresses and traumas of humanitarian work.

extreme poverty or marked by severe inequality contributes to a sense of culture shock. This is not something to be ashamed of, but is a common component of development research and it is important to have coping strategies in place for these stresses. These strategies may include mechanisms for keeping in touch with family and friends, finding ways to relax and recording and reflecting on your experiences and emotions in your research diary. It is important to remember that culture shock generally wears off as you become accustomed to new cultural prompts and cues and the stress of dealing with unfamiliar contexts and cultural cues declines.

When you are participating in research – whether as a lone researcher or during a fieldclass – you should adhere to a 'code of conduct'. This may be a document that is specifically developed for a fieldclass and to which you have to agree to adhere in order to participate. Alternatively, it may be a generic code required of all researchers at your university whether they are conducting their own research or within a group project or fieldclass. In either case, this document will set out the standards of behaviour that are expected of you as a representative or ambassador for the university. Thus, you are expected not only to adhere to ethical practices in research but also to behave in a way that brings credit to the university and does not damage its reputation – and this extends to your conduct during 'personal' time on a fieldclass.

This is not to say that you cannot relax during fieldwork or a fieldclass – these are often incredibly intense and draining academic experiences and you will want, and need, to unwind. It is, however, important to consider the boundaries for relaxation activities, their appropriateness for the local context and culture (as well as the type of fieldwork being undertaken) and how they may impact on locals and/or other members of the fieldclass. For instance, at the fieldsite used by Dan's development geography fieldtrip to Kenya the collective social space is next to the huts (*bandas*) usually occupied by fieldtrip staff and other visitors to the centre, rather than in the middle of the student-occupied *bandas*. In a rural area, with open windows and little background noise, the sound generated by 30 students talking and laughing can be enough to disturb those in nearby huts. In response to this, informal curfews have been used as well as ensuring students move to a campfire lit in the space by their huts before 10pm in order to minimize disturbance to others at the centre.

As well as finding time and space in which to relax and unwind, another important way of dealing with the stresses of fieldclass participation is through your research diary. Keeping a research diary is a key practice during fieldwork. This private document can be used not only to record and reflect on the practices and findings of the day's research activities, but also to reflect more holistically on the research experience, your emotional and physical wellbeing and general reflections and observations. Research diaries are useful on a number of levels – on one level, as tools for reflection and development of research skills (cf. Engin 2011), on another, as a space in which to let off steam and for reflecting on your own emotions in your research and your personal experience and, on yet another level, as a space in which to record descriptive parts of the research encounters – details and observations from travel, interviews and informal interactions. A research diary is an invaluable component of fieldwork, providing a space in which to record emotions, reflections and decisions that would otherwise be forgotten. It is a tool for reflexive learning through private reflections on the successes and failures of your research and the evolution of your project.

Your research diary is also an important research document as it provides you with the space in which to keep detailed notes and records of your day-to-day observations and research findings. Whether you keep your diary as a handwritten document, a computer file, or a combination of the two is down to your preference and the context and location(s) of your fieldwork. Regardless of the medium in which you keep your diary, it is important to keep multiple copies to minimize the risk of losing these data. In terms of the content of your research diary, Burgess (1981, 2002) identifies three key types of content and reflection – your substantive notes on the day's research experience (what you've done, who you've spoken with, where you've been, etc.), methodological reflections and analysis and reflections on your findings and experiences (see also Morrison 2007). By continually recording, reflecting and analyzing your diary, you can help ensure an iterative engagement with your research and assist in developing core themes and narratives in your findings. This practice, along with other activities, can help you get the most out of your fieldwork (see Box 4.4).

Working as a group

Group work during fieldtrips is a common teaching and learning experience. The knowledge generated through group work may be used in individual assessments as well as in group presentations, essays or policy briefs that also form part of your course assessment. Group work is incorporated into fieldtrips for a number of reasons, some to do with safety, cost and logistics (for example when working with local guides and translators), but primarily because group work provides the potential for you to develop deeper engagements with a topic and to produce a more substantial output. Group projects can also promote peer learning and support as well as developing team working, leadership and communication skills. These transferable skills are increasingly in demand from employers and group work during fieldclasses provides you with the chance to develop

Box 4.4 Making the most of a fieldclass and fieldwork

1   Be organized and take responsibility – pay attention to basic details such as dates, timings (of lectures, departures, meals, etc.) and make a note of these: one of the biggest frustrations for fieldclass leaders is students not taking responsibility for their own learning and repeatedly asking for basic information they are already in possession of. Keeping the course leader happy will improve the mood of the fieldclass in general!

2   Do the background/preliminary reading – it is highly unlikely that your fieldclass will be visiting an area that has no previously published academic, popular and press material relating to it. Spending some time reading this before you go will help inform how you develop your research questions and methods, but also help you to understand the things you see and encounter while in the field. If you can access a local or national newspaper for the fieldclass destination via the internet this can help you get a sense of the everyday realities of the area you are visiting and any pressing issues in the region.

3   Keep your research diary – this may seem like a chore but it is a vital component of reflexive research, allowing you to record not only the practical aspects of the day but also the social, emotional and abstract experiences and reflections. Do not leave this for a few days, write it up every day – if you have time during the day to jot a few things down, do so, otherwise spend the evening making notes and reflections – these may be used in your assessment: some assessments are specifically based on research diaries and reflections and even in those that do not you are likely to find the detail and depth recorded in your diary useful in describing and analyzing your findings.

4   Be prepared but be flexible – the best way to enter the field is to have a well-defined research focus and know how you want to execute this. In reality, you are likely to have to compromise and adjust in the field – because a guide turns up late, a key informant is busy or out of town, a minibus breaks down or the rains come early and flood the route to the village. So, be prepared and be prepared to be flexible.

5   Make the effort to participate and contribute to group work and to group debrief sessions – in essence, the old adage of 'the more you put in, the more you get out' applies here. Be prepared and willing to take on roles within a group, to be creative and proactive in pushing forward your research projects and to be willing to speak up in group discussions and meetings.

these skills and to then draw on these experiences as concrete examples of group work under pressurized situations in job interviews.

It is important, however, to remember that simply being in a group is not the same as working as a group or a team. During fieldtrips, you may well find yourself working in a group with people you do not know, have differing types of ability and personality and with whom you may or may not get on. For the success of the research – and your assessed work – you will need to find ways of working effectively together, identifying a common set of goals and a pathway to achieve these, the roles you will adopt within the group to achieve these and how you will work together to these ends.

As well as having to work together in groups, fieldclasses require you to spend a lot of time together in an intense environment – travelling together on buses, planes, trains and minibuses, eating together, socializing together and sharing rooms. Throughout the fieldclass period you will utilize different skills – from motivating others to conflict resolution to general communication techniques. There are no set rules on the ethics and etiquette of living and travelling together on fieldtrips, but these – often very intense experiences – require careful management of power relations, boundaries and behaviour. It is important to remember throughout a fieldclass that your own behaviour can have significant impacts on those travelling with you and that by giving consideration to how your actions affect your colleagues' wellbeing many potential conflicts and flashpoints can be prevented. While we will not develop these dynamics in detail here, there are some useful guidelines and suggestions in Richard Phillips' and Jennifer Johns' (2012) *Fieldwork for Human Geography* (especially pages 91–103).

Practicalities – logistics and red tape

Planning fieldwork can be both immensely invigorating and exciting, but also tedious and frustrating. Doing a good job of planning your fieldwork and ensuring the core logistical building blocks are in place are integral to the success of your research. These decisions are, inherently, informed by other factors including constraints of time and money, as well as personal research interests and the context from which you are engaging in the research (for instance is it to a predetermined destination of the visit due to participation in a fieldclass or working as a researcher on an existing project or is it driven entirely by your own interests?). As Tom Smith's reflections (Box 4.5) on preparing for and carrying out fieldwork in Tanzania demonstrate, careful planning is important not least because you are then better prepared to cope with unexpected developments and complications. In this section, we work through a number of influencing factors before providing a checklist of key considerations in planning your fieldwork.

There are many logistical and practical decisions to be made in organizing fieldwork, and many of the core decisions may be determined by – or may determine – your central research questions (see Chapter 1). Clearly the questions of *what, where, why* and *how* are at the heart of the matter. The decision as to *why* to do fieldwork is linked to your research question (*what*) and methodology (*how*). Deciding *where* to go is one of the foundational decisions for your research. As noted, this may be determined by your involvement in a bigger project or may be driven entirely by your individual interest.

Box 4.5 Thomas Smith: 'Getting there': Organizing remote fieldwork

(extract from field notes in Tanzania, 10 April 2010) Travelling to Rukwa takes 3 days. From Dar es Salaam to Mbeya by coach: 13 hours, but relatively comfortable. We spend a night in a very basic guesthouse at the bus station. The following morning we take another bus for 7 hours to Sumbawanga . . . the road is unsealed and the surface is rutted by vehicle tracks. At one point the coach lurches upwards, sending the passengers crashing into the ceiling. One woman cuts her leg badly on a metal bar that has snapped off from one of the seats. There is a considerable amount of blood and the woman feints [sic]. The bus pulls into a nearby medical centre. She is patched up, and appears to be much better, so we continue.

We spend a night in Sumbawanga. From Sumbawanga to the villages in Ilemba we take a 4WD Toyota which costs us each 10,000 Shillings, expensive by Tanzanian standards. This is the only form of public transport, and the vehicle is packed out with other passengers. The road takes us up gradually to the top of the plateau of the rift valley. We reach a high point and can finally see down towards Lake Rukwa and the summits of hills which line it to the West. The descent road down the rift valley escarpment to the villages is steep and in poor condition. We have some frightening moments, although David [my research assistant] suggests that in wet conditions the road becomes impassable. When we reach the flat ground around the villages my heart rate noticeably lowers.

This research was concerned with environmental education of young people in Tanzania. I had conducted research in an urban area and a coastal town, but, for my final case study, I wanted to reflect on these issues in rural communities. My research assistant, David, suggested a relatively remote village on the opposite side of the country. Why? Because he grew up in this village, his family were still there and the place met our criteria: it was remote and rural.

Organizationally, working in, and planning to go to, remote places to conduct research opens up as many challenges as it does opportunities. The first logistical challenge is simply getting there and having time available to do so. There are hazards associated with getting there and it is important to assess these risks. If you are beginning to feel familiar within a country, research in remote locations can substantially change the rules of the game. In this remote part of Rukwa, different tribes dominated the social makeup, accompanied by ethnic tensions that I had little experience in negotiating. Gender inequalities were more notable, as were age hierarchies, making it considerably more difficult to do research with women and young people. Travel between villages and access to remote farms was challenging. We rented bicycles and used these to travel on the dirt roads to nearby villages (including carrying the bikes over waist-deep river waters). Reaching places off the dirt roads required a considerable amount of hiking up and down the escarpment.

'Doing' research in this remote place was arguably more challenging than conducting research in rural locations close to Dar es Salaam, my field base. Yet these challenges yielded important rewards. Organizing focus groups and interviews was aided considerably by my field assistant, David, who knew many locals and could narrate histories and ongoing environmental issues in the villages. He became my insider guide. While the place was spatially remote, his position within the community made it more accessible, more familiar, then had we visited anywhere else. Many researchers who work in the global south work closely with field assistants and integrating their life experiences and knowledges into the research process can be one way of engaging collaboratively with them. Significantly, the richness of the social and environmental differences between this place and my other study areas had considerable impact on the findings of the project, making the travails of organizing research in a remote location worth the eventual rewards.

The *where* question is double edged – clearly your research focus and question will determine to a great extent the location(s) of study (attempting a study of coastal livelihoods is not going to be particularly successful in Bolivia for example) but in turn, the location may influence the viability of the research (the *what* question). In thinking about *where* it is also important to consider whether you will be able to do the research in this location and, if so, *how* you would be able to do this.

While the common expectation is likely to be that these questions – and the subsequent considerations outlined below – are approached by the lone researcher setting out on a solitary, knowledge-gathering expedition, this is clearly a flawed assumption. Some fieldwork is undertaken with colleagues, while some researchers may travel to the field accompanied by family members. While accompanied fieldwork will present further factors and questions to consider – from securing visas and insurance through to questions relating to hiring domestic help, employment for partner/spouse, access to schools for children – these should not preclude the possibility of accompanied fieldwork. Reflections from academics who have undertaken fieldwork when accompanied by family members demonstrate the viability of this practice, as well as pointing out the potential benefits for the research process (in some situations aiding in the development of rapport and access) as well as being a potential hindrance (restricting movement, preventing or complicating certain forms or periods of data collection) (Cupples and Kindon 2003). Conducting fieldwork accompanied by colleagues also presents a series of questions and both challenges and benefits as factors of age, gender, language, power, race, identity and other factors affect the negotiation of access and building of rapport in the field. Whether you are planning on travelling and conducting your research alone or with others, you will need to consider a range of factors in organizing your fieldwork.

Even in outlining these initial questions, a series of subsidiary questions begins to emerge. Some of these questions require specialist knowledge or information to be addressed while others may be addressed through relatively basic background information and data. Thus, gaining some basic information relating to your proposed research site(s) can often provide a useful starting point to thinking through the organization of your fieldwork. Common sources of background information to an area can include travel guides, websites, news coverage and personal accounts of those who have previously visited the area. As you progress through such general background information, a series of more discrete and defined considerations will emerge. A number of the central questions to be addressed at an early stage in planning your fieldwork relate to your own health and safety and completion of risk assessments (see Chapter 6). As discussed in greater length in Chapter 6, you need to think about the safety and security situation of your proposed fieldsite(s) and factor these concerns into your planning. In situations in which safety concerns are raised, you will need to address these in your risk assessment and may also have to modify your research plans, potentially relocating your fieldwork site, identifying contingencies in case of deterioration of security situation or adapting your daily research practice to minimize risk. As well as considering safety concerns, other contextual factors to consider include the willingness of local communities and elders to provide access and support to your research. Is the area suffering from researcher fatigue? Are there adequate local accommodation, infrastructure and transport options to allow you to work in the area?

Accommodation during fieldwork is important. Depending on the length of time, location and budget for the fieldwork there are various possibilities – local guest houses or hotels, renting a room in a local family's house, renting a cottage/bedsit, university accommodation for visiting scholars. Each option has benefits and costs, both in terms of the economic cost (not only of the room but also in terms of food, laundry, transport to fieldsites) but also in terms of ability to interact and socialize with local communities. Whichever option(s) you pursue, it is important that you have a space in which you feel safe, comfortable and able to both work and relax. Staying with a local family can help you learn more about the local community, cultures and customs through everyday interactions, but can also lead to your being positioned within the community through your hosts. Whichever type of accommodation you intend to use, it is important to have at least your accommodation for the first few nights arranged, not only for your own peace of mind but also because this information may be required for visa forms and landing cards on entry into a country.

The consideration of when to conduct fieldwork is another practical concern. In some situations, this will be critical to the fieldwork, perhaps due to the nature of the research questions or because of ability to access certain areas during religious festivals or due to seasonal climatic variations. For instance, conducting research with Muslim communities or in Islamic countries during Ramadan may be affected by requirements for prayers, fasting and the social ritual and importance of breaking fast. Similarly, conducting fieldwork during the wet season may present serious challenges to travel, access and infrastructure. However, if research is only conducted in the dry season then the seasonal and everyday practices of the wet season may be missed: this 'dry season' bias has been extensively critiqued (see, for instance, Chambers 1983). If your research involves education and schooling, you may need to think about timing your research in relation to school term times and holiday times, as well as with reference to examination periods or other periods when access to schools by researchers is limited (such as in the final semester of schooling in South Africa).

Once questions of where and when are resolved, a series of further considerations emerge. How will you travel to the fieldsite? How will you move around the fieldsite? What are the possible transport options available to you and what are the costs and benefits – in terms of cost, time and safety – of these (see Box 4.6 for some reflections on these concerns)? The consideration of such questions is worth attention: see, for instance, Newhouse's (2012) reflections on the politics of walking in the field and the ways in which this practice shaped knowledge production. If you are planning on using local transport, what options are available to you and how do you identify the 'best' option? If you are planning on buying or hiring a vehicle during your research, what are the legal requirements relating to this – is your licence valid (and for how long) in that country? Do you need an international driver's licence? (If you need an international driver's permit, how long will this take to organize – for UK-based readers, you can find this out here: https://www.gov.uk/driving-abroad.) Must you carry your driving licence with you while driving? Further to this, are you required to carry some form of identification and/or proof of your visa with you at all times? More generally, are you aware of local laws and customs?

Box 4.6 Everyday transport and unexpected concerns

During recent fieldwork in Uganda, Dan travelled between Gulu and Kampala and within both urban areas. To travel between the two urban areas, Dan used local bus services – opting to use the *PostBus*. While this ticket was slightly more expensive than other long-distance buses, the company has a better safety and reliability record. Alternatives could have included hiring a local driver (a faster journey but much more expensive) or seeking a car share with other people making the journey (no control over how long it would take to arrange or over departure times, no control over quality of vehicle or driving).

Within the urban areas, Dan utilized a local taxi – booked through his hotel – for the airport transfers (while a more expensive option, this was more reliable and less stressful than waiting for local minibuses or having to haggle with local taxi drivers on arrival in Kampala in the early hours of the morning) and then relied on *boda boda* drivers for everyday travel. *Boda bodas* are local motorbike taxis that can take the passenger(s) to locations across the city at a low fare (although subject to haggling) and relatively quickly as they can manoeuvre through traffic jams. Travelling by *boda boda* does have its dangers: having seen several accidents, Dan's risk assessment for future trips will include taking a motorbike helmet with him to wear while taking a *boda boda*.

The message here is that local information and knowledge can be vital in helping plan fieldwork, but also in identifying and responding to both logistical and health and safety concerns prior to departure.

As well as checking on regulations regarding driving licences, it is also imperative to check that your passport is valid (remember to check things like the length of time it must be valid for beyond the end of your planned trip and if there are requirements on the number of blank pages) and you have secured any required visas. Remember that securing a research visa may be a much longer and more complex (and expensive) process than a normal tourist visa, so you need to give yourself time to arrange this. You may also need to secure research permissions from government (national or local) departments or be required to hold a 'visiting' affiliation with a local university – often with some financial cost involved. Negotiating such demands can be a complex process involving discussion with government and quasi-government bodies to secure a visa, letters of introduction/permission and various stamps and authorizations. Through the course of such negotiations, your research focus and questions may be reframed and repositioned or even changed and altered as you work through these initial gatekeepers (Turner 2013).

Holding a 'visiting' affiliation at a local university is sometimes required in order to secure a research visa. In other situations, researchers opt to hold such links for various reasons including access to resources and facilities, as a means of building and maintaining local academic networks, to facilitate recruitment of research assistants for their project or because such an affiliation may help them gain access in the research context. If you are required to or decide it is beneficial to have a visiting affiliation and/or to collaborate with a local non-governmental or civil society organization, you first need to find out which organizations are present in the area and who is working in a field of relevance to your research. You then need to identify a contact person, make contact and build rapport with relevant staff and academics and to check in advance what requirements there may be for holding an affiliation. This may be payment of a fee or the securing of research ethics approval for your project from the host institution. In making these decisions, it is important to consider what the potential benefits and complications are – what are the expectations of the host/collaborator? How will this affiliation affect your position in the field? Building links with local NGOs may also provide further options for recruiting research assistants as well as assisting you in building rapport with and an understanding of local communities and making contact with local gatekeepers and leaders. In developing such links, it is important that you consider the reciprocal nature of a collaborative agreement – many NGOs are wary of working with academics who are viewed as imposing significant burdens on the host organization while rarely delivering on their promises of reports and feedback (see Moseley 2007).

Depending on national legislation and the focus of your research, you may also need to secure ethical approval and permission to conduct research from relevant professional bodies, national or provincial government departments, as well as specific institutions (schools, hospitals) and individuals (local district chiefs). Approval of research ethics and risk assessment forms are vital to planning and preparing fieldwork (see Chapters 5 and 6) and take time, so ensure you allow adequate time for these processes to be completed.

At an early stage of planning, it is also important to check on health issues – do you require any inoculations, vaccinations or boosters, do you need anti-malarial drugs or other medication? Will you have access to clean drinking water or will you need to drink boiled or bottled water or take your own water purification supplies with you? Have you ensured that you have an adequate supply of any hygiene products and personal medication that you will not be able to get hold of while in the field (and that it is possible to transport and store this), not only for the planned duration of the fieldwork but also to cover unexpected delays? Do you know what medical facilities and resources there are in the vicinity of your fieldsite? Have you packed a personal first aid/medical kit? Are there any local laws that may prevent your carrying certain medicines (for instance in parts of the Middle East)? Do you have emergency contact details – of local health service providers (perhaps including air ambulance and evacuation providers), of designated contacts at your university, for the local high commission/embassy (have you registered with the commission/embassy?), and of next of kin? Some of these will be of use to you in accessing health care, others of use to people assisting you in the event of a serious incident or emergency.

Following on from this, it is important to ensure that both you and a designated contact at your university (perhaps your supervisor or personal tutor or department health and safety representative) hold copies of your itinerary (including flight details and initial accommodation provider), passport, visa, travel insurance, personal and emergency contacts, etc. Do you also have your banking details with you – your internet banking log-on details, the phone numbers for reporting lost or stolen cards, etc.? Have you informed your bank(s) that you are taking your card(s) overseas? Have you checked that you will be able to use your cards overseas and what fees you will be charged for doing so? You should also identify the preferred local currency and convert some cash into this currency as required. You may also want to have a contingency plan in case your purse/wallet is stolen – something as simple as splitting your cash and cards between two wallets/purses can be invaluable.

Other practicalities to consider may include language training, depending on where and with whom the research will be conducted. It may not be possible to become fluent in the local language, but it is important and incredibly useful to at least be able to muster a few words and phrases as a means of building rapport. You may also need to think about whether or not you want/need to use a local research assistant or translator. While the complexities of these roles and relations are explored elsewhere (see Chapter 5), at this juncture it is important to consider how you would find an appropriate assistant, what skills they would need, how much you would pay them, how long you would need them for and any logistical considerations (for instance, would they need to travel to other places with you and, if so, how do you arrange and pay for transport, accommodation, food). Even if you do not need a translator, you may want to consider having a local assistant – at least in the initial stages of fieldwork – to assist with introductions, building rapport and developing your understanding of the local context, while bearing in mind that their presence and role in your research will affect your positionality and how you are viewed by local participants (see Temple and Edwards 2002).

Once you have identified your fieldsite, it is often useful to make contact with local gatekeepers, either before arrival or once you get to the fieldsite. For short duration fieldwork, ensuring contact with gatekeepers or key informants prior to departure is often vital to being able to hit the ground running. You need to think about how you will position and introduce yourself and also how you will communicate with people (both in the field and at home) while you are in the field. Have you checked whether your mobile phone/SIM card will work overseas and what the roaming costs are for this? Are you able to get –

and what are the costs of using – a local SIM card (do you need proof of identity and/or address to register a new SIM – as in Uganda, Kenya or South Africa)? Have you got a list of important phone numbers, in case your phone is lost or stolen?

The spread of the internet means that email, Skype and social media are increasingly accessible in many areas but coverage is by no means universal. Where there is reasonable internet access, this can be vital in providing supervisory support and interaction during fieldwork, as well as to maintain contact with friends and family at home and as a means of contacting local organizations and participants. But it is important not to take this access for granted so think about alternative plans and arrangements – and make sure you have printed out copies of electronic tickets and bookings.

On a perhaps more mundane level, it is also important to consider what you need to pack in terms of clothing and equipment. In deciding on clothing, it is important to consider the local climate and weather of your fieldsite, what local cultural and religious norms and expectations are in terms of attire and appearance, as well as how different fieldwork settings and encounters may require different styles of clothing or perhaps

---

Box 4.7 Preliminary fieldwork planning checklist

In preparing for your fieldwork, have you:

- Done some background research on destination?
- Gained ethics and risk assessment approval for the fieldwork, as well as gaining any necessary local research permissions?
- Secured necessary tickets, visa, travel insurance, plus local travel and initial accommodation?
- Checked what local laws and customs are, including any need to carry ID with you?
- Considered and made necessary arrangements for banking and money – currency, travellers' cheques, informed banks, record of numbers for lost/stolen cards, etc.?
- Left copies of important documents with designated contact and made additional copies to be kept in own possession?
- Checked health information and sought relevant vaccinations etc. and carrying sufficient personal medication and hygiene products?
- Put in place basic communication plans, for both local and home contacts – including any reporting in and supervisory requirements?
- Paid attention to and addresses safety and security considerations, and registered/made arrangements to register with high commission/embassy in fieldwork destination country?
- Packed appropriate fieldwork equipment and supplies?
- Packed appropriate clothing, toiletries and personal items (including books, music, etc.)?

specialist clothing. At the same time, you need to consider the research resources and equipment you will need – from having sufficient copies of ethics forms and plain language statements (or 'letter of information'), to having pens and other basic supplies through to a laptop computer, digital camera, digital voice recorder and any plug adaptors or power transformers required. Do you also have sufficient batteries and connector and power cables for your equipment and is the correct software and drivers installed on your laptop? Do you also know how to use the equipment and perform basic maintenance functions (such as changing batteries in a voice recorder)?

Negotiating access and field relations

Linked to the multiple considerations in planning and preparing for fieldwork are the dynamics and practices involved in actually accessing the field. Entering the field for the first time is often a stressful and uncertain process, even if you have established links with gatekeepers and potential participants prior to your arrival in the field. It is important that your interactions and practices in the field are culturally sensitive and respectful (see Kawulich 2011; Ndimande 2012). To this end, it is important to develop a knowledge and understanding of legal and moral norms, not only at a national scale (for instance in relation to alcohol consumption or periods of fasting) but also at a local and individual level (see Ntseane 2009). While your understanding and awareness will become more nuanced during your fieldwork, it is useful to think through various everyday practices and protocols before entering the field. The mundane, everyday actions and encounters that are part and parcel of the research experience influence your relations in the field and the knowledge you are able to produce (see Faria and Good 2012). For instance, is your clothing culturally appropriate for the field? Are you greeting people in an appropriate manner: shaking hands may be common in some cultures, but not in others (such as Pokot pastoralists in Kenya). Do you take off your shoes and hat when entering a house? Have you taken a gift or food with you to an interview (in some situations, this may relate to payment for participation (see Chapter 5) but in others it may be a common practice when visiting friends)? Do you accept food or drinks offered by your host? Are your questions appropriately framed?

While background research prior to fieldwork can help you pre-empt many of these practices, others will become evident during your fieldwork as your understanding of the complexities of local cultures increases. At the same time, your relations in the field will evolve. These relations, with participants, gatekeepers, and others, are generally temporary, fragile, complicated and imbued with power relations (see Chapter 3). Relations with host communities and research participants are informed by dynamics of trust and rapport which can be developed both through specific activities and intentional interactions, but also through accidental encounters and engagements. Developing trust is an important process, one that is partial and ongoing throughout the fieldwork period. Trust and rapport will influence and delimit the questions you can ask and topics you can explore with people and it is important to recognize and work within these limits (see Chakravarty 2012). As the researcher, you need to negotiate these relations as you negotiate access and overcome the barriers and uncertainties of gaining access to and being in the field, while reflecting on your own positionality and how your background and training informs your interactions with and views of the field (Nelson 2013).

Of particular concern for many researchers entering the field is the management of relations with gatekeepers (Box 4.8) and how this affects their access to, and the position ascribed to them by, the local community. Given the potential power wielded by gatekeepers, Kaufmann (2002) refers to these individuals as 'overseers' as they can both support and obstruct or deny access, influence not only who speaks with you and about what, but also what you see and hear more broadly. At the same time, the position and reputation of a gatekeeper/overseer can influence how you are seen locally, influencing who may be willing to speak with you. In some instances, a gatekeeper can be a vital intermediary to access marginal or distrustful communities; in other situations, your apparent affiliation with a specific gatekeeper can lead to hostility or distrust from individuals or groups who have a poor relationship with that individual (see McAreavey and Das 2013).

As your fieldwork progresses, your relations in the field will also change with gatekeepers/overseers as well as with participants and the local community in general. Many researchers have talked about a movement along an outsider–insider continuum or their simultaneous movement and positioning along a number of insider–outsider continua (see Box 4.9 for suggested further readings). Movement along these continua relate to the continual renegotiation of multiple identities and identifiers – nationality, race, ethnicity, language, age, experience, personal knowledge – with different groups and individuals. While many researchers argue that becoming more of an 'insider', of aiming

Box 4.8 Working with gatekeepers

The role played by gatekeepers can be crucial to the success of your fieldwork. These individuals wield tremendous power over who you speak with, where you visit, what you see and what you hear. There is an extensive literature reflecting on the complexities of working with gatekeepers and you may find the following sources of use in reflecting further on these issues:

Clark, T. (2011) 'Gaining and maintaining access: exploring the mechanisms that support and challenge the relationship between gatekeepers and researchers', *Qualitative Social Research*, 10: 485–502.

Kaufmann, J. (2002) 'The informant as resolute overseer', *History in Africa*, 29: 231–55.

Kawulich, B. (2011) 'Gatekeeping: an ongoing adventure in research', *Field Methods*, 23: 57–76.

McAreavey, R. and Das, C. (2013) 'A delicate balancing act: negotiating with gatekeepers for ethical research when researching minority communities', *International Journal of Qualitative Methods*, 12: 113–31.

Reeves, C. (2010) 'A difficult negotiation: fieldwork relations with gatekeepers', *Qualitative Research*, 10: 315–31.

Box 4.9 Insider and outsider

Scholars have reflected on the complexities of their shifting insider–outsider positionality and how they have utilized this in different ways. The following articles provide some interesting insights into these processes:

Ergun, A. and Erdemir, A. (2010) 'Negotiating insider and outsider identities in the field: "insider" in a foreign land; "outsider" in one's own land', *Field Methods*, 22: 16–38.

Ndimande, B. (2012) 'Decolonizing research in post-apartheid South Africa: the politics of methodology', *Qualitative Inquiry*, 18: 215–26.

Petray, T. (2012) 'A walk in the park: political emotions and ethnographic vacillation in activist research', *Qualitative Research*, 12: 554–64.

Rubin, M. (2012) '"Insiders" versus "outsiders": what difference does it really make?', *Singapore Journal of Tropical Geography*, 33: 303–307.

Ryan, L., Kofman, E. and Aaron, P. (2010) 'Insiders and outsiders: working with peer researchers in researching Muslim communities', *International Journal of Social Research Methodology*, 14: 49–60.

Tieken, M. (2013) 'The distance to delight: a graduate student enters the field', *Qualitative Inquiry*, 19: 320–26.

---

to achieve 'good faith membership', gives more accurate and nuanced insights into social phenomena, the position of 'outsider' at the start of a research project also has benefits. Being positioned as an 'outsider' means there is little expectation of your understanding local issues and relations. This makes it easier for you to ask the obvious and basic questions that may become more difficult later in the process when people may assume you hold this knowledge. Throughout the research process, your own efforts to position yourself, and to project certain identities dependent on the context and interaction, will encounter the positions foisted on you by local communities and participants. At times the negotiations of these positions can be frustrating, but can also provide insights into the local context and politics.

You may find yourself adopting a range of positionalities during fieldwork – researchers often strategically emphasize or de-emphasize certain identities and affiliations (academic, development worker, NGO link, etc.) as they utilize an 'elasticity of positionality' to reduce the difference between them and their research participants (Rice 2010: 74). While these practices can aid in developing rapport and gaining access for research, there are ethical implications to such practices relating to personal politics, standards of behaviour and the need to ensure informed consent: a set of concerns discussed by Routledge (2002) in the context of research on tourism and development in Goa. Therefore, if you do make use of the fluidity of positionality in the field, it is important to be reflexive and to consider the limits to this.

Isolation, loneliness and mental wellbeing in the field

We always hope that field research, whether as a lone researcher or working with a group, goes smoothly. In planning for this outcome, we often focus on the logistical issues and economic costs of fieldwork and can overlook broader costs relating to our health and wellbeing. Concerns relating to physical health are discussed elsewhere (both earlier in terms of practicalities and the need for travel insurance, vaccinations etc. and in Chapter 6), but it is worth reiterating the importance of these considerations in planning and conducting your fieldwork. In essence, this is a reminder that you need to look after yourself during fieldwork – to eat, drink and sleep enough.

In addition to looking after your physical wellbeing in the field, it is vital to look after your mental health and wellbeing. The intensity of and the conditions experienced during periods of fieldwork are often emotionally and mentally draining. These pressures can emerge through the daily practices of fieldwork, of being witness to traumatic events or testimony or daily encounters with poverty and destitution, as well as through demands and expectations of data collection. At the same time as encountering these causes of stress and disquiet, you are likely to be removed from your usual support networks of friends and family while having to negotiate different societal norms and behavioural expectations. The accumulation of these factors, of academic, personal, ideological, moral, emotional pressures, can result in fieldwork feeling like a lonely and isolating experience. Many researchers experience a sense of loneliness and isolation at times during fieldwork and it is important to recognize these and to put in place mechanisms to support yourself through these times.

From our own experiences of fieldwork, we know how challenging these feelings can be and encourage you to think about possible strategies for coping with and responding to such pressures. Factors such as jet leg and travel tiredness, as well as disrupted sleep patterns, uncomfortable beds and adjusting to new climates can be stressors, so where time and resources allow ensure you provide time for acclimatization and recovery from long journeys before trying to carry out fieldwork. Depending on the location and focus of your research, you may find yourself dealing with traumatic testimonies and upsetting encounters. In such situations, sharing your feelings and emotions can be an important coping mechanism and your research diary can be a useful tool for this, but it may not be an adequate substitute for talking to family or friends. With modern technology, it is increasingly possible to maintain contact with family and friends while undertaking fieldwork and we would encourage you to use letters, email, telephone, Skype and social media to maintain contact with your support networks at home while carrying out fieldwork (see Figure 4.3).

Allied to this, it is important to remember that you are a human being – once you take on the role of doing development research you do not suddenly become a humanitarian superhero with ceaseless energy and positivity. Doing research is demanding, requiring a lot of emotional work to project confidence, overcome shyness and embarrassment, to maintain enthusiasm in interactions with different people as part of the project of a public persona (see Scott et al. 2012). Because of this, fieldwork is tiring and it is important to give yourself permission and time to relax and unwind – whether this is with some

Figure 4.3 During fieldwork, it is important to take time to look after yourself, to reflect on the research experience and to relax and take in your surroundings

exercise, time spent on a hobby, meeting friends, watching a film or any other activity that helps you to relax. In saying this, it is also useful to note that alongside periods of intense activity, there are often periods of inactivity and boredom and that these times can be as draining as the busy times. So, again, it can be useful to think ahead and plan activities or resources that can help at these times. For instance, if you have a laptop with a DVD drive and have access to a power source why not take a few favourite films or TV boxed sets with you to watch at such times? If you are a keen runner, why not make sure you have some running kit with you? (See Box 4.9 for some of our practices in the field.)

Finally, it is important to acknowledge that while mental health issues are often shrouded in stigma and misunderstanding, they are not the taboo topic they once were. Academia is perhaps one of the more supportive environments for those dealing with such concerns and personal tutors, academic mentors and heads of department can provide support in addition to helping you access counselling and other support services. Our experiences suggest that the castigation of depression, isolation and fear in fieldwork that

---

Box 4.9 Coping with the isolation and the mental strains of fieldwork: Some personal examples

We have each experienced the trials and tribulations of fieldwork and have developed our own ways of managing some of the challenges posed by these periods of research. Here are some of the practices Dan has adopted over the years to look after ourselves while in the field.

Dan, reflecting on fieldclass teaching in Kenya: Due to the stress of leading the fieldtrip, as well as the humid nights and uncomfortable bed, I struggle from severely disrupted sleep patterns. Apart from running down my torch battery watching the scorpions moving around the room, I have found the best way for me to get through the sleepless nights is to download a few audiobooks to my iPod and listen to these at night.

Dan, reflecting on a period of intensive ethnographic fieldwork in South Africa: I was in the midst of a project working on citizenship and education in South Africa, spending weeks at a time with schools in a number of urban areas across the country. I would end up spending more than 12 hours each weekday on data collection and archiving and then large parts of each weekend catching up on administrative tasks.

This period of fieldwork was particularly stressful and lonely, so having a couple of series of *The Wire* on DVD provided a vital means of relaxing at a time when I was removed from my usual support network of friends and family.

Irwin (2007) describes confronting is decreasing. Supervisors are increasingly sensitive to issues of culture shock, mental health and personal wellbeing. Irwin's (2007) reflections on the causes and consequences of culture shock while conducting fieldwork in Kenya provide an honest and intimate portrayal of the importance of recognizing and constructively engaging with the mental and emotional strains of fieldwork. As she compellingly notes, these stresses and strains – and her willingness to reflect on them publicly – do not mean she is 'less' of a researcher or 'inadequate'. We would echo this sentiment: fieldwork can be stressful and draining and it is not a sign of weakness to reflect on and respond to this aspect of development research.

What to take from this chapter

You will have seen from the reflections contained in this chapter, and the seemingly endless lists of questions posed, that organizing fieldwork can be complex and time consuming. Ensuring you have enough time and planning far enough ahead so you are not rushing and panicking at the last minute are important ways of getting your research off to a good start. As we noted at the start of the chapter, there are many ways in to the field, each with its own set of challenges and administrative requirements. From the checklists and prompts in this chapter, you will have a good basis from which to develop your own specific list of things that need organizing, booking, doing and reflecting on. In short, from this chapter you should have an informed expectation of what you will need to do before getting to and on arriving in the field. In closing – remember that it is important to be prepared but also to be prepared to be flexible and adaptable to the realities of development fieldwork!

If you want to explore the ideas covered in this chapter in more detail, you may find these further readings useful:

Bell, K. (2013) 'Doing qualitative fieldwork in Cuba: social research in politically sensitive locations', *International Journal of Social Research Methodology*, 16: 109–124.

Lee-Treweek, G. and Linkogle, S. (eds.) (2000) *Danger in the Field: Risk and Ethics in Social Research*, London: Routledge.

McAreavey, R. and Das, C. (2013) 'A delicate balancing act: negotiating with gatekeepers for ethical research when researching minority communities', *International Journal of Qualitative Methods*, 12: 113–31.

Mukeredzi, T. (2011) 'Qualitative data gathering challenges in a politically unstable rural environment: a Zimbabwean experience', *International Journal of Qualitative Methods*, 11: 1–11.

Scott, S., Miller, F. and Lloyd, K. (2006) 'Doing fieldwork in development geography: research culture and research spaces in Vietnam', *Geographical Research*, 44: 28–40.

5

ETHICS IN DEVELOPMENT FIELDWORK

Research ethics are a critical consideration for development researchers (see Laws et al. 2013, especially Chapter 9). This importance lies not only in the necessity of gaining IRB/ethics committee approval for your fieldwork and completion of associated paperwork, but also in a broader set of reflections and responsibilities embedded in our role as social scientists: as Curran (2006: 198) outlines, consideration of the treatment of participants, transparency of methods and reflections on practice are integral to the 'ethical obligation and responsibility that come with the entitlements of being a social science researcher'. These broader considerations are particularly pronounced for development researchers, encompassing both abstract considerations such as 'why is this research worthwhile' and 'how does this research meet my broader moral views on responsibility', through to practical considerations around responsibility and morality in fieldwork as well as the three main tenets of ethical research, namely protecting others, minimizing harm and increasing benefits (Wilson 1993; Israel and Hay 2006).

The importance of these considerations further comes to the fore in the midst of general bemoaning of Institutional Review Board (IRB)/ethics committee members' lack of understanding of the realities of conducting fieldwork in the global south. Despite these lamentations, constructive engagement with these processes is not only an essential part of the research process, but a particularly useful moment to reflect on how and why you are carrying out your research. These questions remain paramount whether you are conducting an independent research project or participating in individual or group research as part of a fieldclass. As institutional policies become tighter, under- and postgraduate student projects are increasingly under scrutiny in these processes (Griffith 2008). As Carapico (2006) observes, while reflecting on the ethical complexities and contradictions of conducting research in the Arab world since 9/11, there are not always easy answers to ethical dilemmas in fieldwork.

To highlight a few of these complexities, and to demonstrate that research ethics are not simply about securing approval by your IRB but a much more fluid and dynamic set of considerations, consider some of the relatively mundane ethical dilemmas we have encountered in our fieldwork:

> *Dan, reflecting on fieldwork on civil society organizations' use of communication technologies to promote citizenship in Uganda*: When I submitted the ethics paperwork for this project, in keeping with common practice I had stated that I

would provide pseudonyms to my interviewees and the organizations they represented. However, during the fieldwork all of my interviewees requested that they and their organizations be named in the writing up of the research findings so as to provide a greater profile for their work. Further complicating this situation was the representative of one organization who requested that they only be named when drawing on positive experiences, but not when discussing the challenges faced by the organization. In this situation, what would you do? Would you abide by the original ethics paperwork or do you respond to the emerging requests of your interviewees?

Dan, reflecting on work on citizenship and education in South Africa: While I was conducting fieldwork in 2009 on the promotion of ideals and practices of citizenship through formal education programmes in South Africa I spent time interviewing teachers and participating in and observing classroom activities relating to citizenship education. At various points during this fieldwork, incidents of corporal punishment occurred – ranging from chalk and chalk dusters being thrown at disruptive students through to caning of a student for misbehaviour – despite corporal punishment having been banned under the South African Schools Act 1996. In this situation, where my ability to conduct fieldwork depended on the goodwill and support of the teaching staff at different schools, should I report these incidents? If so, to whom – the principal, the Provincial Department of Education, the police? Witnessing the challenges faced by teaching staff and hearing about the lack of professional development training and support for alternative disciplinary measure these teachers received, would any benefits arise from reporting these incidents? Knowing that reporting such events could result in someone losing their livelihood and being sent to prison placed enormous pressure on me, leaving aside the implications for fatally undermining my own research.

Chasca, reflecting on her (and other colleagues) experiences of witnessing what would be considered in the UK as animal cruelty: While working in Botswana, I saw a group of children tie a dog that had stolen some food from the house up to a tree. Once they had tied the dog up they began throwing stones at the dog as punishment for stealing the food. In this situation, what should I do? Intervene – because it goes against the ethical standards of animal welfare in British (and other cultures), while aware this may jeopardize my research? Talk to the children about why this is seen as cruel and they should not do it, according to British cultural values? Accept that I am working in a different cultural context and avoid the situation? What would you do in this situation?

This chapter builds on our own research experience as well as broader debates relating to ethical practices in development research to help guide you through your considerations of and reflections on ensuring your own research practices are ethical. We start with a few broad reflections as to what we mean by research ethics and the importance of ethics in research. This discussion leads on to consideration of the formal, legislative rationale

> ## Box 5.1 Negotiating ethical dilemmas
>
> To give further insights into the complexities and dynamic nature of research ethics during a research project, read Morrell, R., Epstein, D. and Moletsane, R. (2012) 'Doubt, dilemmas and decisions: towards ethical research on gender and schooling in South Africa', *Qualitative Research*, 12: 613–29.
>
> In this article, Morrell et al. discuss the challenges they faced when, during research on gender relations in South African schools, they uncovered allegations that a teacher involved in the research project was sleeping with a female student. In this situation, they faced a number of dilemmas associated with commitments to anonymity, how to ascertain the validity of the allegations, how they were situated with local legal and social norms and the potential consequences of such allegations upon the research and the teacher in question.
>
> After reading this article, reflect on how you would have handled this situation and your thoughts as to how the research team dealt with it.

for compliance with ethics procedures before exploring how these concerns are embodied in the research process. We then discuss and reflect on a range of specific ethical and moral considerations in fieldwork, including friendship in the field, negotiating relations with research assistants, payment and reciprocity for participation and dealing with unexpected encounters in the field.

These discussions should provide you with insights into ethical dilemmas and practices while also instilling an understanding that research ethics are not only of concern during the research design phase and in securing IRB approval. Rather, you should recognize that ethics infuse the entire research process and require a continual and reflexive engagement in the same way as you will implicitly carry out dynamic risk assessment practices throughout your fieldwork. This ideal is captured by Pearson and Paige's (2012: 75) observation that while 'ethical guidance from the researcher's institution and academic mentors may be required to conduct the project, the real ethical decision-making is often subtle yet spontaneous' as you interact and engage with everyday life in the field (Price 2012).

Glyn Williams' reflections (Box 5.2) on negotiating unexpected ethical dilemmas during fieldwork in India illustrates this point, reflecting both the importance of careful ethical consideration in project design and an ongoing reflexivity on emerging ethical concerns. With this argument in mind, we offer reflections on our personal experiences and provide guidance on ethics codes, practices and documents while wishing to avoid any sense that we are trying to reduce complex ethical negotiations of research to a series of simple solutions: as Bourgois (1990: 43) notes, ethical engagements 'are too complicated and important to be reduced to unambiguous absolutes or even perhaps to be clearly defined' (see also Fujii 2012). You should, therefore, use our examples and reflections to develop your own understanding and tools to address the ethical concerns arising from your own research practice.

Box 5.2 Glyn Williams: Ensuring ethical encounters in research: A postcard from India

I begin with an ethical dilemma that I faced when undertaking a research project on poverty and participatory governance in India (Williams and Meth 2010; Williams 2012). The team I was leading was holding a series of user-engagement workshops, presenting our research findings to audiences of local academics, senior politicians and civil servants. In our presentation to participants, we criticized Kerala's Kudumbashree programme – an innovative scheme that attempts to link poverty alleviation with female political empowerment – arguing that it was unintentionally excluding the poorest and most socially marginalized women. Our presentation sparked lively debate over our findings and the current state of Kerala's local government institutions. We were happy with a job well done. A few days later, however, a senior opposition politician approached a member of our research team and asked whether she would write something for him on the difficulties Kudumbashree was facing. How should we respond?

The university ethics approval process this project underwent, and through which I guide students undertaking their own research, didn't address dilemmas such as this. Formal approval processes often focus on data collection and are driven by procedural issues such as demonstrating that research participants have given informed consent or ensuring that data are managed in accordance with clear protocols. These are important issues, but taken in isolation they can leave you with a truncated view of what 'ethical research encounters' might mean. Particularly for research in the global south, following 'correct procedure' could still reproduce very unequal power/knowledge relationships in which researchers 'extract data' from Southern subjects. For over a generation, post-development scholarship has rightly criticized these inequalities (Escobar, 1995) and 'alternative' development practitioners have aimed to overcome them through participatory research methods (Chambers 1994). As other postcards in this volume demonstrate, good ethical practice means asking additional questions: When might our role as researchers blur with that of concerned citizens, activists or friends? When might we have an ethical duty not simply to follow a protocol, but to intervene actively to cause social change? Conversely, when might we have a duty to put our own urge to act on hold out of respect for values we may not share?

If ethics approval processes don't help here, there are other sources of support. For the researcher wanting to know how to conduct themselves ethically, one starting point might be social science ethical guidelines (such as in anthropology: http://www.theasa.org/ethics/Ethical_guidelines.pdf) or learning from the direct experiences of others (see Lunn 2014, for a collection of PhD researchers' own experiences). These are useful not in giving rules to be followed slavishly, but rather in anticipating the broad range of ethical questions raised by the research process, from the framing of a project's objectives through to dissemination of its results. The best preparation here is perhaps to think and talk with others (friends, colleagues, supervisors, etc.) about what your responses to these ethical challenges would be. This won't pre-empt everything that the research process may bring, but it will help you to recognize the ethical questions that face you and to make reasoned responses to them.

With this in mind, how did we address our dilemma? We decided *not* to talk to the politician, largely because we felt we would become pawns in a game we did not control (c.f. Bénit-Gbaffou, 2010). As 'international researchers', our status would have been used to discredit the Kudumbashree programme, but it was unlikely that our research would 'shape policy' in a meaningful way: it would have also positioned Indian members of our team as open critics of the outgoing government (a cost not borne by me as the only foreigner in the team). We felt that the workshop was the right place to pose questions about an important government programme that had complex effects on gender relations and poverty, but we didn't want our role as critical researchers to be overshadowed by becoming enlisted within someone else's political agenda. Ultimately, however, it's not our particular 'solution' that matters, but rather to learn skills of situated ethical judgment.

The importance of ethics in research

Research ethics are integral to good social science research practice. Price's (2012: 3) maxim that 'we ultimately have the well-being, dignity, and rights of those who participate in our research with us as our primary goal' should inform all of your research endeavours. These concerns are most commonly associated with methodological concerns relating to informed consent and doing no harm but should also be considered in relation to broader questions of power, social justice, development and human rights. For those working in development-related research such concerns are often paramount. At times, ethical dilemmas are relatively easily dealt with; at others they are far more complex and can involve actions that are simultaneously subjectively ethical and unethical (see for instance Dhanju and O'Reilly's (2013) reflections on handling a life-and-death situation during research in India). Many of us still take inspiration from Bourgois' (1990: 43) critique of North American anthropology when he argued that projects addressing human rights issues and socio-political injustices were essential in order to meet:

> [A] historical responsibility to address larger moral issues because their [anthropologists'] discipline's traditional research subjects – exotic others in remote Third World settings – are violently being incorporated into the world economy in a traumatic manner that often includes starvation, political repression, or even genocide.

Unfortunately, historical experience shows that such concerns with ethics and responsibility in research are far from universal.

Historical abuses of power and ill treatment of research participants have driven the enshrining of codes and practices of research ethics in legislation at multiple levels. While the specific content and spirit of legislation varies between territories and from institution to institution a set of common concerns underpins these policies. These common concerns emanate from the findings of the Nuremburg War Crimes Tribunal, established to investigate and prosecute the abuses perpetrated in Nazi concentration camps during World War Two, which included the conducting of medical experiments and tests on prisoners without their consent. The rulings of the tribunal enshrined a set of principles that underpin research: respect for participants, preventing harm and safeguarding of research participants and gaining informed, voluntary consent prior to participation (Hall 2009). These principles have evolved and been expanded over time, initially in the biomedical field (due to the nature of the atrocities examined at Nuremburg), but more recently within social science and humanities research (for a detailed summary history of the evolution of ethical review legislation and practice, see Dyer and Demeritt 2009). These developments are manifest in the increasingly extensive and detailed ethics guidelines and review processes demanded by funding councils, philanthropic organizations, universities and other institutions. These processes emphasize informed and voluntary consent, confidentiality of data, avoidance of harm to participants and the disclosure of the independence or otherwise of research (Hall 2009).

Consequently, researchers are now required to adhere to various ethical guidelines as established by national governments and/or by professional or academic institutions. For instance, researchers in the US are obliged to meet the requirements of the National Research Act (1974), as implemented by US Department of Health and Human Services through 45 CFR 46 (Price 2012). While this legislation is framed primarily in terms of biomedical concerns and research, it covers all research in the US involving human participants including work in the social sciences and humanities. This legislation enshrines three key ethical principles in research that underpin the decision-making practices and processes of ethics review boards, namely respect for persons, beneficence and justice (Price 2012). Elsewhere, such all-encompassing legislation is less common. In these contexts, researchers are bound instead by institutional and other sources of ethical oversight.

The influence of biomedical sciences on the evolution of research ethics within the social sciences has triggered a range of criticisms of IRBs for being inflexibly positivist in their approach to research and transferring quantitative methods-based assumptions into qualitative research (for instance Tolich and Fitzgerald 2006; Dyer and Demeritt 2009). Despite these concerns, it is important to recognize the importance and validity of the ethics review process and use this as an opportunity to address issues of inequitable power relations, potential exploitation and working with potentially vulnerable groups – all key concerns within development research. Indeed, IRBs often pay increased attention to development-related research proposals because of these concerns.

In these situations, rather than responding by simply telling the IRB what you think it is that they will want to hear, you should emphasize contextually appropriate ethical practices (Tolich and Fitzgerald 2006; Hammersley 2009). More commonly referred to as *situational ethics*, this is a process by which you outline the ethical concerns pertaining to the context and focus of your research and justify the measures taken to address these in ways that are suitable for, and recognize, local custom and culture (see Flyvberg 2001). A good way of thinking through this process is to consider Bell's (2013) discussion of her continual negotiation of ethical practices during research in Cuba and to reflect on how you could justify these within an ethics review document (see also Box 5.3).

Box 5.3 Dealing with the positivist approach of an ethics committee

As you approach your ethics review documents and committee, the reflections provided in the following papers as to the challenges and difficulties some academics have faced in securing permission for qualitative research – and ethnography in particular – from committees grounded in positivist methods and working from a set of requirements influenced by the biomedical roots of the expansion of ethics committees, may provide some useful pointers:

Hemmings, A. (2006) 'Great ethical divides: bridging the gap between institutional review boards and researchers', *Educational Researcher*, 35: 12–18.
McAreavey, R. and Muir, J. (2011) 'Research ethics committees: values and power in higher education', *International Journal of Social Research Methodology*, 14: 391–405.
Murray, L., Pushor, D. and Renihan, P. (2012) 'Reflections on the ethics-approval process', *Qualitative Inquiry*, 18: 43–54.
Tolich, M. and Fitzgerald, M. (2006) 'If ethics committees were designed for ethnography', *Journal of Empirical Research on Human Research Ethics*, 1: 71–8.

Ethical development research

As approaches to development and development research have changed, moving away from instrumentalist, externally imposed, top-down approaches to more collaborative, iterative and participatory approaches informed by post-colonial theory, the design and benefits of research are subject to increasing ethical scrutiny (see Brydon 2006). Ultimately, a core ethical consideration for development researchers is to ensure – as far as possible – that their research is not 'parasitic', that it is not an exercise in data and knowledge extraction that is then used solely to advance theory and promote an individual career (see Binns 2006). Indeed, researchers such as Kiragu and Warrington (2012) have argued that development researchers should go beyond the ethical norm of 'doing no harm' to a more radical position of specifically 'doing good' and ensuring research contributes to a quest for social justice.

The historical experience of exploration, colonialism and imperialism contributes to the added importance of these research ethics debates in development research.

Historical travelogues and reporting created a discourse of non-Western societies as barbaric, uncivilized and backwards: they were inferior 'others' to be observed, recorded and 'civilized'. For social anthropologists, among others, the fascination with the 'Other' coalesced around observing cultures and learning about people and places. For missionaries and other types of philanthropist, their engagements with the 'Other' were intended to 'civilize' and 'uplift' societies to Western styles of living. These narratives foreground current development research. Histories of exploitation, expropriation and subjugation continue to inform how researchers are viewed in the field as well as how we, as researchers, need to think about the field and our research practices. These engagements draw on a range of historical concerns that continue to resonate with contemporary research practices. Critical reflection on development research highlights how power inequalities between the privileged (white) Northern researcher and the (black) Southern researched serve to replicate historical inequities and manifest a new form of extractive industry – knowledge extraction – in which knowledge and data are 'mined' from countries and communities in the South and appropriated and exported by Northern scholars who use this knowledge to advance their own careers (see Leadbeater 2000) (see Figure 5.1).

While such broader ethical dilemmas (Box 5.4) are often beyond the procedural focus of IRB paperwork, it is important that you consider such issues as your research practices will also form part of this evolving narrative. Your research will not only be informed by previous experiences and popular memory, but will also function as a new

Figure 5.1 During fieldwork, everyday actions and practices can take on new ethical dimensions and be imbued with multiple power relations that require careful reflection and consideration

Box 5.4 Broader ethical questions

We have outlined why it is important that you give due consideration to the broader ethical context and framing of your research. To make this a bit more accessible and relevant to your specific research, think about your responses to these questions and the rationale for them:

1   *Is the research well designed?* Poorly designed research suggests a lack of respect for participants and the demands placed on their time. Throughout the research process you are a guest. Your participants are your hosts, sharing their time and expertise with you – it is therefore vital to ensure your research is well designed so as to be as efficient and effective as possible, to minimize the disruption to your participants' lives and to inconvenience them as little as possible.

2   *For whom is the research worthwhile?* Not all research can/will have beneficial impacts on development practice or on the host community/individual participants, but this needs to be recognized and understood from the outset. Often academic research is beneficial for the academic's career but has no benefit to the community in which the work is based or may only benefit a certain part of that community. This is a perennial concern for many development researchers – should development research be aligned more closely to activist research? Is this the nature of academia?

3   *Who has been involved in designing the research?* Postcolonial theorists would argue that development research should be designed and developed by the participants in the research, that they should lead (or at least have a significant input into the research design process). While this ideal may be possible in longer term projects, short-term student projects are unlikely to have the time or resources to allow for this process to occur. However, you should consider the measures that could be taken to provide the local/host community a means of inputting into the research design and (re)shaping this as the project progresses.

4   *What benefits will be delivered for the local community?* The benefits accruing to a local community from your research may arise from your findings and any implications they have for policy and practice. You may also make other contributions during your fieldwork, either through the sharing of skills or training, acting as an activist academic or even simply in terms of the financial contribution made through your interactions with the local economy or through payment for participation. In reflecting on these concerns, you may also want to consider to which sections of the community these benefits will accrue, how will they affect local dynamics and relations (reinforce or challenge power relations for example) and how might they affect how you are positioned while conducting your research.

encounter added to this canon of experience that will influence how future researchers are welcomed, supported or rejected. Our point here is twofold. First, that historical exploitation and unethical practices mean there is a particular need to ensure your research is conducted to the highest ethical standards. Second, that your own practice in a community will leave a legacy that will shape any future research interactions. In essence, there is a double imperative for you to ensure that you have reflected carefully on the ethical conduct of your research, both in ensuring that your research ethics meet the requirements imposed by your institutional review board but that they are also situationally appropriate (Flyvberg 2001).

Ethics review procedures and paperwork

It is increasingly common for all research activities to be subject to ethical review. If your research involves human participants, it is almost guaranteed that your research proposal and design will be subject to IRB assessment prior to your fieldwork. The exact structure, flow and requirements of ethics approval procedures will vary between countries, universities and departments. Some institutions use similar processes for qualitative social science/humanities research as for biomedical work, requiring you to develop strategies for dealing with emotional distress and psychological harm should this arise (see Corbin and Morse 2003). Other institutions have a much lighter touch approach to qualitative social science research. The best advice here is to find out what procedures and forms are required at your institution as early as possible so that these can be an integral part of the research design process.

Getting hold of the outline IRB forms and paperwork (Box 5.5) early in the research design process allows you to reflect on your use of methods and practices throughout the research design process. You will also have a clear idea as to the specific procedures in place for providing ethical approval (i.e. who will sign off on your ethics forms). If you are working on a non-sensitive topic with non-vulnerable groups, it may be that your supervisor is able to approve the ethics paperwork. In other situations, perhaps where you are working on a more sensitive topic, with vulnerable groups or on a project with the potential to cause emotional distress, your proposal may be subject to an escalating set of ethics review boards depending on the perceived level of ethical risk associated with the research. In some institutions, all ethics proposals are reviewed by a departmental, faculty or even university committee (in some countries, there is also a growing trend towards including non-academics on ethics review boards).

If your research involves partners from other universities and/or fieldwork overseas you may need to secure ethics clearance from multiple review boards: for Dan's work on citizenship education in South Africa he needed to get the research design and protocols approved by IRBs at two universities as well as by three provincial education departments. Understanding what procedures you need to go through will help to clarify the process and reduce the stress involved in completing this paperwork.

It is important to remember that IRBs are not in place to prevent your working with vulnerable groups or on sensitive topics. Their role and function is to ensure that when

Box 5.5 Common key documents

Ethics form: Document required by IRB procedures to be completed prior to the undertaking of research involving human subjects. This document usually requires the researcher to provide a basic outline of their research and how they will conduct the work for approval by supervisor/ethics committee before fieldwork can begin.

Plain language statement/letter of information: A document to be left with the research participant(s) that outlines in simple terms the nature and focus of the research, the sponsoring institution, any sources of funding, your contact details plus those of your supervisor and the ethics committee.

Informed consent form: A document for your records that the participant signs to demonstrate their agreement to be involved in the research. This document will contain the same basic outline of the research as the plain language statement/letter of information.

this kind research is undertaken, it is carried out in an ethical manner and with appropriate practices and safeguards in place. Due to the nature and focus of development research, you may well find your work being subject to an enhanced review – often under the pretext of your work involving 'vulnerable' groups.

Definitions of 'vulnerable' groups vary greatly between institutions, but will generally include children and youth, elders, those in state institutions (such as prisons) and those with learning and behavioural difficulties. Of more frequent relevance to development researchers are definitions of 'vulnerability' relating to language ability and comprehension (i.e. if participants' first language is not that of the researcher and in which the research will be conducted), migration status, socio-economic situation (i.e. absolute poverty) and other factors leading to heightened power inequalities (cf. Perry 2011).

If your work involves such groups, this should not prevent you from carrying out the research but requires greater reflection on practice and detail on how you will address additional ethical concerns. For instance, if you are conducting a survey on housing and health with an isiXhosa-speaking community, one way to address concerns regarding literacy, understanding and informed consent may be to provide pre-translated plain language statements and consent forms for respondents and to use local isiXhosa-speaking research assistants who can translate the survey questions (as Dan did in his work with Niamh Shortt in Imizamo Yethu, Cape Town (Shortt and Hammett 2013)).

The purpose of the ethics review paperwork – which will likely comprise a research ethics form/statement plus supporting documents such as plain language statement/letter of information, consent form and, in some cases, outlines of interview schedules or survey questionnaires – is to establish that you (the researcher) have adequately engaged with and integrated ethical issues into the design of your research. Common areas of interest to ethics committees include your explanation and justification for the selection of groups

Box 5.6 Further sources of ethics guidance

Guidance and briefing notes on research ethics with relevance to development research can be found through numerous professional/academic associations' and funding bodies' webpages. We include a few links below that you may find useful.

We would also recommend utilizing the University of Nebraska-Lincoln Ethics Centre's online sociology ethics resources to explore many of these issues in more detail, including a series of case studies relating to research ethics and decision making:

http://ethics.unl.edu/ethics_resources/online/sociology.shtml#case_studies.

Selected sources of further guidance:

- *American Anthropological Association*: www.aaanet.org/cmtes/ethics/CoE-Fieldwork-Dilemmas.cfm; www.aaanet.org/cmtes/ethics/Ethics-Resources.cfm

- *American Association of Geographers*: http://www.aag.org/cs/about_aag/governance/statement_of_professional_ethics (see also the ethics statement from the Latin America Specialty Group of the AAG: http://www.community.aag.org/lasg/home/)

- *American Political Science Association*: ww.apsanet.org/imgtest/ethicsguideweb.pdf

- *Australian Government and Australian Research Council*: Australian Code for the Responsible Conduct of Research: http://www.nhmrc.gov.au/guidelines/publications/ r39

- *British Sociological Association*: http://www.britsoc.co.uk/about/equality/statement-of-ethical-practice.aspx

- *Developing Areas Research Group (DARG) of the Royal Geographical Society-Institute of British Geographers*: www.gg.rhul.ac.uk/DARG/ethical.html

- *Economic and Social Research Council (ESRC)* – one of the UK's main funding bodies for social science research: http://www.esrc.ac.uk/about-esrc/information/research-ethics.aspx

- *Office for Human Research Protections*, US Department of Health and Human Services – the central government agency in the US tasked with providing guidance on and ensuring ethical conduct of research involving human subjects. The Section 45 guidance provides the framework for gaining research ethics approval in the US:

    www.hhs.gov/ohrp/humansubjects/guidance/45cfr46.html

- *Social Research Association*: http://the-sra.org.uk/sra_resources/research-ethics/ethics-guidelines/

to be involved in the research (are they vulnerable?), how participants are recruited, what topics and questions will be raised (are they seen as sensitive?), what methods will be used during the project, what measures are in place to ensure participants know what they are involved in and how any information provided will be used, and what safeguards are in place to ensure anonymity and confidentiality of data (these considerations should also include any requirements for public archiving of data, a practice often required by funding agencies). (See Morrow et al. (2014) for useful reflections on this concern.) Your ethics paperwork should demonstrate to the reviewer(s) that you have thought carefully about the potential ethical concerns and implications of your work and the strategies you have employed to ensure best practice and the upholding of ethical standards in your research in ways that are appropriate to the context in which you will conduct your research.

Box 5.7 Plain language statement/letter of information

A plain language statement is intended to provide potential research participants with a clear and simple outline of the research you are doing, why you are asking them to be involved, what their involvement would entail and what you will do with the data from the research. To achieve this, the document should be short (c. one or two pages only) and written in clear, accessible language and free from technical jargon.

Remember – you need enough copies of this document for all participants to keep a copy.

Your document should include:

- the title of the project
- an invitation to participate
- an outline of the purpose of the research
- an indication as to why you are asking them to participate
- an outline of what will happen if they agree to participate
- an outline of any foreseeable risks involved in the research (including an outline of how the data will be handled and used)
- who is doing the research
- who is funding the research
- if there are any benefits to the participant for taking part
- what you will use the data from the research for
- a reminder that participation is voluntary and a clear outline of what you will do with any data should they withdraw from the study
- contact details for you (the researcher) as well as the chair of your ethics committee and/or your supervisor.

Two examples from Dan's research are included here.

Research Project: Community health and social tensions in informal settlement upgrade projects in Cape Town: A study of Imizamo Yethu

What is this project?
This project explores the health and social wellbeing outcomes of informal settlement upgrade projects in Cape Town. It will consider how local, national, and international organisations are involved in the design and implementation of these projects. We are interested in how residents' views of their own health are related to the quality of their houses. We are also interested in how new, formal housing in informal settlement areas affects community relations and residents perceptions of community and belonging.

Who funds this project?
This research is funded by a small research grant from the Hayter Travel and Field Research Awards scheme (University of Edinburgh, UK).

Why are you being contacted?
You have been contacted because you are a resident in the case study site and we are interested in your views on local informal housing upgrade projects.

How is the research being conducted?
Involvement in this research is entirely voluntary and you may withdraw at any time and/or refuse to answer any questions you are uncomfortable with. Data will be collected through the completion of a questionnaire with a research assistant with an expected duration of 40 minutes. These data will be collated by the researchers, coded and entered into a secure electronic database. No personal identifying information will be stored with this data – your responses will be entirely anonymous. You may stop your participation at any time during the process.

Data from this project will be securely stored and may be used in future academic projects. Data from this project will only be used for academic purposes by the principal and co-investigator named below.

No personal risks are envisaged from participation, and consideration to research ethics has been given in the design of this project. Research data will be destroyed after five years.

Who is responsible for this research?
Researcher 1 (POSITION & UNIVERSITY AFFILIATION). Contact: email address
Phone – in South Africa on NUMBER, in the UK on NUMBER.
Researcher 2 (POSITION & UNIVERSITY AFFILIATION). Contact: email address
Phone – in South Africa on NUMBER, in the UK on NUMBER.

If you have concerns with the conduct of this research, please contact the Chair, Institute of Geography Ethics Committee, University of Edinburgh (CONTACT EMAIL ADDRESS).

Research Project: Education, Citizenship Formation, and Democratisation in South Africa

What is this project?
This project is a self-contained research project considering how notions of citizenship are expressed in high-school education. The questions posed in the course of this work will explore how government formulates the idea of citizenship in the post-apartheid nation and how this official discourse is informed by the wider project of nation-building and managing racial history and multiculturalism. Questions will also consider how challenges to this conceptualisation of citizenship are encountered and negotiated in citizenship education materials and practices.

Who funds this project?
This research is funded under a small research grant award from the UK's Economic and Social Research Council.

Why are you being contacted?
You have been contacted because of your involvement in citizenship education in one of a number of capacities – politician, civil servant, policy maker, academic, school principal, school teacher, community activist. We are collecting data from this range of informants in order to further explore the questions outlined above and to develop a variety of perspectives on the issues involved.

How is the research being conducted?
Involvement in this research is entirely voluntarily and you may withdraw at any time and/or refuse to answer any questions you are uncomfortable with. The collection of data will involve individual (one-to-one) interviews with an expected duration of 1 hour. These will be digitally audio recorded. You will have control to pause or stop the recording and/or the interview at any time.

Interview data will be transcribed and analysed and respondents assigned pseudonyms in any material produced from this data. Schools and institutions will also be given pseudonyms and information which could identify an institution will be removed. Data from this project will be securely stored at the University of Edinburgh and may be used in future academic projects. Data from this project will only be used for academic purposes by the principal and co-investigator named below.

No personal risks are envisaged from participation in this research, and consideration to research ethics has been given in the design of this project. Data from this research will be destroyed after five years.

Who is responsible for this research?
Principal Investigator: RESEARCHER 1 (POSITION & AFFILIATION)
Contact: EMAIL ADDRESS
By phone – in South Africa on NUMBER in the UK on NUMBER.

Co-Investigator: RESEARCHER 2 (POSITION & AFFILIATION)
Contact: EMAIL ADDRESS

If you have concerns with the conduct of this research, please contact the Chair, Institute of Geography Ethics Committee, University of Edinburgh (EMAIL ADDRESS).

Informed consent

Informed consent is a core component of ethical research practice. In simple terms, this means that research participants 'must fully understand what it means to participate in terms of risks and benefits' (Corbin and Morse 2003: 348). As a general rule, you must always secure informed consent from research participants, ideally in written form using a consent form and plain language statement. In securing this consent, it is vital that you explain that participation is voluntary and that participants may withdraw from the research at any time or refuse to answer any questions they are not comfortable with. However, there are times when gaining this consent in written form is not possible or appropriate, presenting you with additional ethical hurdles and challenges. Participants may be reluctant to sign a consent form due to illiteracy (in the language used by the researcher), distrust of official paperwork, a fear that documents may be used for an ulterior purpose or concerns over potential repercussions should their involvement in your research become known. In these situations, you may be able to gain verbal consent but you need to have ratified this process with your IRB before commencing your fieldwork.

Anticipating such challenges is part of the process of reflecting on the ethical design and conduct of research and should be included in the paperwork required for an ethics review. While written consent may be the 'gold standard', if you can explain in your ethics forms why this may be inappropriate and that there is justified reason for pursuing an alternative method of ensuring informed consent, this should satisfy an IRB as you are showing an awareness of potential limitations and offering justified alternatives in line with accepted good ethical practices.

Other challenges to securing informed consent can arise from particular methods, notably prolonged ethnographic research. During such work, you may be immersed within a community for a prolonged period of time and engage – to varying degrees of depth – with a large number of people. In this situation, can you ensure appropriate and informed consent from everyone you encounter? Is it feasible or necessary to gain written consent from everyone encountered? Are there other levels of consent that may be gained that are more appropriate to the context and encounters? In other situations, you may find that individuals are willing to speak to you 'off the record' about a particular topic or concern. In these situations, it is highly unlikely that you will be able to get written consent, indeed even using the information provided in your analysis and writing may prove problematic.

Working towards the 'gold standard' for informed consent, in order to ensure and to demonstrate that you have secured this from your research participants, you are likely to be asked to produce and use two core documents: a letter of information/plain language statement and a consent form. While aspects of these documents are very similar, they perform different functions – one is for your participants to know what they are agreeing to and to retain for future reference and the other is for you to be able to evidence that you secured informed consent from your participants. Further details of the typical content of these documents can be found in Boxes 5.7 and 5.8.

Box 5.8 Consent forms

A consent form provides you with a record demonstrating that research participants have voluntarily agreed to participate in the research based on a clear explanation of the project, as well as any specific permissions related to the exact nature of the project. To achieve this, the document should be short (c. one or two pages only) and written in clear, accessible language and free from technical jargon. As can be seen in the following example, it is increasingly common for these forms to take the style of a tick list, with participants initialling each statement to indicate agreement to it, before signing the end of the form. The form should also include your contact details and those of the chair of your ethics committee and/or your supervisor.

As you can see, the form above is simpler and easier to understand and gives more precise indication of what permission has been granted for. The form below is from Dan's project in Imizamo Yethu. You can see how much more difficult it would be to understand a consent form with this much information on it. The content in the form is similar to the plain language statement included in the text box above but translated into the dominant language of the township. (The English language version is included here.)

CONSENT FORM

Title of Research Project

Please initial box

1. I confirm that I have read and have understood the information sheet for the above study. I have had the opportunity to consider the information, ask questions and have had these answered satisfactorily. ☐

2. I understand that my participation is voluntary and that I am free to withdraw at any time without giving any reason, without my rights being affected. ☐

3. I understand there is no payment or compensation for participation. ☐

4. I understand that I can at any time ask for access to the information I provide and I can also request the destruction of that information if I wish. ☐

    ☐

5. I agree to take part in the above study.

6. I agree for the interview to be audio-recorded (recordings will be securely stored in digital format and deleted 12 months after the completion of the project)   Yes: ☐   No: ☐

7. I give permission for the transcript of my interview/research to be used for research purposes only (including research publications and reports)   ☐

8. I understand that such information will be treated as strictly confidential. I understand that I have the right to anonymity. I assign copyright of my transcript to the researcher, who may quote the transcript with strict preservation of anonymity.   ☐

_____   _____   _____
Participant Name   Date   Signature

_____   _____   _____
Researcher(s)   Date   Signature

If you wish to contact the researcher, NAME can be reached at:
CONTACT DETAILS

If you have any concerns with the ethical conduct of this research you can contact the Ethics Committee chairperson:
CONTACT DETAILS

Ucwangciso Phando:
Ngezemfundo, Ulwakhiwo-bumi no Lawulo Lwesininzi eMzantsi Afrika

Umphandi: RESEARCHER 1 (POSITION & AFFILIATION)
Uqhakamshelano: EMAIL   Ph – (South Africa) NUMBER

Umphandi: RESEARCHER 2 (POSITION & AFFILIATION)
Uqhakamshelwano: EMAIL   Ph – (UK) NUMBER

Le project inika ingqalelo iziqhamo kwezempilo nezentlalo-ntle kuphuhliso ematyotyombeni ase Imizamo Yethu. Imibuzo eyakubuzwa koluphando iyakugxininisa kumkhomba-ndlela notshintsho kwezempilo, nangayo indlela ekhangeleka ngayo xa uthelekisa nonikezelo lweendawo zokuhlala. Imibuzo iyakuphinda inike ingqwalasela kwindlela uphuhliso oluchaphazele ngayo intlalo yoluntu, kanjalo namaqumrhu lawo achaphazeleka kunikezelo lweendawo zokuhlala banxulumana kanjani noluntu lwaloondawo neenkokheli zabo.

Ukuzifaka koluphando-nzulu kona kungokuzithandela, apho unako ukurhoxa nangaliphi na ixesha okanye ungavumi kuphendula namnye umbuzo oziva ungakhululekanga nguwo. Uqokelelo lwezinkcukacha iyakuba kukuqokelelwa kugqitywe loo mibuzo ngumncedisi lowo wophando ngokwexesha elilinganiselwa kwimizuzu engamashumi amane. Lamaxwebhu ayakulungeleliswa ke ngabaphandi abo, agcinwe ngokhuseleko kumathala eenkcukacha kangangeminyaka emihlanu phambi kokuba atshatyalaliswe okanye abenokusetyenziswa kwiiprojekti ezikumila kunje zexa elizayo. Wena uchaphazelekayo akukho nalinye inqanaba koluphando apho unokwaziswa ngaphandle kwemvume yakho. Kungekho naxabiso limiselekileyo kolu phando.

Azikho nezigrogriso zoluntu ezithe zabonakala kothe wazibandakanya koluphando. Ngokunjalo ithe yanikwa ingqwalasela nemeko yobuzwe kwizalathiso zale projekti. Le miqulu iyakushunqulwa ngohlobo lwakhona iyekufakwa koovimba abangafuni kwanto engathi ibhentsise iincukacha zoluntu.

Ukuba ngaba unemvakalelo noba yeyaluphi uhlobo ngoluphando, ungaqhakamshelana no: uSihlalo, Institute of Geography Ethics Committee, University of Edinburgh
        (EMAIL ADDRESS)

Olu phando alunaxabiso liqinisekisiweyo ntonje ziincukacha ngophando kuphela.

Mna ....(igama) ndiyavuma ukuthatha inxaxheba koluphando-nzulu lunngasentla, ndiyiqonda neyokuba ndinakho ukurhoxa naninina. Ndikwavuma nokuba noshicilelo lweencukacha luyakugcinakala ngokhuseleko ze lisetyenziswe kwiimfuno zexa elizayo.

Signed: Date

Umphandi Oncedisayo:

Research Project: Education, Citizenship Formation, and Democratisation in South Africa

Researcher: RESEARCHER 1 (POSITION & AFFILAITION)
Contact: EMAIL ADDRESS   Phone – (South Africa) NUMBER.

Researcher: RESEARCHER21 (POSITION & AFFILAITION)
Contact: EMAIL ADDRESS   Phone: (UK) NUMBER.

This project considers the health and social well-being outcomes of informal settlement upgrade work in Imizamo Yethu. Questions posed in this research will explore recollections of health indicators and changes in these and how they are seen as relating to housing provision. Questions will also consider the ways in which upgrade projects have affected local community relations, and how different organisations involved in providing housing have consulted with local communities and leaders.

Involvement in this research is entirely voluntary and you may withdraw at any time and/or refuse to answer any questions you are uncomfortable with. The collection of data will involve the completion of a questionnaire with a research assistant with an expected duration of 40 minutes. These forms will be collated by the researchers and securely stored in a database for five years before destruction and may be used in future academic projects. You will not be identified at any stage of the research process. There is no commercial value to this research.

No personal risks are envisaged from participation in this research, and consideration to research ethics has been given in the design of this project. Questionnaire data will be coded and entered into an electronic database that will not contain any personal identifying information.

If you have concerns with the conduct of this research, please contact the Chair, Institute of Geography Ethics Committee, University of Edinburgh (EMAIL ADDRESS).

This research has no commercial value and is for research purposes only.

I. .... (print name) hereby consent to participation in the above research project and understand that I may withdraw at any time. I consent that recordings of interviews may be securely stored and used for future academic purposes.

Signed: Date:

Research Assistant:

Broader ethical considerations

A range of broader ethics concerns are of particular relevance to development research. Previous chapters have identified the historical background to development research which informs contemporary practices and experiences (Chapter 2), as well as reflecting on questions of power and the politics of development research (Chapters 3 and 4). Inequalities in power, privilege and wealth are often rendered very starkly in development research experience, but these inequalities should not lead you to overlook the agency and power retained by hosts and participants. You are reliant on their goodwill, as can be made evident in refusals to participate, bartering for payment or the offering of very scripted or partial narratives. A number of related considerations are discussed below. However, this discussion is not exhaustive and you may also want to reflect on moral and ethical obligations relating to delivering on promises made to 'give back' to host communities/organizations, on how to credit the support provided by hosts, translators, guides and research assistants in your work (something that is often overlooked, leaving these individuals largely invisible in research outputs (see Schumaker 2001; Molony and Hammett 2007)) and on how to ensure appropriate use of research data in your analysis and outputs.

Positionality and boundaries of friendship

Negotiating the boundaries of work and personal life can be a continual fieldwork challenge, particularly in policing the blurred boundaries of friendship in the field. While conducting fieldwork, you will hold multiple roles and subjectivities and encounter a range of emotions and emotional work that lead to ethical conundrums in forming and maintaining friendships (and even relationships) in the field (Whiteford and Trotter 2008; Hall 2009; Kay and Oldfield 2011). These interpersonal dynamics can be tremendously powerful and have significant implications for your research, particularly if a perception emerges of an exploitation of friendship or a relationship in order to advance your research agenda. These concerns are particularly pronounced in development research, given histories of sexploitation within colonialism and the construction of narratives of eroticism, sex and sexuality through the colonial gaze (for further engagement with these issues, see McClintock 1995; Stoler 2002). Given this historical colonial sexualized gaze, some consider sexual relations in the field to be unethical, while others – responding to feminist and postmodern critiques – have called for a more critical and reflexive engagement with sex and sexuality in the field that engages with issues of power, desire and the ways in which knowledge is constructed (Cupples 2002; Rubenstein 2004; Irwin 2006; Grauerholz et al. 2013).

Dilemmas around sex(uality) in the field arise frequently, both in terms of how you seek to position yourself and view others, but also in the ways in which you are viewed as positioned in turn. Underpinning these gazes are a range of historical and contemporary power and other relations, as explored in depth by many scholars including Cupples (2002), Malam (2005) and Vanderbeck (2005). This is a delicate and complex topic, one that is incredibly personal but also – in this situation – with potential professional implications. How you enter the field – alone/accompanied, single/in a relationship – and your personal moral compass will influence your position and desires within the field. You must also be aware of how you are positioned in turn, how your appearance, relationship status, gender, etc. will lead to ideas about sex(uality) being ascribed to you (Groes-Green 2012). These need reflection and consideration, both in terms of their meanings for your research but also for your personal life and ethical practices: as Groes-Green (2012) outlines, their participation in the nightlife of their research participants aided their research by altering their ascribed positionality but triggered another set of ethical concerns.

Your negotiation of sex(uality) in the field will involve not only your own personal morals (and experiences of broader cultural norms) but also professional consideration of ethical research practices as well as understanding and respecting local boundaries, values and norms (see Markowitz and Ashkenazi 1999). At times these may be legal strictures, for instance in relation to legislation on homosexuality and at others these may be sociocultural norms, for instance with regards to unmarried couples sharing a bedroom. You need, therefore, to be supremely sensitive to local expectations and practices and to respect these boundaries. Layered on to these concerns are issues relating to positionality and the (ab/mis)use of power and negotiation of expectation. On one side, these concerns relate back to the colonial erotic/exotic gaze (Cupples 2002), on another, they relate to the performance of gendered identities and stereotypes – particularly of masculinities (Vanderbeck 2005), but also in terms of your position within the field and the risks of sexual encounters stigmatizing you within your 'host' community or being (perceived as) a misuse of power and status. Ultimately, your personal (including sexual) behaviour in the field must always seek to avoid harm and must never involve deceit or exploitation (Irwin 2006).

As a researcher, you are a human being with emotions and feelings. At the same time, you may well be in a position of privilege amid a complex web of power relations. It is imperative, therefore, that your actions remain ethical and that you reflect on your positionalities and the boundaries these place on your relations while in the field. You must therefore reflect on how you differentiate between your multiple roles and subjectivities during your fieldwork and ensure that these are clear to those you are working and socializing with. This is not to say that you can never conduct research with friends or become friends with your participants, but they are ethical considerations that must be given attention. Such concerns, including reflections on the potential repercussions and censure if allegations are made that you have exploited positions of privilege or power, are discussed at length elsewhere (for instance Kulick and Wilson 1995; Lewin and Leap 1996; Markowitz and Ashkenazi 1999; Grauerholz et al. 2013). These concerns are increasingly important as social media sites and mobile technology mean that friendships increasingly extend across spatial and temporal boundaries in ways that may, among other things, reveal personal information about participants to each other as well as revealing personal information about the researcher (Hall 2009).

Reciprocity in research

The issue of reciprocity and 'giving something back' has become an increasingly prominent concern in development research (see Harrison et al. 2001). These concerns are manifest on numerous levels, including the post-colonial-inspired call to design research questions in dialogue with researched communities, ensuring research findings are accessible to and are useful to multiple audiences and the question of whether or not to provide payment or gifts to research participants. Across all of these dilemmas 'issues of reciprocity and debt to informants come to the fore and are agonized over by most researchers' (Robson 1997: 66).

In determining how you will deal with the question of reciprocity in research, particularly with regards to whether or not to pay or provide gifts for participation, you will go through a 'negotiation of research ethics, local expectations, power and wealth inequalities, guilt and personal discomfort' (Hammett and Sporton 2012: 496). While Lloyd-Evans (2006: 158–9) suggests that 'researchers should value their participants' inclusion in a research project by offering an incentive or gift of some kind' we would suggest that this is too glib a response. Your decision on whether to pay for participation needs to be made based on a series of considerations including whether this will encourage people to participate and, if so, which groups and why? Are these the appropriate groups for your work? Might payment be seen as inducement rather than encouragement? Does the motivation for participation have the potential to affect your findings? (See Figure 5.2.)

Moving further with these considerations, you will need to consider what the current expectations are within the research site. Is there an expectation for payment, either due to precedent or because participants recognize that they are helping you to further your career (see Kaufmann 2002)? How will your decision fit to previous experience and what precedent might it set for future researchers? Are there concerns that payment for participation in research is leading to a commodification and routinization of research participation (and payment) and rehearsal of participants responses (Leadbeater 2000; McKeagney 2001; Cook and Nunkoosing 2008)? Will you pay all participants equally? Would you, for instance, provide a bigger payment/gift to an elite informant than to an impoverished participant? An elite informants' time may 'cost' more in absolute terms, but the relative cost to someone struggling to survive on a subsistence livelihood may be much greater (see Meth 2003a, 2003b; Van Blerk 2006; Porter et al. 2010).

You should also consider what your motivations are in making this decision, and whether or not you believe that payments and/or gifts mitigate or re-inscribe power and wealth inequalities (for contrasting views see McDowell (2001) and Meth (2003b) and a critical reflection between these views, Ansell (2001)). How do such decisions fit within broader debates concerning development research as a new form of extraction and colonial exploitation?

If you do decide to offer some form of payment or recompense, how do you determine who has access and the opportunity to participate in the research? This can be a particular concern in resource scarce settings, with concomitant issues of (perceptions of) favouritism, local gatekeepers determining who can/does benefit and who does not, potential risks to community goodwill and support for your presence and research if only a few benefit. In these situations, you may consider a donation to a community project or collective development need or activity. However, while this action may address some ethical concerns, it poses new questions. Will such an action affect potential participants' willingness to be involved? Will participants accept that non-participants in the research

Figure 5.2 When carrying out research, it is important to recognize the importance of being clear on any expectations and practices of reciprocity. In some situations, the provision of refreshments may be required or expected, at other times the sharing of information and research findings may be considered. Here, a local non-governmental organization (NGO) provided refreshments to elder women who were interviewed by students during a fieldtrip and the students provided their research findings back to the host NGO

Box 5.9 Whose decision is it?

In reflecting on the question of payment for participation during a fieldclass to rural Kenya, Hammett and Sporton critically reflect on the way in which this decision was reached and suggest that '[E]thics decisions during fieldwork . . . cannot solely be the domain of the researcher(s) and the privileging of their own ethical attitudes. Ideally, this negotiation of research practice should have involved local community members, translators/guides and local elites in order to have reached a contextually suited decision founded upon broader participation' (Hammett and Sporton 2012: 500).

To what extent do you agree with this statement? Why do you reach this conclusion? Were you to adhere to the sentiments expressed in this statement, how would you reconcile this with/integrate this into the ethics review process?

stand to benefit equally from the payment? If a community donation is made, who determines what this is spent on and who will benefit? These considerations will involve your engagement with local power hierarchies and relations, topics for consideration particularly if your work concerns issues of power, gender and equality (see Box 5.9).

In a similar vein, many researchers have expressed concern about whether or not an individual recipient of a payment for participation will make 'appropriate' use of this. We put scare quotes around 'appropriate' because these questions revolve around a set of (moral) judgements as to how resources *should* be used and implicates a set of concerns around who can make such judgements and raises questions of agency, (neo)colonialism and paternalism that require reflection. For some researchers, their preference is to provide donations or gifts of essential items, food, phone credit or fuel rather than cash donations, for others, a cash payment is preferred and for yet others, the decision remains to not provide payment or gifts at all. These decisions can be very challenging, both while carrying out individual research and while involved in a fieldclass. Harriet Lowes' (Box 5.10) experiences of these dilemmas while involved with an undergraduate fieldclass to rural Kenya not only highlight these complexities but also how ethical decision making continues throughout and beyond the lifetime of a fieldwork period.

In making these decisions it is important to find out about and reflect on how your practice fits into previous experience of researchers (is there a precedence to be followed) and the local moral economy, networks of reciprocal obligations and local social relations and rituals (see Bollig 1998; Hammett and Sporton 2012). In certain contexts, requests for payments and gifts may be made but without the expectation or obligation that they will always be met, but rather are 'playful' and a component of local social rituals (Twyman et al. 1999). At times, participants may make requests of researchers that are within local social norms but may be seen as unethical according to the norms and training of the researcher (Corbin and Morse 2003). It is important that you begin to think through these concerns prior to your fieldwork, not only because you will need to outline your decision and rationale in your ethics forms but also so that you are prepared to handle these often emotive and contested encounters in the field.

Working with assistants

Working with research assistants and translators raises a series of ethical questions beyond determining and agreeing an appropriate rate of pay (as well as expectations in terms of duties, hours and times of work, need to travel (and how this will be paid for) and when paid (how often, in advance or arrears)). There are a number of issues to consider, not least whether or not you will need an assistant at all. If you do, then you need to think about contextual factors, ethnicity, gender, age, language abilities, educational level and other factors in determining who would be appropriate for your project. Whoever you end up working with, you should discuss with them their perspectives on the issues being researched to ensure they are comfortable with working on the topic and that they are clear as to the necessity of ensuring that their views do not influence the work being done (Temple and Edwards 2002). Taking the time to ensure your guide/assistant/translator understands your research and their role is a crucial part

Box 5.10 Harriet Lowes: Negotiating payment and reciprocity on a development fieldclass

My experiences researching mobility and gender issues during a development geography fieldclass to rural Kenya in March 2014 triggered a great deal of personal, moral and ethical reflection. One of the biggest dilemmas we encountered was: how do you pay participants for their time, if you even pay them at all? Until we discussed it in a lecture prior to going to Kenya, I had given little to no thought to paying for interviews, I assumed that it would be a negotiation that I didn't have to be a part of. We discussed the ethics around paying for interview participation, having read Hammett and Sporton's (2012) paper, in the safety of our lecture theatre and the decision seemed an easy one: we wouldn't make payments to each interviewee, but would contribute to the community development fund to try and meet a collective need, as had been done in the past.

After nearly two days of travelling we reached the field centre and met our guides and explained our research topic and what kind of participants we were looking for. His immediate response: 'so, in the morning before interviews we will go and buy sugar for participants, you have to give them sugar'. Tentatively we tried to explain we had decided that we would give to a development fund and wouldn't be paying for interviews. No, he said, we had to give them sugar or no one would talk to us. Every guide insisted on this with each group. So, instilled with the fear that we would return from Kenya without data, and so nothing to write a research project on, we went back to the drawing board. The decision lay totally in the hands of the students, which provided a learning experience (although I think all of us, in a way, wanted to be told what to do as we were struggling to come to a solution that satisfied all 30 of us) and insight into some of the issues that need negotiating in the field during development research.

Our final decision was that we would give a small token of gratitude to participants for their time (soap, not sugar, as it transpired the guide's suggestion of sugar was tainted by vested interest), as well as contributing to a community development fund after returning to Sheffield. The most important thing to highlight is that the process of negotiating ethics and reciprocation for interview participation is mutable and heavily influenced by the fieldclass context. It has also been a process – in this case – that is not limited to the field and has continued since leaving (with the setting up of fundraising activities from Sheffield to send back to Marich Pass).

Whether to offer payment for participation was an ethical question that in many ways characterized this fieldclass. There was a strong desire to remain consistent while simultaneously offering relevant compensation. Ensign (2003) and Cook and Nunkoosing (2008) agree that payment should have the sole aim of compensating for time and expenses; however, in this case we placed an emphasis on the positive impact that human collectivity can have, and maybe should have (Cloke 2002), on the research community. The community development fund contribution was seen as a more holistic method of reciprocation for interview participation that did not solely aim to compensate for time and hospitality, and was intended to ensure that the research undertaken becomes less of a self-indulgent process and has more practical, useful applications at a local scale. Although we tried to come up with the best solution possible to suit the very specific context within which we were working, it's clear that there is no single, universal solution in determining how to compensate for participation.

Box 5.11 Siân Parkinson: Reflections on the role of the translator in cross-cultural research

While conducting field research during a fieldclass to Nepal, the significant role and influence of a translator in fieldwork was underlined on several occasions. We were conducting collaborative research on female empowerment with women from disadvantaged backgrounds and were aware that this was a sensitive research topic. The sensitivity of the topic enhanced the significance of the translator's role in the research process: she was tasked with translating questions that dealt with potentially sensitive topics, as well as personal and sometimes emotional responses.

Recognizing this, we began by establishing a positive and open relationship with our translator. We sat together and discussed at length our research topic, prior to starting fieldwork, so that she understood our research aims and the academic context in which they were situated. This process was vital in ensuring that the meaning of our research was fully understood by our translator and meant that she was able to critically engage with our research throughout the research process. Essentially, our translator provided both services of translation and as an external research critic, where from her own perspective as a Nepalese citizen she was able to offer us a deeper understanding into the cultural context of our research.

A key methodological issue associated with translation is whether language and its meanings are translated accurately and whether the research is able to capture the essence of the participant's responses. Language is socially acquired and therefore its meaning is dependent on the social and cultural context in which it is produced. The fact that our translator was familiar with the culture of our research area meant that she was able to provide our research with greater contextual grounding, permitting for a more in-depth analysis of responses.

Furthermore, the importance of fully explaining our research and its academic context to the translator was emphasized on a couple of occasions where our translator took the initiative to negotiate dialogues, later providing us with details and justifications as to why. The expertise of our translator meant that she was able to fully engage with the interview process, making decisions to reorder the interview questions, how to respond to answers and decide when was appropriate to translate, based on her own understanding of the situation. This experience of using a translator during cross-cultural field research emphasizes the numerous ways in which translators can influence and contribute to the research process.

Box 5.12 Assistant or overseer?

Kaufmann (2002) provides a vivid description of a range of challenges and questions posed by his ethnographic research in Madagascar being rigorously overseen by a local host and informant. Questions of power, identity, money and politics are clearly evident. After reading Kaufmann's article, reflect on how you might have dealt with such challenges and what alternative practices might have circumvented these. Would these have been possible and what other challenges might these have caused?

(Source: Kaufmann, J. (2002) 'The informant as resolute overseer', *History in Africa*, 29: 231–55)

---

of the process, as Siân Parkinson's reflections (Box 5.11) reveal from her experience of fieldwork during a postgraduate student fieldclass to Nepal. In this situation, developing the translator's understanding of the research meant they could support Siân in further developing her research questions and practice while in the field.

When hiring someone to work with you in your research it is important that you make clear at the outset that it is a trial period with no commitment to a length of employment. This is important to allow you to get to know your translator/assistant and to reflect on how comfortable you are with their work, whether they are steering the work in different ways to those you want to pursue or if their positionality or personality affects the interactions with your research participants (see Box 5.12).

Working with a local guide or assistant can affect your positionality as well as how potential participants view and locate your research. This can arise from your assistant's position within the local community: are they aligned to a particular group, marginalized from some sectors of the community, potentially distrusted because of ethnicity, level of education, previous conduct? For instance, Molony explains how working with a female research assistant was vital for accessing researching female traders in Tanzania due to religious and cultural norms, but this assistant was then accused of being a prostitute because of her employment by a white, male researcher (Molony and Hammett 2007). Drawing on his extensive fieldwork experience in East Africa, Thomas Molony's reflections (Box 5.13) on working with a number of research assistants highlight not only the logistical but also ethical decisions involved when working with assistants.

Allied to the dilemmas and concerns outlined in Molony's reflections are the intricacies of how to identify and access potential research assistants. Often this is done through a local university or NGO, at other times by word of mouth from previous researchers and in other contexts they are located through local community contacts. A number of questions arise through this process. If an assistant is sourced through a local university or NGO who appoints or approves them? If they are brought in from outside the community, might they be seen as a threat – perhaps someone who will report back to authorities about them (cf. Turner 2013)? If they are from within the community, are there

Box 5.13 Thomas Molony: Personal reflection on the role of research assistants

Language is crucial to any research. This is especially so for research conducted in a language that is not the researcher's mother tongue. Ideally the researcher becomes as competent as possible in the required language(s). The very act of learning a language can help the researcher to become accepted by some of those who speak it. It also helps the researcher to better understand the society that s/he is researching. Where and when the required language is learnt depends on a number of issues, including time, money and the availability of native speakers. In some situations, it is most practical for a researcher to learn the language where the fieldwork is to take place. It is often the case, however, that a researcher is not fluent in a language by the time s/he reaches the field. This can be where an interpreter or translator is used.

There has been a conspicuous absence of theory concerning the role of research assistants. Research assistants – those, often locals, who assist with research – are rarely addressed in 'how-to' research training guides that students may read before undertaking fieldwork. As Lynn Schumaker (2001: 12) has observed of anthropologists' finished texts, research assistants are generally invisible and, 'like wives', often receive a token measure of gratitude in the preface. Careful consideration of the role of any possible research assistance is integral to research design and planning *before* entering the field. This should extend to the ethical reviews that many researchers are expected to undertake prior to conducting fieldwork.

The need for research assistance is, of course, context specific, but the following questions are still important to consider as a starting point for all research that may rely on the assistance of others – whether the research is undertaken by a local with many years of experience or it is being conducted by novice:

Why is a research assistant required? As interpreter/translator? (Can I learn the language myself?) As a 'fixer'? (Do I need a guide? Can I build my own networks?) For his/her academic knowledge? (Is employment required? Might an academic act as a filter?) As a companion? (Does s/he need to be involved in my work?) As a factotum? (Are these tasks that one person is expected to fulfil or should I get to know different people with different strengths?) The assistance that is required will change, and depends on how much time the researcher can spend in the field.

Efforts need to be made into locating the best person for the job. *And it is a job*, since the research assistant is usually paid – although, in some cases, the payment may be in kind. The crux of the issue of research assistance is: What is the understanding between the researcher and the assistant over what the research involves, and who is expected to do what? Is there a contract for this employment? Is it a 'gentlemen's agreement' or a more formal contract? Who determines the terms of the contract? How might others become involved if the contract is breached?

Some aspects of the working relationship with a research assistant can be anticipated and should be written into any contract – if a contract is deemed necessary. But unforeseen events do arise and this is usually when the nitty-gritty of interpersonal relationships come into play. Unexpected circumstances are often the time when the unspoken power dynamic between the researcher (the employer, who is generally wealthier) and the assistant (or employee, who usually needs the work) is exposed.

Many successful working relationships take place between researcher and assistant, but this should not ignore the possibility of accidents and suchlike that may occur during fieldwork. The researcher usually has personal insurance, but this is sometimes not the case for local assistants: Should s/he be insured? If so, who pays for the insurance, and what does it cover? Contractual issues can reduce some of the excitement of doing research in new places, and may take some time. They certainly force both parties to discuss the research and the practical, day-to-day relationship between researcher and assistant. For example, is a time set aside – perhaps at the end of each working day – for both employer and employee to voice concerns about all aspects of the research?

The aims of the researcher and the assistant, linked in the short term, can be very different in the long term. For the researcher, the relationship can be a brief association within a long-term personal strategy (ultimately, in many cases, of career advancement). Some assistants who have regularly worked with a number of researchers may be used to this relationship and may have a similar strategy. Other research assistants can have different expectations. All parties should speak frankly about their hopes for the working relationship and for the research before they work together. This will raise many issues that are particular to the individual research project. Perhaps the one question that best gets the ball rolling and which may alter any working relationship between researcher and assistant is: What role, if any, should the research assistant have in the production of research outputs?

concerns that the knowledge they gain during your research might become public knowledge or be used against participants? Whoever you hire, and wherever they are from, it is vital that clear boundaries and ethical expectations are set out from the outset of fieldwork, including issues of confidentiality and protection of data and participants.

It is also important to consider the skills and proficiency needed for an assistant to help with your research to ensure the data produced are reliable and accurate (Molony and Hammett 2007). These will depend on the methods used, the theme addressed and the types of participants involved in the research (Boxes 5.14 and 5.15). In understanding the experience and abilities of a research assistant, you can ensure that these are appropriate for your research design while also considering whether to draw on their existing skills to develop new aspects of your research or whether your assistant may benefit from any forms of skills development and training you may be able to provide. You should also consider whether your research may have an emotional or other impact

Box 5.14 Who to hire?

Reflect on the following scenarios:

- You are working on a project about male and female sexual health in both rural and urban areas of Senegal. What factors do you consider in deciding on whether to utilize local research support? Assuming that you do decide to use local research assistant(s), what are some of your considerations in determining who to hire?

- You are a female European student working on a project with village elders in rural Tanzania relating to HIV/AIDS. What are the potential challenges and benefits of hiring a local, 24-year-old female assistant? What are the potential challenges and benefits of hiring a local, 50-year-old male assistant?

Box 5.15 Questions to consider when hiring research assistants

- Will your research need or benefit from an assistant or guide or translator? If so, why?

- What are the expectations of their role – in other words, what will you need them to do?

- What skills will be required to deliver on this role – language skills, research methods skills/experience, driving licence?

- For how long will you need them?

- How much are you planning on paying them? What further costs may be required (travel, accommodation)?

- Are there any factors to consider in terms of positionality and cultural norms in relation to the kind of person you are looking to recruit for the position?

- How might the presence of an assistant/translator affect your research – perhaps in terms of negotiating access, building rapport?

- What are the potential costs and benefits for an assistant in working for you?

- What groups are you seeking to engage and interact with and who is then best equipped to help with this?

- How will you acknowledge their role in your research? What involvement will they have in the writing up and presenting of findings?

on your assistant (for instance if working on human rights abuses) and how you could provide support in these situations.

Another important consideration for a translator or research assistant is payment. Initially, this may relate to how much to pay them: if you recruit through a university or non-governmental organization then it is likely that they will have a defined level of payment for such work. If, however, you source someone through a local community or other contacts you will need to negotiate a rate of pay directly with the individual. In such circumstances, it is useful to get an idea of what local wages are and to offer a fair and appropriate rate of payment. As Molony notes earlier, it is also important to think about other costs such as insurance, travel, food and so on.

An important practice to implement is to document all payments made, whether as 'wages' or 'expenses', to your translator/assistant. One of the easiest ways to do this is to purchase a receipt book and have your translator/assistant sign to confirm receipt of all payments made to them. We give this advice based on experiences of problems – our own and those encountered by some of our students – with translators and assistants claiming not to have been paid. This has been very much in a minority of occasions, but it is something easily and best avoided.

Box 5.16 Further exploration

To explore some of the issues relating to employing translators and research assistants these articles may be of use:

Araali, B.B. (2011) 'Perceptions of research assistants on how their research participants view informed consent and its documentation in Africa', *Research Ethics Review*, 7: 39–50.
Molony, T. and Hammett, D. (2007) 'The friendly financier: talking money with the silenced assistant', *Human Organization*, 66: 292–300.
Sanjek, R. (1993) 'Anthropology's hidden colonialism: assistants and their ethnographers', *Anthropology Today*, 9: 13–18.
Turner, S. (2010) 'The silenced assistant: reflections of invisible interpreters and research assistants', *Asia Pacific Viewpoint*, 51: 206–19.

Anonymity/confidentiality

In addition to gaining informed consent, you must also pay careful attention to the safety and wellbeing of your participants. In some situations, this may relate to the provision of safe spaces for interviews and focus groups, at other times, this may relate to minimizing the potential for emotional harm and distress. In other encounters, this attention may be related to the longer term safety and security of your participants and the information they have provided to you. Keeping research data safe and secure is a concern for all researchers in order to adhere to confidentiality pledges and ensure best practice.

Failure to do so can lead to participants being put at risk if interview or other data can be traced back to them. In some situations, this may cause embarrassment or tensions and problems within a community. In other contexts, such failings may place your participants in physical danger from state security forces, militia, criminal organizations or other parties. Such concerns are of particular concern when working in authoritarian states, illiberal spaces within otherwise liberal states or on sensitive or potentially controversial topics which mean it can be undesirable or unsafe to speak about certain topics or events (see Clark 2006; Koch 2013).

Therefore, you need to think about how you will handle and store data and paperwork so as to minimize the risk of these data and participants' details being accessed by third parties. You should consider how and where you will store materials, especially digital materials: USB pens are useful and portable but can easily be lost, individual computers are less likely to be lost but are more prone to theft, online and server-based storage options such as Googledrive or Dropbox provide benefits against the loss or theft of the storage device but may be vulnerable to data theft through hacking and also suffer from limitations on privacy as well as challenges regarding the ability to and security of data uploading. After deciding which storage device(s) you think are most secure, two further issues arise, namely, ensuring you have a back-up of your data and protecting or encrypting your files and storage devices. You should also give consideration to the secure storage of hard copies of both raw and processed data. These concerns are particularly important in conflict and post-conflict situations or when dealing with politically sensitive topics (for a compelling personal account of such encounters, see Gentile 2013). These concerns are high on the agenda of organizations such as Human Rights Watch. As Anna Neistat's reflection (Box 5.17) outlines, these considerations really are a matter of life and death.

A relatively common concern for development researchers relates to the use of transcription services to produce written versions of interview recordings. If you use a third party to transcribe interview or focus group data or to input survey data, you need to include this in your ethics paperwork and practices. In these situations, you need to ensure that the service provider has a rigorous policy on confidentiality and data disposal that meets the ethical requirements of your research and that they adhere to this. You should also have your own data disposal procedures in place, usually in keeping with your institution or funding body's guidelines on data storage and disposal. These procedures will include the length of time for which data are held upon completion of the project, how they should be stored and then the procedures for secure disposal of both hard-copy and digital data. Complicating these practices are increasing demands from funding bodies for research data to be made available through data repositories for use by other researchers. These pressures, as well as legal issues (e.g. obligations to report certain forms of criminal activity) and demands made by state agencies or through freedom of information requests for access to research data, underline the importance of explaining to participants that, while you will endeavour to provide anonymity and confidentiality, there are certain limitations to this (see Morrow et al. 2014).

In writing up your findings, there are a number of common steps you should take if you are seeking to provide anonymity to your participants. The primary practice here is the assigning of pseudonyms to participants – this should not only be practised in any

Box 5.17 Anna Niestat: Safety of witnesses as a primary concern

In the course of any research project, we conduct dozens, if not hundreds, of interviews with victims and witnesses of human rights violations and, at times, with the perpetrators as well. In all cases, we try to get as close as possible to where the violations occur, which sometimes means working in battlefields or in 'closed' countries or regions in which the authorities do not welcome our presence. In addition to methodological and logistical challenges, such work requires close attention to security concerns. We need to ensure not only the safety of our staff, but also the safety of witnesses and those helping us with our research: translators, drivers and local consultants.

One of the first steps toward ensuring the safety of witnesses is choosing a right location for an interview. It is always a difficult choice between the interviewee's dwelling (a home, a tent in a refugee camp), which provides the necessary privacy, but which could also draw unnecessary attention to the person, and a public place, which ensures the necessary anonymity and is less likely to be under surveillance.

At the start of an interview, we explain how the information is going to be used, indicate the potential security implications and make sure that the interviewee shares information voluntarily and with informed consent. We always respect requests for anonymity and would often avoid using not only names but also other details that might allow the identification of the interviewee in our publications.

Given the sensitivity of our research, we take special care when choosing translators, drivers or any other local staff with whom witnesses and victims may be in contact. We make it very clear that translators have to be discreet about what they see or hear during the interviews. When hiring support staff, we always seek recommendations and check their background to ensure that our research and the safety of our interlocutors won't be compromised. We also take measures to protect our local staff and their families should they face repercussions as a result of their work with us.

All our researchers are trained in communication and information security. A simple phone call through an insecure line or an unencrypted file may put someone's freedom or even life at risk. Before each field mission, we discuss in detail which method of communication would be the safest and most efficient: local or international mobile phones, satellite phones and internet, secure emails or other.

We also go to great lengths protecting the information we collect, such as interview notes or recordings, photographs and documents. In high-risk locations, we dispose of paper notes as soon as possible and keep the information electronically on encrypted laptops or remote servers. We act under the assumption that our possessions may be searched and they should not reveal the identities of the people we met with or the locations we visited.

Being security conscious during a research mission requires training, experience, and time, but it is a crucial component of our work. It allows us to ensure that the high-quality research that we produce is never done at the expense of anyone's safety.

written reports and papers, but also in your storage and handling of the data. In some cases, this may be sufficient, but you should also think about whether you may also need to change the names of places, institutions and organizations so as to provide a more robust level of anonymity. During Dan's work on citizenship and education in South Africa, the names of teachers and schools were changed but not the suburbs and townships in which these were located. As the research was relatively non-controversial, such a level of anonymity was considered adequate although those with detailed local knowledge may have been able to identify a number of the schools involved in the work. Depending on what you outlined in your ethics paperwork and agreed with your participants, your practices in data storage, handling and reporting need to adhere to associated best practice.

Morality and ethics in research practice

Fieldwork often throws up unexpected ethical dilemmas. Consider, for instance, how you would deal with a situation in which your research uncovers evidence of bribery and corruption: do you report this and, if so, to whom and with what potential implications for your own work and safety? Would this depend on the extent of these activities and whether such practices were relatively 'low level' and endemic to the extent they are an 'accepted' part of local practice? Alternatively, how would you deal with a situation wherein you are asked for a bribe? This may be to 'buy' your way out of a speeding ticket or to facilitate the processing of a research request. While there may seem to be an obvious 'ethical' decision to be made in these situations, you will have to consider the context and reality of your situation in determining your response. In some situations, you may feel that paying a bribe may be necessary to ensure your personal safety, in others, the exposing of corruption may place you or your research participants in danger. We cannot give a clear-cut answer as how to act in such situations. Rather, and as our reflections below indicate, ethical decision making during fieldwork is often complex, contested and can be simultaneously ethical and unethical (see Dhanju and O'Reilly 2013).

In this section, we want to share with you some encounters we have faced during fieldwork – either while conducting our own research or while leading student fieldtrips. In so doing, we intend to show you how ethical conundrums emerge during fieldwork and that you need to adopt an ongoing, reflective approach to research ethics above and beyond the paperwork exercises conducted prior to departing for the field. In presenting these experiences, we do not give you answers as to how to respond to them, but rather pose a series of questions for you to engage with and to reflect on how you might deal with such encounters.

On a fieldtrip to rural Kenya, two small groups of students, a university staff member, two local guides and local driver had travelled to a pastoralist community to conduct interviews. During the day one of the groups interviewed a woman with a very sick infant who urgently needed hospital treatment. The mother was unable to pay the cost for a motorbike taxi to take her to the local hospital, a mere seven kilometres away, and was waiting for her husband to return home in order to transport the child to receive health care. The students were unsure what to do – should they continue the interview or leave?

Could they offer to ask our driver to take the mother and child to the hospital? Could they offer to give the woman the money for the cost of a motorbike taxi to get her and her child to the hospital?

During fieldwork on citizenship education in South Africa, Dan spent time observing history and life orientation lessons in a number of schools across the country to see how teachers translated textbook citizenship ideals into meaningful content for high school students from a range of socio-economic backgrounds. During recesses Dan spent time talking with staff and students. On one such occasion, Dan was talking with a male teacher when a female teacher who was on playground duty rushed over to get the male teacher's help – a large fight was breaking out between some of the older students and she needed help to break it up. Without thinking about the ethics of the situation, Dan joined in the

---

Box 5.18 Further examples of ethical dilemmas in development research

Phillipe Bourgois, a leading American anthropologist, provides a vivid and personal account of a number of his engagements with ethical concerns and dilemmas during a number of fieldwork projects in Latin America. While few researchers will encounter ethical and moral conundrums in such visceral ways during their fieldwork, Bourgois's recounting of his experiences of conflict and violence and subsequent dilemmas regarding potential uses of his data, threats of his exclusion from academia and so on provide some fascinating and provocative insights into these debates (see Bourgois, P. (1990) 'Confronting anthropological ethics: ethnographic lessons from Central America', *Journal of Peace Research*, 27: 43–54).

If you are considering working in a conflict area or on a topic relating to conflict or violence, you may also find this collection of essays useful: Cramer, C., Hammond, L. and Pottier, J. (eds.) (2011) *Researching Violence in Africa: Ethical and Methodological Challenges*, Leiden: Brill.

With resonance to some of the later considerations raised in Bourgois's work, David Price's reflections on how intelligence agencies seek to access and make use of anthropological research raises a further set of ethical dilemmas: Price, D. (2002) 'Interlopers and invited guests: on anthropology's witting and unwitting links to intelligence agencies', *Anthropology Today*, 18: 16–21.

In another context, Jelke Boesten (2008) reflects on an unravelling of research in the field amid a set of ethical and moral challenges. In this instance, research on HIV/AIDS interventions in Tanzania with a carefully considered participatory approach was brought to a premature end amid acrimony and conflict. Boesten's reflections on these experiences reminds us of the ongoing negotiation of field relations and fieldwork ethics and the multiple layers to these: see Boesten, J. (2008) 'A relationship gone wrong? Research ethics, participation, and fieldwork realities'. NGPA Working Paper Series No. 20, London School of Economics.

teachers' efforts to break up the fight. Once the dust had settled, Dan realized that, while this instinctive reaction may have been helpful to the teachers, it also posed dilemmas about his positionality and a blurring of the lines of his role in the school.

What to take from this chapter

This chapter has provided a brief background to research ethics and the importance of these in designing and conducting research. There is a tendency to view ethics forms and procedures as a bureaucratic burden and hindrance to academic research. However, the importance of a rigorous ethics process is not only the ethics paperwork produced, but the thought process involved as you scrutinize your research plans and methods. These reflections contribute to ensuring that your research is viable and adheres to practices that will minimize the chances of your research causing harm to participants. In proceeding through the ethics process, you will also have developed key documents, including a consent form and plain language statement, which will be used during fieldwork.

More broadly, we have provided a set of reflections and thoughts on wider ethical dilemmas in fieldwork, drawing out some of our own, personal experiences of fieldwork to demonstrate how unexpected ethical and moral conundrums are part of the fieldwork experience. As was noted at the start of the chapter, we have not sought here to provide a series of easy solutions to ethics concerns, but instead attempted to guide you through some of the common procedures and experiences from our own fieldwork. In so doing, we do not claim that our way of doing things is the best or right way.

If you want to explore the ideas covered in this chapter in more detail, you may find these further readings useful:

Griffith, D. (2008) 'Ethical considerations in geographic research: what especially graduate students need to know', *Ethics, Place and Environment*, 11: 237–52.
Iphofen, R. (2009) *Ethical Decision-Making in Social Research*. Basingstoke: Palgrave Macmillan.
Israel, M. and Hay, I. (2006) *Research Ethics for Social Scientists*. London: Sage.
Lunn, J. (ed.) (2014) *Fieldwork in the Global South: Ethical Challenges and Dilemmas*. London: Routledge.
Mertens, D. and Ginsberg, P. (eds.) (2009) *The Handbook of Social Research Ethics*. London: Sage.
Thomson, S., Ansoms, A. and Murison, J. (eds.) (2012) *Emotional and Ethical Challenges for Field Research in Africa: The Story behind the Findings*. London: Palgrave Macmillan.

6
RISK AND FIELDWORK

Identification and assessment of risk relating to fieldwork, and mechanisms to reduce these dangers, are integral parts of the research design process. Increased stringency of legal responsibilities – both with regards to criminal and civil liabilities – have resulted in universities and other institutions implementing increasingly robust procedures for risk assessment practices. For those researchers travelling overseas, particularly to remote locations in the global south, these pressures are increasingly evident in the scope of pre-trip paperwork required. In this chapter, we outline the importance of risk assessment practices and then proceed to reflect on a range of risk areas and possible practices that can help minimize the residual risk involved in fieldwork. These engagements are important not only with regards to your safety and wellbeing but because your approach and reaction to risk and danger will influence your data collection and everyday field practice (Kovats-Bernat 2002).

In providing this guidance, we hope that you do not get caught up in the popular perception that health and safety regulations and risk assessments are designed to stop you doing things. In our encounters with safety services/health and safety staff through training sessions and individual communications, they present a consistent message: risk assessment is not designed to prevent work that involves a degree of risk; it is intended to support and facilitate safe working practices through the identification and adoption of strategies that reduce the severity and likelihood of harm. This ethos comes through in Shane Winser's reflections (Box 6.1) based on her experiences as Geography Outdoors Manager for the Royal Geographical Society – a role in which she supports the training and best practice of fieldwork and research. As she notes, risk assessments are about demonstrating a substantive engagement with the risks of fieldwork and recognition of making sure that what you *do* in the field keeps you and others safe.

To illustrate this approach in practice, we can reflect on Dan's experience of conducting a risk assessment for an undergraduate fieldclass to Kenya in the immediate aftermath of the Westgate Shopping Mall attack in September 2013. Dan had begun updating the risk assessment from the 2012 fieldtrip in early September 2013 (ahead of the March 2014 fieldclass) and identifying various action points and details that needed updating. The importance of making sure this was completed in detail and with appropriate consideration for inherent dangers of fieldwork in Kenya had already been underscored by the challenges faced by emergency services responding to the fire that destroyed part of Jomo Kenyatta International Airport in Nairobi on 7 August 2013. Then, on 21 September 2013, al-Shabab

Box 6.1 Shane Winser: Managing risk as part of development fieldwork

The challenges of doing fieldwork in unfamiliar places, where local cultural norms are outside a researcher's previous experience, can be scary both for the fieldworker and their supervisors back home. Not everyone will embrace the experience but all concerned should have a reasonable expectation that the researchers should return with no lasting damage to their physical or mental health.

The Royal Geographical Society (with IBG) encourages university students to undertake field research in remote and challenging environments by providing funding through its grants programme and with advice, information and training though Geography Outdoors: the centre supporting field research, exploration and outdoor learning. As such, it has a moral responsibility to ensure researchers do not cause or come to harm and, in an increasingly litigious world, support higher education institutions (HEIs) that may feel increasingly nervous about allowing students to take part in development fieldwork either in supervised groups or on their own.

Fear of litigation and UK health and safety regulations have led to an increasing emphasis on risk assessment documents as a means of demonstrating due diligence by HEIs. Thus institutional health and safety requirements are often seen as an unnecessary bureaucracy rather than a means of keeping researchers safe from avoidable harm. Risk assessments should be seen as tools to demonstrate that researchers have appreciated the significant risks associated with their fieldwork and have proactive measures in place to reduce the likelihood of these occurrences.

Tick boxes that list the risks to be considered are not sufficient. The key is good preparation and training, where the competence of individuals to participate in a specific fieldwork project can be evaluated. Here the key role of the supervisor is to make a professional judgment about the suitability of the researcher to undertake their intended fieldwork. If the student does not have the necessary competencies, the supervisor will need to encourage the student to modify their plans until these are within their capabilities.

Risk assessments need to reflect what people will actually *do* in the field to keep themselves and others safe. These actions must be specific to the individual piece of fieldwork and practical in their application. The focus needs to be on the significant and foreseeable risks: If things do go wrong, and they often will, a good emergency response plan should make access to help less stressful and more efficient and might well save a life. But remember the old adage: prevention is better than cure!

Resources:

The British standard for overseas expeditions and fieldwork. BS 8848 (2007 + A1: 2009): a specification for the provision of visits, fieldwork, expeditions, and adventurous activities outside the UK. This specifies the operational requirements for those responsible for organizing overseas activities and student placements abroad.

Off-site safety management course for those involved in planned visits in the UK or overseas. Spread over two days, it offers an in-depth look at the safety management issues involved in planning, managing and evaluating local visits, fieldtrips, residentials and exchanges: www.rgs.org/OSSM.

Universities & Colleges Employers Association (2011) UCEA/USHA Guidance on Health and Safety in Fieldwork. Published jointly with USHA, this document provides a framework for establishing policies and procedures that enable staff, students and other participants in higher education institutions to undertake fieldwork safely.

terrorists attacked the Westgate Shopping Mall in Nairobi. Even though Dan's fieldtrip would spend only a few hours in Nairobi, the Westgate Mall attack had focused the attention of students, their parents, academic staff and safety services staff on the safety of the Kenya fieldclass – and on the risk assessment document in particular. Although the existing version was thorough and robust, a number of additional practices were identified and incorporated into the document and fieldclass procedures, including scheduled reporting in activities and improved strategies for monitoring FCO travel advisory updates during the fieldclass, in an effort to further mitigate potential dangers and enhance the capacity of the fieldclass teaching team to respond to a crisis event.

Why risk matters

We encounter risk in our daily lives and constantly take steps to avoid the hazards and dangers we face in navigating everyday needs to minimize the chances of loss, injury or death. These efforts are simple forms of self-preservation, from the mundane actions involved in crossing a road safely, through not consuming food or drink in a laboratory, to avoiding the various 'dumb ways to die' identified by Australia's Metro train company (http://dumbwaystodie.com/). When you are representing your university or another institution, either as a staff member or a student, these everyday actions become one part of a broader set of health and safety expectations and obligations. These obligations are enshrined in criminal and civil liability, requiring you to act in ways that protect your own health and safety and that of others.

The legislative context varies between countries, but the common principles remain largely consistent. For UK-based researchers, fieldwork (defined as work conducted 'for the purpose of teaching, research or other activities while representing the institution off-site' (USHA/UCEA 2011: 7)) is covered by various legislation including the Health and Safety at Work etc. Act (1974), the Management of Health and Safety at Work Regulations (1999) and the Corporate Manslaughter and Corporate Homicide Act (2007), as well as being subject to the laws of any countries in which fieldwork is undertaken. The criminal liabilities resulting from this legislation relate to the responsibilities of both institutions and individuals for the wellbeing of the staff member(s) or researcher(s) as well as students and the general public who may be affected by your actions. Researchers based elsewhere are subject to similar legislation: useful guidance for health and safety legislation in the USA can be found here: https://www.osha.gov/; for Australian-based researchers such details can be found here: http://www.worksafe.act.gov.au/health_safety; and for Canadian researchers guidance is provided here: http://www.ccohs.ca/.

Risk matters in fieldwork planning. First, it is important to ensure your own health and wellbeing, as well as anyone else involved directly in your fieldclass or fieldwork. Second, it is important because you hold certain responsibilities – linked to criminal and civil liabilities – to act in ways that do not place others at risk. As an independent fieldworker, you are responsible for taking reasonable care in your activities, including ensuring that you have a sufficient risk assessment in place for your fieldwork. If you are a participant in a fieldclass, you retain responsibility for your own conduct, while also observing and following appropriate instructions provided to you by fieldclass staff and reporting to them any problems or questions.

Awareness of risk and the identification of ways to avoid risky behaviour and minimize risk are important throughout the research process. These considerations will inform the design of your research (from identifying research sites to decisions over modes of travel and transport operators) through the period of fieldwork (your actions and behaviour in the field) and into the writing up and sharing of research findings (your storage and use of data and efforts to ensure the safety of your research participants). As a result, your engagement with risk assessment extends beyond the completion of a risk assessment document prior to fieldwork; the practice of dynamic risk assessment carries on throughout your research. Dynamic risk assessment means that you continually assess for unforeseen risks or changes in risk levels and adapt your actions and activities accordingly to control these threats.

In the midst of thinking about how best to manage risk and danger in your fieldwork you should also reflect on how you will handle other people's fears and stories of dangerous places. On the one side, this may be seen in terms of reassuring an (over)anxious parent, partner or friend that the place(s) you are going to and the things you will be doing are not as dangerous as they fear and you will act in such a way as to further reduce these dangers. Often this reassurance is required to counteract negative media coverage and poor destination images of many areas of the global south (see Ferreira 1999; Cornelissen 2005; Nelson 2005).

On other occasions, you may encounter family, friends, researchers, travellers and expats regaling you with cautionary tales of the dangers (from wildlife, climate, crime, road safety, etc.) of the place you will be visiting. In these situations, you will need to consider how accurate these stories are and what likelihood you think there is of these happening: in some instances, the warnings of places to avoid will be accurate and useful, on other occasions, the tales will be overblown anecdotes with little foundation in reality. It is down to you to identify and utilize the useful tips, bearing in mind your perception and tolerance of risk may well be very different from the person talking to you. You may want to compare anecdotes with government travel advice and/or seek counsel from supervisors or other students you know who have travelled or worked in the area you will be visiting. This can allow you to both get a sense of the veracity of the danger being highlighted and a chance to talk about your own fears or concerns in a supportive and constructive manner.

The next section of this chapter addresses your engagement with risk and risk assessment in planning fieldwork, followed by a discussion of dynamic engagements with risk assessment during fieldwork. The discussions presented here are not exhaustive. There will be other factors and risks that we do not address here and there are a range of additional useful resources relating to risk and danger in fieldwork that we would encourage you to engage with. It is important to remember that as well as the national legislative framework your university will have its own health and safety policy on fieldwork that you must adhere to. It is possible, therefore, that you will need to provide details of risk assessment and management (including consideration of threats due to (in)security, climatic extremes, endemic health issues and crisis events) as well as consideration of emergency response and management.

Assessing risk in preparing fieldwork

Risk assessment is an integral part of research design and, as Simon Jones (Box 6.2) outlines in his reflections of many years of fieldwork and field teaching, is an activity that needs to be conducted rigorously. One of your primary considerations in designing your research is *where* you will do your fieldwork. At this early stage, you will need to be aware of any potential threats to the stability of your proposed fieldsite. For instance, is your proposed destination the site of ongoing conflict or subject to any travel advisory notices? Most countries now provide online country-specific travel advice. You can find the UK's Foreign and Commonwealth Office guidelines here: https://www.gov.uk/foreign-travel-advice; the US's Bureau of Consular Affairs guidance here: http://travel.state.gov/travel/travel_1744.html; the Canadian government's advice here: http://travel.gc.ca/travelling/advisories; and Australia's 'Smart Traveller' pages here: http://www.smartraveller.gov.au/.

If there are travel advisory notices in place you will need to consider whether you are still willing to travel to a (potentially) unstable area. If you are, then you will need to discuss this in detail with your supervisor and safety services/health and safety personnel at your institution to explore whether such travel would be approved, subject to a satisfactory risk assessment and justification for such travel as well as being able to secure appropriate travel insurance. Many insurance providers exclude travel to areas subject to 'do not travel' notices so you would need to check your policy carefully to make sure you would still be covered if travelling to such areas. There are a number of specialist insurance providers that may be able to provide cover, subject to a premium. In addition, you will need to reflect on how the context of (in)security will influence your methodological approach and data collection practices (see Kovats-Bernat 2002; Haer and Becher 2011).

Completing your risk assessment form can be a useful way of helping you think through the intricacies of your research design. While there is no single, universal template for these forms, it is likely that you will be asked to respond to a series of issues and questions

Box 6.2 Simon Jones: Risk assessment and incident management in fieldwork

Fieldwork has defined the teaching of natural sciences and those who teach it. It's common to receive the offer to 'carry your cases on the next trip', to be asked 'where are you next going on holiday?' and have to deny the wardrobe cliché of the corduroy jacket (with leather elbow patches).

The growing diversity of fieldwork in higher education means that fieldwork is no longer confined to the natural sciences and reflects the academic trend to incorporate experiential student-centred learning into an enriched curriculum. Fieldwork is thus a vivid highlight for many students, remembered long after numerous lectures and seminars have blurred into one, then faded. Consequently, stout walking boots, corduroy and Kendal Mint Cake are no longer sufficient to sustain a decent day in the field . . .

I have organized more fieldwork events than I care to count, ranging from full days local to the university to extended international expeditions in remote parts of the world. Preparation for fieldwork is an art! The need to recruit to the module, accommodate timetable disruption, secure the teaching team and confirm travel logistics demands sophisticated project and budget planning. In addition, there will be an unpredictable gamut of health issues and dietary requirements within each year's diverse group of students to manage.

While crises very rarely happen during fieldwork (when such acute and challenging circumstances arise, managing the situation is rarely borne individually), it is the risk assessment paperwork against which your actions will be judged. Hence, the increasingly systematized and potentially formulaic tasks of completing the annual risk assessment proforma for the field course, a background administrative task in a time-pressured working environment, carries inherent risks if ignored or completed inadequately.

In reflecting on my preparing for fieldwork, I carry years of investment in first aid courses and expeditionary medicine qualification, mental ill health workshops and risk assessment courses. I have actively engaged with the Royal Geographical Society, British Standard 8848 systems, private medical training courses and pored over field medicine handbooks, field course textbooks and HEA workshops. I carry a comprehensive first aid kit that would not look out of place on a TV hospital drama and have on occasions had to plunder its contents. However, completing the risk assessment proforma with the student cohort has always been the start of the field course and it happens in the lecture theatre. Shared discussion of risk, the variety of environments to be encountered, cultural differences and the behaviour expected and an appreciation of health and safety in the field enables the students' buy-in to the risk management. Additionally, hidden resources are more easily discovered among the students that may include local language skills, first aid qualifications, mountain leadership qualifications and the like.

From a basis of sharing an appreciation of the risk all face, actions required during a crisis and how such events may be recovered from, anticipation can actually be heightened and expectation managed.

Box 6.3 Safety guidelines for fieldwork

Specific guidelines and requirements for risk and safety during fieldwork will vary by institution and legislative context. The list of resources below provides a range of examples of safety guidelines, checklists and protocols from universities around the world. These should provide a sense of the kinds of guideline and practice you may be expected to adhere to in designing and conducting your own fieldwork (see also the sources listed in Shane Winser's contribution above):

Duke University:
   http://www.safety.duke.edu/SafetyManuals/University/Z-SafetyGuidelines Fieldwork.pdf
University of Amsterdam:
   http://www.falw.vu.nl/en/students/regulations/safety-in-fieldwork/index.asp
University of Exeter:
   http://www.exeter.ac.uk/staff/wellbeing/safety/guidance/fieldwork/
University of Texas at Austin:
   https://www.utexas.edu/safety/ehs/fieldguide/field_guide.pdf
University of Western Australia:
   http://www.safety.uwa.edu.au/topics/off-campus/field-work-remote
University of Western Sydney:
   http://www.uws.edu.au/__data/assets/pdf_file/0020/7058/ Fieldwork_Safety_ Guidelines.pdf

(for guidance on common questions and issues see the guidelines signposted in Box 6.3). Initial information requested is likely to be mainly logistical, asking for your itinerary, contact details while in the field – for you plus any other travellers and your next of kin, details of your passport, visa, travel insurance, initial accommodation and similar logistical details, as well as a brief justification for the trip. You may also be asked to detail any vaccinations or medication requirements and health conditions.

Further requests for information may include outlining contingency plans and emergency communication and management procedures you have put in place. In addition, you should detail any reporting/checking in procedures you have put in place, as well as arrangements for remote supervision during the fieldwork period. These may include a rough schedule of email correspondence, Skype meetings, telephone calls or other communication.

The bulk of the risk assessment form will comprise your detailing of the various hazards foreseen for the fieldwork, a level of risk associated with these, details of the potential consequences from the identified risk, your control measures (how you will reduce the level of risk) and the residual risk left once the control measures have been put in place. The quantifying of risk level is subjective and can be measured in a number of ways. One of the most common scoring systems is based on a matrix of high, medium and low

Table 6.1 *Risk-level matrix*

	Rating and score	Description
Severity	High (3)	Potential for death or severe injury/illness
	Medium (2)	Potential for injury or illness resulting in absence from work
	Low (1)	Potential for minor injury requiring first aid treatment
Likelihood	High (3)	Likely to occur in time, hazard permanently and consistently present
	Medium (2)	Likely to occur in time but hazard intermittently present
	Low (1)	May occur in time with hazard being infrequent
	None (0)	Hazard removed or rendered impossible
Overall risk rating	High (6, 9)	Priority risk, requires further control measures to reduce risk level
	Medium (3, 4)	Lower priority risk, should consider additional mechanisms to further reduce risk
	Low (0, 1, 2)	No further action required

scores for the severity of harm and likelihood of occurrence, with the risk rating based on a multiplication of the two scores (see Table 6.1).

When you complete your risk assessment form, you should demonstrate both an awareness of the range of potential/likely risks you will encounter during your fieldwork and how you will seek to minimize these. In outlining these control measures, you need to demonstrate that you are meeting the legal requirement to control for a risk 'so far as is reasonably practicable' (USHA/UCEA 2011: 20). For instance, if you know that while conducting fieldwork in Uganda you will travel around Gulu or Kampala by *boda boda* then there is a relatively high likelihood of being involved in an accident resulting in injury or death, giving this hazard an initial score of 9. In an ideal world, you might say that you would use a private taxi to travel around as a means of reducing/removing this danger. Assuming that is not possible due to cost, you can take certain control measures such as only using *boda boda* during daylight hours and taking your own motorcycle helmet to wear when riding a *boda boda*. These actions would reduce the risk rating while providing specific justification for continued practice.

Your risk assessment form will ask you to detail potential hazards linked to a number of categories, commonly including health, transport, legal, crime/security, political instability/terrorism, extremes of climate or topography, hazards associated with fieldwork activities, accommodation and any other hazards or considerations (for a further list, see Box 6.4). Attempting to cover all potential risks within these categories is beyond the scope of this book, but what follows are some common sense considerations.

Health issues can be considered in relation to various concerns. Are you travelling to an area subject to endemic or epidemic diseases? If so, how do you mitigate the dangers of these? In some instances, this relies on vaccinations and/or prophylactic medication such as anti-malarial drugs. In other cases, conducting or avoiding certain activities may be required – for instance, covering bare skin and using insect repellent or avoiding swimming in rivers or pools. General knowledge and education about local health dangers

Box 6.4 Common hazards – a checklist

- Travel to domestic departure point
- Travel to international destination (mode of transport, transport provider – e.g. are all airlines you are using IATA accredited)
- Travel from international arrival point to and around vicinity of the fieldwork location
- Personal health concerns and ability to access health care services
- Access to clean food and water in fieldwork location
- Communication options – for general contact, supervision/reporting in and in emergency
- Environmental and climatic factors and extreme weather
- Threats from political instability or terrorism
- Crime and safety – in relation to personal wellbeing, loss of cash/cards/passport/valuables, loss of data
- Hazardous flora and fauna
- Lone working
- Equipment and manual handling
- Methods-related risks
- Unsafe accommodation
- Dangerous activities
- Risky behaviour

is integral to these considerations. You should also ensure that you carry with you any personal medication you may require, including a reserve supply to last a few additional days in case of delays in travel or loss of luggage. In instances where medication may require storage under particular conditions, you will need to account for how you will achieve this while on fieldwork (see Figure 6.1).

Other health concerns can include access to safe and clean food and water. In such situations, you may need to think about how you will access clean water: buying bottled water is one option, but you may also think about water purification techniques. Consideration should also be given to your ability to access health facilities, whether for first aid treatment through to accident and emergency facilities. Common measures you can adopt here include ensuring that you are carrying your own first aid kit (which may include a 'sharps' kit containing sterile surgical and dental supplies), knowing the contact details for local medical facilities and national emergency contact numbers (which may include flying doctors).

Climate and topography can present a number of risks, ranging from exposure to sun stroke to altitude sickness. In responding to these risks, you should think about appropriate clothing, rehydration techniques, use of sunscreen, adapting working patterns to avoid the worst of the heat or storms and so on. You should also ensure you are aware of the symptoms of relevant medical conditions and have in place contingency plans for

Figure 6.1
Health and safety issues can arise from a range of
dangers, including handling equipment or carrying out
tests on local water sources (as in this picture). Within
your risk assessment you should note these dangers
and how you would mitigate them (in this instance, by
adopting careful hand-washing practices, sterilizing
equipment and not carrying out the water-testing
procedures in food preparation areas for example)

evacuation if necessary (for instance in case of altitude sickness or pulmonary oedema).
You may also be working in areas where there are cliffs, caves, quarries, forests,
quicksand, rivers or oceans, each of which presents its own set of risks and you should
consider how you will mitigate or avoid these through your everyday actions in the field.

Concerns about crime and personal safety may inform how and when you move around
in or avoid certain areas. You may want to think about how to spread the risk posed by
losing your wallet or purse or its being stolen. For instance, can you split your money
and bank cards into different 'stashes' and ensure that they are kept secure separately?
It is always useful to have a separate note of the lost/stolen contact number for your bank
cards. You should also have a copy of your passport and visa page. If your passport is
lost or stolen this can help tremendously in getting new documents issued and in some

countries in which you are required to carry identification with you at all times, you may be able to carry these copies rather than your passport.

As you can see from the limited number of reflections and questions provided above, a comprehensive risk assessment can be a lengthy document and one that takes time to complete. You can find an example template of a risk assessment form used for development fieldwork in Box 6.5. It is likely that your institution has its own risk assessment paperwork and you should complete their version of the paperwork: we include an example here to help you think through key issues and questions. You should ensure you give yourself plenty of time prior to your planned fieldwork to get this paperwork completed and approved. The protocols and control measures set out in the document should inform your daily practice in the field. However, fieldwork is a dynamic experience and the context within which you are doing your research can change rapidly. As a result, you will need to maintain a dynamic risk assessment throughout your time in the field.

Dynamic risk assessment and management during fieldwork

The political, social and physical environment in which you conduct fieldwork can alter rapidly as political protests rapidly escalate or a natural disaster such as a hurricane or tsunami strikes. Such episodic events may require you to conduct a dynamic risk assessment of the emergent and shifting risks that results in a sudden and dramatic change in your fieldwork practices. It is good practice to register, where possible, with your country's consulate or high commission in your fieldwork country, particularly if you are working in a country or region prone to instability (such details are readily available online). You can also register for country-specific travel advisory updates by email from the Foreign and Commonwealth Office (www.gov.uk/foreign-travel-advice).

Continually reflecting on levels of risk and ways of reducing this should not be confined to such episodic events, but should rather play an important role in your everyday research practice. A common concern during fieldwork relates to harassment and petty crime. As you become accustomed to the field, you may recognize certain spaces and times when these threats are more pronounced and adopt practices to avoid these, either by travelling at different times or by alternative routes or adopting another form of transport. Other practices may include efforts to modify appearance through changing clothing styles and not carrying items that make it clear you are a visitor to reduce these attentions to a degree (see Bell 2013).

Dynamic risk assessment is an integral component of fieldwork but you should not use this as a reason for not having adequate contingency plans in place. In many instances, these plans are relatively straightforward, involving slight adaptations of time, location or transport mode. In more serious situations, contingency plans should include consideration of evacuation plans and emergency communication procedures. In such situations, being registered with the relevant consulate (see above) is important as well as ensuring that you have a clear list of emergency contact numbers for in-country and home-country support. These may include contact numbers for your consulate and for emergency medical evacuation providers, as well as designated contacts in your university and/or department (for instance your supervisor, head of department or institution's safety

Box 6.5 Example risk assessment form

Risk Assessment Form
Department
University

Name:

Course (if relevant):

Module code (if relevant):

Dates of travel (outward and return):

Fieldwork & Risk Assessment

Fieldwork is practical work carried out by staff or students of the University for teaching and/or research in places which are not under University control, but where the University is responsible for the safety of its staff and/or students and others exposed to their activities. Across the Univeristy, but particularly within the Department of Geography, employees and students travel extensively for fieldwork purposes.

The majority of this fieldwork will occur within areas deemed safe by the Foreign and Commonwealth Office. For proposed work in areas deemed as unsafe, the Department required a very robust risk assessment to be completed before consideration is given to approving such a trip.

All proposed fieldwork must be covered by a valid risk assessment before commencement. This requirement is rooted in the Health and Safety at Work Act, 1974 (HSAWA). For employees, HSAWA requires you to take responsbility for your own safety and to co-operate with your Institution's health and safety arrangements: for fieldwork this means that you MUST complete and have approved a Risk Assessment for the trip.

The Head of Department must, under HSAWA and recognised in civil and criminal law, exercise a 'duty of care' towards staff and students and is responsible for risk assessments (the appropriate regulations state that they are "to make a suitable and sufficient assessment of (a) the risks to the health and safety of employees.... and (b) the risks to the health and safety of persons not in his/her) employment". The completion of a risk assessment for fieldwork is required under this commitment.

Personal and Travel Details

Please provide the following details	Details here...
List the main activities on your fieldwork	
Give the location(s) – country, regional, town/village.	
Dates of travel (a trip itinerary must be attached):	
Passport details of traveller(s):	
Confirm your passport valid for appropriate period of time after travel and with required number of blank pages:	

Is a visa required for your travel? If so, please give visa details.	
Do you need any other letter of permissions or introduction? If so, please give details.	
Provide next of kin details for traveller(s):	
Provide travel insurance policy details (ensure is comprehensive, covers all trip activities and includes repatriation/evacuation cover) (include company, policy number, emergency evacuation number etc)	
Confirm that you have checked the FCO travel advice – www.fco.gov.uk and it is safe to travel (date last checked).	
Provide justification for the trip if it involves any areas subject to travel advisory notices from the FCO.	
Flight details (airline, flight number, departure and arrival time):	
Airport transfer arrangements:	
Accommodation details while in-country (name, address, phone, duration of stay):	
In-country contact details for traveller(s):	
Emergency contact details for use by traveller(s) in emergency – local:	
Emergency contact details for use by traveller(s) in emergency – UK:	
Contact details for relevant in-country High Commission (UK and your national embassy)	
Contingency plan for emergency:	
Do you have any health conditions the Department should be aware of? If yes, please provide details on a separate sheet in confidence.	
What immunizations are required?	
Confirm that you have adequate personal medication, required immunizations and personal first aid kit.	
Details of supervisory arrangements (for directly, indirectly and non-supervised fieldwork) and frequency of checking in via email, telephone etc.	

Hazard Detail specific hazards foreseen within each category	Initial Risk Level	Potential consequences (Detail potential outcomes of hazards)	Minimise risk by: (What control measures will you take to reduce the level of risk?)	Residual risk
Disease (e.g. malaria, rabies, other infectious diseases)				
Climate & weather (e.g. exposure)				
Hazardous flora and fauna (e.g. venomous creatures, predators, poisonous fungi)				
Physical safety (e.g. personal attack, abuse, assault)				
Transport and vehicular (e.g. local driving conditions, excessive driving hours, road-worthiness of vehicles, remote or hazardous terrain, check validity of license & insurance)				
Food and drink (e.g. safety of local water)				
Terrain (e.g. quicksand, cliffs)				
Methods-related (e.g. lone working, interviews in private spaces)				
Security (e.g. theft)				
Accommodation (e.g. security, emergency procedures/fire risk)				
Local customs (e.g. religious practices, dress codes)				
Security of data and prevention of harm to participants				
Economic (e.g. loss of bank card, theft of cash)				
Legal (e.g. specific local laws and customs, alcohol prohibition)				
Political stability & Terrorism (protests, civil unrest, terrorist activities)				
Final assessment of overall risk level:				
Assessor: Date:				

services). Having a clear note of your next of kin or other designated contact person can be useful for emergency responders in case you are involved in a serious incident. One accepted method for this is to have an ICE (in case of emergency) number in your mobile phone – this number being for a local designated contact who knows who you are, can provide some assistance locally and who can activate further contacts to next of kin, university or other support providers.

In addressing these concerns, you should also consider how reliable communication networks will be in your fieldsite (i.e. can you call for help if you need to), ensuring your travel insurance will cover evacuation and medical treatment costs and ensuring you have the capacity to cover any emergency transport or other costs.

Doing no harm – risk to others

As noted in Chapter 5, the principle of doing no harm underpins social science research. As well as an ethical consideration, this concern also deserves attention within risk assessment practices. It is rare for risk assessment forms to contain a specific category for 'security of data and prevention of harm to participants' but we would encourage you to add this to the 'other hazards' section of your form. The risks in this case relate to your safety and security, as well as of your research assistants, drivers, translators and participants if your presence is noted and/or data are stolen, lost or requisitioned by a state agency or other interested party and anonymity and confidentiality compromised. Government interest in research data is not unheard of: Gentile (2013) offers a useful summary of different ways in which the state may become interested in your work, how this may manifest itself and influence your research and your safety and potential ways to try to mitigate such issues.

While in the field, you will need to navigate the local political context and associated concerns with surveillance and state interference. Your actions in the field will, to an extent, be informed by these attentions. Depending on the situation and level of state interest you may opt to 'play dumb' and carry on, try to befriend the authorities or to bribe them (not that we are condoning this) or to change the emphasis or popular story about your research to try to reduce the apparent interest of your work to the organs of the state (Gentile 2013; also Clark 2006). As well as influencing your actions in the field, you should also consider your data storage and handling practices. At a basic level, you should ensure that names are not attached to recordings or transcripts and that your metadata containing identifying information is encrypted or password protected. You may also need to think about alternatives for data storage – if you are working in particularly sensitive areas perhaps using a secure server to host your research materials is necessary, rather than saving them to a computer or memory stick.

The risk posed to participants and research co-workers may not only occur while you are in the field but also continue after you have left the field. Depending on the political and security context in which you conducted your work just the knowledge that someone participated in your work may place them in danger (for an accessible set of reflections on conducting research during insurgency and fears for the safety of research participants see Pettigrew et al. 2004). It is important, therefore, to ensure robust practices relating to anonymity and confidentiality in your analysis and writing. If you are concerned that including a piece of information or quote may place an individual, group or community in danger, leave it out.

What to take from this chapter

Fear should not pervade your research design or practice. However, it is important to give adequate attention to questions of risk and danger throughout the research process. To put it bluntly, you can only write up and reflect on (dangerous) contexts if you survive them. To survive them you need to consider the specific local context and potential risks presented by working in this environment and to identify and adopt strategies or control measures that reduce risk to a low level.

Risk assessments, as with ethics (see Chapter 5), should not be viewed simply as a bureaucratic hurdle to be jumped or as a single, static one-off consideration. Instead, you can use the completion of a pre-fieldtrip risk assessment as an opportunity to reflect on the intricacies of your research design, from logistics to methods to ethics. An ongoing, dynamic approach to risk assessment throughout your fieldwork will not only help ensure your safety but can also provide useful reflections on your data collection and fieldwork practices.

If you want to explore the ideas covered in this chapter in more detail, you may find these further readings useful:

Belousov, K., Horlick-Jones, T., Bloor, T., Gilinskiy, Y., Golbert, V., Kostikovsky, Y. et al. (2007) 'Any port in a storm: fieldwork difficulties in dangerous and crisis-ridden settings', *Qualitative Research*, 7: 155–75.

Hecht, T. (1998) *At Home in the Street: Street Children of Northeast Brazil*. Cambridge: Cambridge University Press.

Kenyon, E. and Hawker, S. (1999) '"Once would be enough": some reflections on the issue of safety for lone researchers', *International Journal of Social Research Methodology*, 2: 313–27.

Lee-Trewick, G. and Linkogle, S. (eds.) (2000) *Danger in the Field: Risk and Ethics in Social Research*. London: Routledge.

Nordstrom, C. and Robben, A. (eds.) (1995) *Fieldwork Under Fire: Contemporary Studies of Violence and Survival*. Berkeley, CA: University of California Press.

Sharp, G. and Kremer, E. (2006) 'The safety dance: confronting harassment, intimidation and violence in the field', *Sociological Methodology*, 36: 317–27.

Part II
COLLECTING AND ANALYZING DATA

7

INTERVIEWS AND FOCUS GROUPS

Interviewing has become one of the most commonly used research methods in social science research. In part, this is because interviews are seen as being an easy way to find things out, a continuation of the everyday practice of conversation or 'conversations with purpose'. The apparent effortlessness of interviews that we witness broadcast on radio and television programmes – be these with expert talking heads on Al-Jazeera's *Inside Story* or celebrities on *Parkinson* or *Letterman* – further reinforce the perception of interviews as an easy and straightforward means of gaining answers from knowledgeable others to the questions we have.

In reality, conducting a good interview or focus group is far from an easy option. They are challenging encounters that require careful planning and preparation, as well as energy and concentration to conduct well. Interviewing is tiring, using mental and nervous energy in preparing and waiting for the interview, concentration and focus during the interview while you listen to responses and constantly think ahead to further questions and restructure the interview to maintain the narrative, as well as adapting and responding to unexpected moments and responses and then the efforts of writing up reflective and contextual notes and transcribing recordings.

This chapter provides you with guidance in the use of interviews and focus groups, from epistemological concerns and the preparation for interviews and focus groups to conducting an interview or focus group and recording both verbal and non-verbal data. In prompting you to engage with this process as a whole, we hope you will realize that these methods construct data through the interaction between the interviewer and the interviewee (Roulston et al. 2003: 645). We hope our guidance is useful, but would strongly urge you to practise your interviewing skills prior to your fieldwork as: 'Interviewing is a skill, one that's not readily learned in the classroom' (Corbin and Morse 2003: 347). Pre-fieldwork practice will help you to develop the skills in preparing for interviews as well as the skills needed to develop rapport, to handle interviews sensitively and to record verbal, non-verbal, contextual data accurately (see Figure 7.1).

Why use interviews and focus groups?

Interviews and focus groups can be used to elicit a range of information. Most commonly, interviews and focus groups are used to develop detailed, *subjective* understandings

Figure 7.1 Carrying out interviews during fieldwork can be a complex process, potentially involving a guide or translator, building trust and rapport with respondents and taking notes both of the interviewee's responses and the micro-geographies of the interview context and location

drawing on people's knowledge, memories and perceptions. For those with a social constructivist or naturalist approach to knowledge the targeted, insightful and detailed responses generated through interviews to explore subjective understandings, experiences and perceptions are of great interest. These methods are best suited to the development of qualitative data to explore 'how'- and 'why'-type research questions through detailed, multi-layered insights into topics such as identity, social issues and citizenship. Interviews and focus groups are less useful for positivist approaches where concerns would be raised over the objectivity of the data generated.

In order to develop good verbal data through interviews and focus groups you need to understand how the answers sought in these conversations speak to your research questions (see Longhurst 2010). Depending on your overarching research questions and approach you will need to decide on the type(s) of interviews you will use. From this understanding, you then need to identify the questions you want to ask and who you want to speak with (and how you will recruit them) based on an engagement with both theoretical and empirical background materials. These steps are important to ensure you are well prepared: a poorly prepared interview is unlikely to produce useful data and will waste your time and that of your research participant.

Types of interview

Various types of interview are used in social science and development fieldwork, each with its own strengths, weaknesses and purposes. While some projects use one type of interviewing, others may strategically use different styles of interviewing at different points in the research process or with different participants to elicit a rich and varied set of responses. Which type(s) of interview you decide to use will depend on the type of research question being addressed, the information and data needed to answer this question, and the types of people you will interact with.

Structured interviews follow a rigid, predetermined set of questions and prompts. They are generally used for finding out factual information (often with key/expert/elite informants from whom you want facts and figures) or as part of survey-style research (to allow you to develop comparisons and analysis between interviews). Structured interviews provide a relatively high level of comparability and the possibility to develop comparative analysis. However, this approach does not allow any freedom to follow up on unexpected responses or to explore meanings and understandings. Structured interviews can also be quite staccato conversations with little rapport and engagement.

Unstructured interviews are approached as free-flowing conversations that have a predetermined starting point but with further questions developed during the interview based on the interviewee's responses. These interviews are challenging to conduct and often lack focus and clarity of purpose, with responses veering away from the topic of interest. In these interviews, you act as an active and focused listener and having asked the interviewee to talk through their engagement with an experience, idea or event you can ask follow-up or clarifying questions during the narrative or at the end (see Corbin and Morse 2003). As there is no consistency to questions across interviews, these are not suited to comparative analysis.

Semi-structured interviews are generally conversational in tone and partially structured, roughly based on a set of predetermined questions to direct a fluid discussion around the central topic (see also Longhurst 2010). The interviewer and interviewee determine the flow of the conversation, exploring responses and emergent topics and tangents of relevance. You engage in active listening throughout, adapting your questions and the interview flow to address salient responses. This type of interview allows for a degree of comparative analysis and is well suited to exploring understandings and perceptions.

Narrative/oral history interviews are used to develop experiential understanding and can be used to explore the generation and outcomes of policies and practices in development. These interviews work from a general narrative question and move on to narrative and specific probing around events and silences or absences. In these interviews, you do not interrupt the speaker's narrative, but listen actively and ask follow-up and probing questions at the end of the narrative. While these interviews take time, both to build the requisite rapport beforehand and to conduct, they can lead to things being said that would otherwise remain unspoken. These interviews are subjective recollections of history and the biases and silences they contain help you to understand the multiple and continual constructions of self and society. Narrative interviews are therefore of limited use for finding out 'facts' or 'truth' but are a useful tool to develop insights into local

Box 7.1 Oral history repositories

Narrative/oral history interviews have been used as a research tool for those conducting biographical research, but can also be used to understand political and developmental challenges over time (see for instance Portelli's (1991) work on how memories of a striking worker's death had a significant impact on social and political relations in Italy). You can find useful further readings as well as oral history archives through the following links:

African Oral Histories Online:
   http://www.africanoralhistory.com
British Library:
   http://www.bl.uk/oralhistory
Centre for Popular Memory,
   University of Cape Town: http://www.popularmemory.org.za/
Collection of Oral History Recording Database,
   National Archives of Singapore: http://a2o.nas.sg/cord/public/internetSearch/
National Library of New Zealand:
   https://natlib.govt.nz/researchers/oral-history-advice
Oral History Australia:
   http://www.oralhistoryaustralia.org.au/index.html
World Bank Archives:
   web.worldbank.org

subjective understandings and experiences and how these shape interactions, identities and relations, not least though relating to the personal and the political. Oral history projects also often produce public archives of these testimonies: Box 7.1 provides some examples of a number of these that are relevant to development research.

Ethnographic interviews are similar to *unstructured* interviews in many ways, lacking a predetermined set of questions or narrative. They are ad hoc discussions that occur in the field that gradually build up a set of information around a topic. They are not simply conversations, but are prefaced by an explanation of the research and gaining consent for the interview to happen. Such interviews can add detail and nuance to your understanding but are challenging to conduct.

There are many other forms of interview that you may also find useful. These include *problem-centred* interviews, which begin with a conversational entry before leading on to general and specific prompting around a problem or scenario. This approach uses an interview guide to give a common theme to the interviews but is used with flexibility to explore follow-up questions in more detail (Flick 2009). *Focused* interviews use a stimulus – a film clip, a newspaper story, a policy document – at the start of the interview during which you ask the respondent non-directive and then increasingly structured questions in order to understand their subjective reading of or response to the stimulus

(Flick 2009). This approach to interviewing could also use ranking exercises or concept cards (see Sutton 2011) through which participants can develop and attach stories, meanings and descriptions as material objects that may help them conceptualize and explain their views and ideas.

Preparation

Recruitment

The first step in recruitment is to identify the appropriate types of person, group or organization to approach based on your research questions. You then need to consider how you will set about sampling. In many cases, this will involve purposive sampling, aimed at ensuring you have respondents relevant to your research, or snowball sampling, particularly if you are working through a local gatekeeper.

Once you have identified appropriate individuals or groups to approach, you will need to negotiate access and contact. In some situations, this is relatively straightforward, involving an email, telephone call or face-to-face conversation and invitation to participate. In others, notably when seeking to access elites, this process can be far more challenging and can require working through gatekeepers (secretaries, personal assistants) as well as the embedded and enacted power structures that surround elites (McDowell 1998; Bachmann 2011; Mikezc 2012). In such situations, personal contacts or a powerful sponsor to support your approaches and facilitate introductions can enable and/or speed up these processes significantly (Rice 2010).

Questions and flow

The success of an interview is largely dependent on the questions asked and how these are linked together. A poorly structured interview or one with irrelevant or poorly framed questions is unlikely to produce good-quality information. In approaching your interview, you need to think through a series of questions yourself:

1   Why ask this question (what relevance does it have to my theoretical frame/my overarching research questions/will it contribute to the research)?
2   What is the substance and focus of the question (how will it help answer my research questions)?
3   Why is this question important to ask (is it simply repeating another question/is it relevant to my research)?
4   Why ask the question in this way (what type of response is desired (open, closed) and is it phrased in a way that is understandable and unambiguous)?
5   Why ask this question at this point in the interview (does it fit to the flow of the interview and relate to the questions around it)?

Overall you need to consider how a flow of questions will best help you discover what it is you want to find out. The first level of this questioning strategy is to think about the structure of the interview and the timing of different kinds of question. For example, you may want to start with broader questions before narrowing down to very specific questions or you may want to begin with something specific and then open the questions out to address the bigger picture. There is no single rule as to the structure of an interview, but thinking through the relationship between your research questions, interview questions and data sought will help you work out the best flow and structure.

Linked to consideration of the structure of your interview are decisions over the types of question and if/where these fit into your interview. Depending on the type of interview being conducted and the kind of information sought, you may decide to use different types of question during an interview. These may include unstructured questions that invite an open response to a broad reflection or issue, semi-structured questions that ask for an open response to a concrete topic or a more defined response to an open topic or structured questions in which both the topic and response are defined. You may also decide to use specifying questions to refer back to a specific event or response in order to gain deeper insight into that issue. In some cases, you may decide to use leading questions in order to transition into new topics or to probe for more depth, but you need to be careful in how you use these questions so that you do not bias responses. Whichever type of questions you are using, it is important to ensure they are worded and framed appropriately for local sociocultural norms.

Ensuring that your questions are clear and accessible is vital. You should therefore only ask one question at a time: asking a question with multiple sub-clauses/sub-questions is confusing so ask a series of simple questions instead. Similarly, questions with contradictions or negatives in them are often confusing. The overall message here is that even if your questions are probing and challenging, you should find simple and straightforward ways of asking them.

In most interviews, you will want to gain some background biographic and/or demographic details from your respondent. Asking a series of closed questions at the beginning of an interview may provide useful conversational starting points and a narrative entry into the interview. Alternatively, they can be disruptive and may damage rapport if they address personal questions (such as age or income) before the participant is comfortable discussing such areas. If such questions are left to the end, however, there is a danger they will get discarded if you run out of time. Working out how (and when) to incorporate these questions can be tricky and you may prefer to use a short questionnaire provided to the participant prior to the interview to collect these data instead.

Conducting interviews

Carrying out interviews can be stressful and intimidating, particularly if you lack confidence or are generally quite shy (see Scott et al. 2012). Your own demeanour and actions, your performance, during an interview can have a strong bearing on the data produced. As Jude Murison's reflections (Box 7.2) on 15 years' working in central Africa identify, there is a big difference between 'any data' and 'good data' and poorly conducted interviews are highly unlikely to furnish you with 'good data'.

Box 7.2 Jude Murison: The patience of an interviewer

'I did not ask you! I asked him!' As the woman spluttered the words with a mixture of hiss and spit, her pointed finger moved from the direction of one man's face to the other. I had unfortunately ended up visiting rural farmers on an NGO project at the same time as someone from the communications department of a major funder seeking 'sound bites' to use with promotional material. While my usual technique is to spend days observing the daily lives of people followed by one-to-one in-depth interview or focus groups discussions preceded by an explanation of who I am, the work I do, and the (un)likely impact of the research, time was clearly of the essence as photographs were being snapped snapped snapped by a camera with a huge lens and no request of consent from the people whose pictures were being taken. As I stood there, like most researchers, and probably like most human beings with an ounce of humility about them, loaded thoughts raced through my mind, 'respect and humanity would be nice', 'has she never interviewed people before', 'I hope these people do not place me in the same category as her' and so on.

A friend of mine once said, go onto a street in your hometown with a clipboard, telling people you're doing research and asking how much money is in their bank account, how many people do you think will tell you the truth? Yet that is what we assume when we do work in developing countries, that somehow we take advantage of the power dynamics whereby the poor think by answering our questions they are helping our research, or perhaps there will be a return – foreigners are, after all, likely to be donors. And as Westerns we take advantage, throwing in a soda or piece of soap as a 'thank you'. Which brings to the main point I always make in discussing research – it's easy to get 'data', but it takes skill to get 'good data'.

Sometimes the best interviews happen though unexpected means: In 2000 I interviewed a refugee settlement commandant who'd been with the refugee ministry since the 1970s. His bosses had given permission for the interview, but despite this he was hesitant to respond. When I asked questions about the party in power in the early 1980s, he responded to my reference to the UPC (the Uganda People's Congress) as if I'd publicly said some secretive code word. The next time I met him I stayed one week at the refugee camp, now defunct since the majority of refugees had returned home. I did my research and apart from cordial conversation, I didn't ask him any questions for fear of triggering his paranoia. On the last day I went to say goodbye, and happened to mention that in another camp, I'd seen the signature of an MP in the visitors' book from the early 1980s. I recalled the date and person to him. You were the camp commandant then, do you remember that day? 'Do I remember? Do I remember? The man wanted me killed . . .' and in great detail he recounted the day and the expulsion of the Banyarwanda from Uganda. I tried to scrawl down some notes and as I did, he started to speak more slowly to assist me in getting down his words. Here was a story he wanted to tell.

What have I learned from 15 years working in Rwanda, Uganda, Burundi and Congo? There is a huge difference between obtaining 'any data' and 'good data'. The latter requires respecting people, being patient and looking for the details that others may have overlooked.

Implicated in Jude's reflection is the need to take care, to show respect and develop your position within the field. Thus, ahead of an interview you need to consider appropriate choices of location (see discussion below) and clothing. If you are interviewing a senior government official or company executive then smart, business-type attire would be advised, whereas if you are interviewing *zabbaleen* in Cairo or *palero basurero* in Manila

---

Box 7.3 Andrew Maclachlan: Reflections on interviewing during an international development fieldclass to South Africa

As an undergraduate human geography student at the University of Sheffield, I travelled to South Africa as part of a fieldclass group to undertake research hosted by local non-profit organizations. The trip itself generated many challenges regarding development research. Of particular concern, and the biggest issue encountered by our group, surrounded interviewees' perceptions of us, a group of young, well-educated Western students. Our research entailed speaking with and interviewing beneficiaries of a charity who had suffered serious life setbacks resulting in homelessness. For the beneficiaries to be greeted by a group of Western students wanting, within a very limited time period, to investigate their lives seemed rather intrusive: at times we questioned the morality of this. We also sought to develop different research skills to mitigate these concerns and provide us with fieldwork findings.

Initial discussions with beneficiaries failed to produce any meaningful research. After reviewing our practices, an alternative method was employed: instead of approaching beneficiaries as a group of four students, we subdivided into pairs and used a more conversational approach in introducing ourselves and building rapport. The potential research participant was then invited to be part of our research, with clear explanation of the benefits. We found this approach to be more successful in increasing participation.

In a similar manner, group conversations were found to be an effective tool for our research. This consisted of researches forming a group circle on the floor and inviting residents to join in a discussion. Many residents would simply sit and listen to the conversation for a period of time and then join in when they felt they could contribute. This method was particularly effective as other participants would encourage additional residents to join in, creating inter-resident dialogue. At the start of one these situations, the research group was asked if they wanted seats, opting, however, to join residents on the floor enabled barriers and hierarchies to be mitigated to an extent through situating both parties on a level playing field. In these situations, we didn't use a Dictaphone or take notes as we did not want the group members to feel they were being examined. Instead, at the end of the conversation the researchers reviewed the discussion and generated detailed notes. While this is not effective as direct recordings, it provided in-depth research within a limited time period.

such business attire would likely be inappropriate. Similarly, your mode of address and body language may vary between interview encounters as you adapt the formality of your speech, phrasing of questions and style of interaction to suit the individual encounter and to put your interviewee at ease. The negotiation of these dynamics requires a degree of cultural sensitivity and self-reflexivity to realise when interview dynamics are working or not, as Andrew Maclachlan's experiences (Box 7.3) during an undergraduate field-class to South Africa demonstrate. Andrew's experiences show how individual and group practices can have a significant bearing on the success – or otherwise – of research interviews.

Your performance during interviews reflects your engagement with broader sets of power relations, whether you are in a position of power and privilege (as is common) or when this is inverted (as may be the case in elite interviews). Conducting interviews with elites poses a number of specific challenges including access, openness and focus; you may well have to adopt different strategies for such interviews compared with other interview respondents (see Harvey 2011). Regardless of the power relations within an interview, you need to be aware of and adhere to cultural norms and practices of behaviour and etiquette in order to develop rapport and trust (Mikecz 2012). Allied to these concerns is awareness that how you (and your assistant/translator if you are working with one) are positioned by your respondent – including your nationality, age, ethnicity, gender, politics, worldview and identity – will influence what information is revealed and how this is presented. While there may be little you can do to influence certain of these markers, you may want to consider how you present your identity in the field and the amount of personal information you reveal – personal history, shared experiences, personal anecdotes, world-views – as these issues can influence how people respond in an interview. (In many situations, the sharing of experiences and ideas can be tremendously beneficial for building rapport and opening up dialogue, as Mary Suffield found during a fieldtrip to Kenya (Box 7.4). In her experience, developing more conversational dialogue and sharing personal experiences helped develop deeper responses to interview questions.) These concerns raise a range of ethical and other issues, including gender- or race-of-interviewer effects on interviewee's willingness to reveal information in interviews, which are addressed in more detail in Gunaratnam (2003), Garton and Copland (2010), Razon and Ross (2012) and Takeda (2012).

It is important to remember that interview data are subjective. Responses and recollections may be partial, inaccurate or biased and you need to reflect on and critically analyze your data accordingly. It is therefore important to respond carefully to requests for clarification and guidance from interviewees and to provide a reassuring and supportive environment in which nervous or shy respondents are able to develop confidence in their participation (see Irvine et al. 2013).

Another common issue in interviews can be interviewees providing answers that they think you want to hear or that position them in a specific way that will gain your approval. To mitigate these challenges, often implicitly expressed in comments from interviews stating 'I hope you got what you wanted/expected' or 'I hope those were the answers you wanted' at the end of an interview, it is important to build rapport before and during an interview and to reiterate that an interview is not a test of knowledge or logic and there are no 'right' answers. You may find techniques such as including emphasizing shared

Box 7.4 Mary Suffield: Sharing information – the two-way process of interviewing

There is only so much that preparation can achieve when conducting development fieldwork for the first time. Open-mindedness and a willingness to deviate from interview schedules and carefully planned methods are necessary to create an environment in which interview participants become empowered, providing richer, more accurate data. Having previously only conducted interview research within familiar cultural contexts, my consideration of reflexivity was limited through unfamiliarity with local culture and context. Conducting research as a small group (four students) during a development geography fieldclass resulted in our strongly adhering to interview schedules and being reluctant to improvise during research encounters. This induced dualities of power, with a clear dominant interviewer and submissive respondent. Instead of an interactive dialogue, an extractive process ensued.

Reflections on the success of each interview in the field carried substantial importance in the progression of our interview technique. It quickly became clear that reciprocity should be actively endorsed, enabling knowledge to be gained not only for our benefit, but also the participant. Questions were invited in both directions, creating a more conversational atmosphere and reducing the monopoly of control over the knowledge exchange that we had previously held. These modifications developed our relationship with the respondent, allowing for empathy and, when appropriate, a sense of humour to be involved. The resulting conversational tone meant that any questions considered to be judgemental or inconsistent with the context could be comfortably challenged by the participant, providing them with additional power. Rather than maintaining a focus solely on our research interests, the participant's own curiosities were explored, ranging from supermarkets to famous footballers' religious beliefs. This was beneficial to both parties. The control the participants had over the knowledge they gained redistributed power dynamics and we benefitted through gaining avenues of exploration previously not thought of.

Interestingly, a demographic pattern emerged: we found that women were more inclined to participate in conversational interviews. As our research group were all female, this may have resulted from a higher level of identification (Scheyvens and Leslie 2000). However, our differences were what truly animated reciprocal discussion. Marital status was regularly discussed; when we exposed a difference in status to interviewees, a wider discussion developed. Although post-colonial theorists may suggest that this form of sharing knowledge may reinforce power discourses of Western culture as superior, the reactions received when sharing our experiences did not reflect this criticism. Female participants were not apprehensive in expressing their opinions on our lifestyles, often adopting a sense of humour in their own criticisms, suggesting that reciprocity motivates participants to speak their mind.

I recognize that sharing information is not unproblematic. While answering the questions of participants, advantages that we experienced were revealed. Sensitivity in responses was key so as not to challenge the social structure by undermining those who hold power within the community. Maintaining positionality from the participants while creating a more informal atmosphere was demanding, especially when questions inducing feelings of discomfort were asked.

characteristics and experiences, using humour and reassuring respondents about their views and competencies within the interview setting can help build rapport and alleviate reticence and shyness (Scott et al. 2012).

Building trust and rapport can be vital if you have to handle sensitive issues and emotional responses. For those dealing with sensitive topics, it is vital to consider how you will broach these questions and how you will deal with responses to these. Consideration should be given to the appropriateness of particular questions and the framing of these. Returning to the underlying mantra of doing no harm, you should begin by thinking about whether questions that may evoke emotional responses on sensitive topics are necessary. More broadly, apparently innocuous research questions can elicit emotional responses and you should be prepared to think about how to continue with a discussion on this topic, whether to take a break, to change topics or to continue with related questions: see Box 7.5 for an example of such an encounter from one of Dan's research projects.

To further illustrate and develop some of these considerations, it is useful to reflect on the following anecdote and extract from Dan's research on race, identity and social

Box 7.5 Emotional issues in interviews

During an interview Dan conducted in Cape Town with a female teacher, her response to a question about memories of anti-apartheid protests evolved into a discussion of her family's experience of apartheid and her disclosure that a family member had committed suicide due to the experience of oppression under apartheid. As this revelation was made, the teacher – in an emotional state – explained that she had never revealed this to friends or colleagues but that the interview had provided an unexpected cathartic moment to talk about this event. In this situation, Dan allowed the interviewee to move the conversation forward at her own pace and direction, providing her the space to deal with the emotions evoked and to steer the discussion as she felt most comfortable.

In this situation, how would you deal with this and support the interviewee? What other ways could you handle the conversation? What emotions and feelings might this have generated for you – the researcher? How you would deal with these?

change in the Cape Flats – an area of Cape Town formerly designated for the 'coloured' and 'Indian' residents of the city. During an interview with one teacher, the discussion turned to the ways in which experiences of apartheid framed his engagements with politics and race in the post-apartheid period. From previous experience, Dan was unsurprised that the teacher began to talk about personal experiences of protests and violence but this narrative was more of an oral history, providing a deeper and more personal story:

> I started out teaching in Manenberg, I was there during the 1976 riots . . . I remember in '76, I approached a venue in Manenberg – it had been looted and policemen were inside. They were waiting – they had heard that people from Gugulethu, the 'black' township were coming to loot. And we were standing there and waiting and as these guys approached from Gugulethu we warned them, 'don't go in there, the police are waiting for you'. They just ignored us completely and went into the bottle store, and we saw bodies flying as the police shot at them from inside. That was first experiencing death right in my face . . . Three days later I was picked up by the security police . . .
>
> While I was at Hewat Training College, we were in what is now Klipfontein Road. We were at the bottom of Klipfontein marching towards Rondebosch, which was a 'white' area. So what the cops did, they covered the 'white' area to keep us out. Now the tactic of the police was that they would shoot the main person in that crowd. They would identify the leader – he's got a white shirt or a red shirt and a sharp shooter would kill that guy and everyone else would disperse. On this particular day, one of my colleagues at Hewat – he had on a red shirt and we were marching towards Rondebosch and he was shot right in front of me: blood spattered on my face. A sharp shooter had taken him out on the order of a 'coloured' sergeant. That evening, we went to the youngster's home. His father was there and we spoke to him and the father was waiting for his son to come home. The son never came home. It was the same sergeant who had given the order to shoot. He had killed his own son. He committed suicide, I think, a month or two later.

This interview remains one of the most challenging of Dan's career, taking the narrative into deeper and more personal reflections on traumatic events than he had anticipated and beyond the level most development researchers might expect to encounter. In this situation, how do you prepare for and react to such stories in terms of minimizing the potential for emotional harm? With such a powerful and personal story, how do you ensure the participant is willing for you to write about this material? With such a compelling and sensitive narrative, how do you try to ensure its accuracy – do you seek to triangulate data or do you accept that the value of oral history testimonies as being partial? In this situation, Dan allowed the interviewee to dictate the direction and pace of the reset of the interview and then at a later date revisited and clarified with the interviewee that they were happy for the content to be used. A number of factors contributed to the depth of this interview: in part it was the rapport and trust built up during ethnographic fieldwork and in part it was because it was part of a strong narrative the interviewee was passionate about and wanted to share. The possibility to develop this rapport and share the narrative in such detail was also aided by the common language spoken by the interviewee and interviewer.

The language used during the interview is another important factor implicated in the production of knowledge. If you are able to conduct interviews in the local language, this can often provide richer data as it allows respondents to think aloud and conceptualize ideas in their preferred language. When this is not possible, you may be able to conduct interviews in English or another colonial language, accepting that this may limit nuance and expressive detail. Alternatively you may need to conduct cross-language interviews using a bilingual research assistant or translator, with associated complications to be considered in terms of knowledge production relating to their positionality and the accuracy of the translation and capturing of nuance (see Box 7.6 for further readings on this concern; and perhaps reread Chapter 5 of this book).

In the midst of these discussions and considerations, you should not forget the basic logistics and practicalities. Do you know where, exactly, you are meeting your

---

Box 7.6 Working with translators

Identifying and working with a translator is a vital decision. You may find the following sources useful for further insights into the factors needing consideration in such a decision:

Lopez, G., Figueroa, M., Connor, S. and Maliski, S. (2008) 'Translation barriers in conducting qualitative research with Spanish speakers', *Qualitative Health Research*, 18: 1729–37.

Maclean, K. (2007) 'Translation in cross-cultural research: an example from Bolivia', *Development in Practice*, 17: 784–90.

Ndimande, B. (2012) 'Decolonizing research in postapartheid South Africa: the politics of methodology', *Qualitative Inquiry*, 18: 215–26.

Temple, B. (2002) 'Crossed wires: interpreters, translators, and bilingual workers in cross-language research', *Qualitative Health Research*, 12: 844–54.

Temple, B. (2006) 'Being bilingual: issues for cross-language research', *Journal of Peace Research*, 2: 1–15.

Temple, B. and Edwards, R. (2002) 'Interpreters/translators and cross-language research: reflexivity and border crossings', *International Journal of Qualitative Methods*, 1: 1–11.

Temple, B. and Young, A. (2004) 'Qualitative research and translation dilemmas', *Qualitative Research*, 4: 161–78.

Wallin, A.-M. and Ahlström, G. (2006) 'Cross-cultural interview studies using interpreters: systematic literature review', *Journal of Advanced Nursing*, 55: 723–35.

Williamson, D., Choi, J., Charchuk, M., Rempel, G., Pitre, N., Breitkreuz, R. and Kushner, K. (2011) 'Interpreter-facilitated cross-language interviews: a research note', *Qualitative Research*, 11: 381–94.

interviewee? Do you know how to get there and how long it will take you? If you are driving, is there somewhere to park (and do you have change for the parking meter)? Depending on where you are meeting an interviewee, do you need to have some form of identification with you or a letter of invitation? This is often the case if visiting government departments or central offices of major NGOs or governance institutions – as Dan has learnt from almost missing out on interviews with senior officials in the South African Department for Foreign Affairs when he had to plead with the building's security officials to be allowed in for the meeting after he had left his passport and driver's licence behind.

Facilitating focus groups

Focus groups are one-off meetings of a group of people, usually between five and ten participants, who engage in a discussion that is facilitated and directed by the researcher (Hopkins 2007; Jakobsen 2012). These groups can be drawn from pre-existing collectives, which can mean there is already trust and rapport, but may preclude openness on certain topics in case opinions expressed are carried beyond the focus group setting and affect ongoing relationships. Alternatively they might be drawn together specifically for the discussion, with resultant challenges of developing rapport and productive group dynamics, although the relative anonymity may also encourage people to speak more openly and freely (Longhurst 2010) (see Figure 7.2).

Figure 7.2 Focus groups can generate useful data for development research, but attention must be paid to how group size, composition and your role as moderator influence the data produced

Focus groups are not suited to gaining biographical or factual details. Instead, we use them to gain group-level, socially grounded insights into everyday life and developing data on issues, subjectivities, perceptions and reactions, possibly in response to specific prompt materials, ranking exercises or simply the facilitator's questions. It is not only the verbal content of the discussion that can be of interest, but also the social interactions between group participants and the dynamics of presenting and defending opinions (see Warr 2005; Farnsworth and Boon 2010; Jakobsen 2012).

Group size matters: when designing research using focus groups you should think about the size and composition of the focus groups. Larger groups are challenging to manage and record (and transcribe) and can be prone to being dominated by a subsection of the group. They can also provide a greater sense of anonymity and freedom for individual contributions. Smaller groups are easier to moderate and analyze, but can reduce participant's willingness to speak openly as they feel under greater scrutiny (Hopkins 2007; Jakobsen 2012). In terms of group composition, you need to bring together a group that is appropriate to your research while considering and mitigating potential power dynamics within the group, perhaps relating to age, gender, topic, etc. This consideration should take into account local conversational and sociocultural norms and community hierarchies as these can have an important bearing on who can speak, on what, when and in front of whom.

It is also important to think about the location of the group discussion. Certain spaces may be logistically more or less appropriate, but there may also be sociocultural significance to these and power dynamics or norms that influence how willing groups and individuals will be to speak out in these locations. Focus groups are viewed by some as important tools for allowing marginalized groups a space in which to speak and to be heard, but great sensitivity is needed to group composition, location and timing to facilitate such engagements (see Madriz 1998). Timing should also be considered. Will holding the focus group at a certain time or on a certain day preclude certain participants due to livelihood or employment needs or religious observance? Are there national holidays, religious festivals or observances that may affect when you should hold a group discussion? You may also find that local, national or international events of relevance have an influence on focus group discussions (see Hopkins 2007).

Regardless of when or where your focus group takes place, you role remains the same: you are a facilitator and moderator, keeping your inputs to the minimum needed to steer the discussion to relevant topics. A key danger with focus groups is that you, as the researcher, engage in a direct question and answer process with individuals within the group. Instead, the benefits of a focus group arise from the communication and interaction between focus group members to generate ideas and explore issues, perceptions and understandings. In essence, once you have asked the opening question(s) you play the role of an eavesdropper into the group discussion and gradually become involved in steering the discussion while being flexible and objective. (Box 7.7 provides some additional ideas as to how to creatively manage and direct focus group discussions.)

A particular skill required is in creating and maintaining an open and safe space for discussion and (dis)agreement without interrupting the inputs and initiative of the participants (see Madriz 1998). There are three main ways in which this can be achieved

Box 7.7 Decentring the researcher in focus group discussions

The positionality of the researcher and other participants can constrain the ways in which individuals speak within a focus group setting. There are a number of strategies available to you in trying to overcome these challenges, including these three options drawn from Jakobsen (2012):

1 Asking focus group participants to complete evaluative or ranking exercises and to record their discussion and rationale for the ranking(s) produced
2 Asking participants to first discuss with the person sitting next to them their response to a prompt or question and then to report these discussions and rationales explored back to the group as a whole for further discussion.
3 Setting up the focus group discussion as a debate between different viewpoints that must be presented and maintained by sub-groups, before shifting the discussion to ask what participants own views are on the topic.

Further ideas can be found in:

Hennink, M. (2007) *International Focus Group Research: A Handbook for the Health and Social Sciences*. Cambridge: Cambridge University Press.
Jakobsen, H. (2012) 'Focus groups and methodological rigour outside the minority world: making the method work and its strengths in Tanzania', *Qualitative Research*, 12: 111–30.

– first, through formal direction (in which you set out an agenda and then simply start and end the discussion), second, through topical steering (through which you introduce new topics and questions during the flow of the discussion) and, third, by using steering dynamics (involving a higher degree of active engagement and direction as you invite and draw in shy personalities while including but restraining dominant and outspoken participants) (see Gibson 2000; Crang and Cook 2007; Hennink 2007). It is also important to reflect on both the group dynamics and interactions between focus group members, including changing positionalities and mediation of dominant or collective voices and how your positionality may affect the dynamics and responses of the focus group (see Warr 2005; Belzile and Öberg 2012). Thus, the complex array of 'talk' and partial accounts that are produced, often in disjointed ways, mean that your analysis of this material needs to consider both content and form of discussion while drawing out how different kinds of interaction contribute to meaning and consensus making.

The role of technology

Technology plays an increasingly important role in how you *do* and you *record* interviews. While most research interviews are conducted in person, this is not always possible. In these situations, various technologies can be used to facilitate an interview. The telephone interview is the most established alternative and is widely recognized as a valid and viable means of conducting interviews. This approach does mean that you will miss out on visual cues and various other interpersonal aspects. Pauses and silences must be handled with care. Gauging whether these are because someone has finished their response, is thinking or is unclear about your question can be difficult. Recording these interviews also requires specific equipment and you will need to ensure you have this available to you (see Cook 2009; Holt 2010). If you conduct telephone interviews it is useful to reflect on these compared with any face-to-face interviews you complete and to recognize and respond to any differences in depth, style and process as these will inform the data collected and your analysis thereof (see Sturges and Hanrahan 2004; Irvine 2011).

Telephone interviews do offer tremendous logistical benefits, reducing travel and other logistical burdens. There remain various drawbacks to telephone interviewing, including low response rates to cold-calling, the restriction of potential respondents to those with access to a (fixed-line) telephone, and the financial cost of the phone calls, particularly for overseas research (for further discussion, see Block and Erskine 2012). One solution to the cost issue has been the emergence of voice-over-IP services such as Skype. The spread of the internet means that using Skype for interviews is increasingly feasible, providing the same logistical benefits as telephone interviews but without the same costs incurred. Further benefits from using Skype and similar platforms include the possibility of recording interviews using software downloaded on to a computer and that video-conference calls can involve multiple groups across geographically disparate locations at the same time (see, for example, Gratton and O'Donnell 2011). Video-calling also allows for visual and other non-verbal cues and gestures to be noted and recorded (Hanna 2012). However, challenges remain with access, poor bandwidth, dropped calls and other technical issues. If you do utilize these technologies, be aware that other challenges of developing rapport, handling silences and other cues in interpersonal interactions remain and will need to be mitigated.

Recording interviews and focus groups is a critical consideration as good-quality recordings are the basis for detailed and accurate transcripts and analysis. Before going to an interview, make sure you have your recording equipment with you, that you know how it works and that it is working (the batteries are not flat and there is enough capacity for your new recording(s)). When setting up for an interview, give careful consideration to where you position the recording device, not only to gain a clear and audible recording of the conversation but also how prominent the device is in the interviewee's line of sight: there is something quite intimidating about having a microphone positioned directly in front of you. Before you start recording, you need to gain your participant's permission and should explain to them why you want to record the interview and how the recording will be stored, used and disposed of.

It is good practice still to take notes during an interview that is being recorded. These notes could focus on the environment and surroundings of an interview, reflect on body language and demeanour of the interviewee, key issues or questions to return to later in the interview and reflections on your own practice (if you are working in a group during a fieldclass, you may be able to devolve different aspects of note taking to different group members). If an interviewee declines for an interview to be recorded, you may be able to/need to negotiate alternative note-taking practices. Conducting an interview while trying to take accurate notes of responses is very difficult as you will be trying to do this while also thinking ahead to the questions you want to ask next, identifying comments or topics to follow up on, as well as trying to note body language and other non-verbal indicators.

You would be well advised to take some time as soon as is feasible after finishing an interview to write up your notes and reflections on the encounter. It is also good practice to back up your recordings at this stage and to enter the details of the interview into a database, capturing salient details so that you can navigate your way through your data during the analysis and writing-up periods of your research.

Micro-geographies of interviews and focus groups

Although interviews and focus groups are primarily focused on gaining verbal data, there is more to be gleaned from these encounters. The space within which an interview is conducted can provide important background and contextual information. A poorly chosen location can also undermine the interview, particularly if you use a space in which the interviewee feels uncomfortable or intimidated. 'Front spaces' such as offices and staff rooms inscribe particular power privileges and often encourage discussions to focus on more 'formal' or 'official' matters, whereas interviews in 'backstage' areas, which are often more hidden and more difficult to access such as particular clubs or social spaces, can uncover less formal content and encourage discussions that diverge from the 'official' line (Mikecz 2012).

It is often useful to invite the interviewee to identify a location for the interview as this allows them to select a space in which they will feel relaxed and comfortable and, as a result, more open and honest with their responses to your questions. In some situations, you may find that conducting an interview while walking can further disrupt power dynamics, mitigate concerns with the spatialization of power and encourage more open and reflective responses to questions (Kuntz and Presnall 2012). This style of interview can also provide a useful introduction to – or even to develop multi-layered narratives about – a fieldsite or location. George Barrett's reflections (Box 7.8) on the differences experienced by changing interview location and dynamics, including utilizing a 'mobile interview' approach, demonstrates how changing the space within which an interview is conducted can profoundly change the dynamics of and data generated by these interactions.

It is important to remember, however, that respondents – elites in particular – may also use decisions over spaces for interviews as a means of exerting power and control,

Box 7.8 George Barrett: The unexpected interview

The teacher sat there, staring at her feet, trying her best to respond to the multitude of questions she was being bombarded with. We, a group of five students, had crammed ourselves into a small, dimly lit school office, frantically extracting any data possible for our fieldwork project. Some perched on tables; I was sitting on the floor behind a frequently opened door. The teacher was clearly uncomfortable at the prospect of being surrounded by overseas undergraduates who were taking it in turns to interrogate her.

It was my first experience of doing development research and, through this encounter, I quickly learned the importance of interview structure and environment when researching under significant time constraints.

On this occasion we were fortunate: the teacher was associated with the non-profit organization (NPO) we were working with to carry out our research. As our first interview was proving problematic, another student suggested we drew this interview to a close and that she and I talk to the teacher again later that day. Our lunch was spent planning an interview focusing on school drop-out levels, the original intent of our research project. This second interview, involving the teacher and two students, took place at the teacher's own office at the NPO's headquarters.

The difference provoked by the change in group size, environment and interview structure was stark. In contrast to the reserved figure interviewed earlier, we were now met with an enthusiastic interviewee: posters were taken out of cupboards, photos pulled from drawers and tales of the crime pupils experienced on journeys to/from school (which discouraged their attendance at school), were recounted at length. 'I wish I could show you where the children come from', the teacher said with considerable verve. Conveniently, the bus we travelled on was outside and, after negotiating with the fieldclass leader, we were shortly being guided around the local area by the teacher.

In previous lectures, the lecturer had advocated the use of walking interviews. While we had the fortune of a bus to save our legs, the notion of interviewing on the move added a totally new dimension to the data collected. The interviewee's points were now being supplemented with a spatial context. The interview had evolved from the confines of the stationary, room-based environment to productive, serendipitous research. The power imbalances inherent in static interviewing appeared to be countered. From an earlier distinction between the researcher and researched, the participant now had greater control over the research process. Consequently, far richer data were collected. The interview was no longer merely a data collection activity, it was now a more participatory research experience.

In the short space of time I spent with the NPO I not only learnt of the importance of interview structure and research environment, I also gained an understanding of the flexibility and adaptability required when interviewing for development research. As it turned out, the best data were collected when least expected.

to withhold information and close of topics of discussion. It is also possible that interviewees will suggest spaces for an interview that are impractical or ill suited to the interaction due to issues of safety or simply because of background noise preventing your being able to capture a clear recording of the conversation.

By allowing the interviewee to select the site for the interview, you may also be able to develop further insights into their responses through observation of the surroundings. Observation of the context, or micro-geography, of the interview can provide information about the participant's identity construction and project of social identities (Elwood and Martin 2000; Sin 2003). By observing what posters, pictures and other decor are evident, you can begin to think about the materiality of identity. Considering why certain spaces are chosen or rejected may also give you insights into meaning and everyday practice, just as seeing the proximity (or distance) of places and resources can give a sense of reality and understanding to participants' responses (for more detail, see Elwood and Martin 2000). During a number of projects involving interviewing school teachers in South Africa in their own classrooms, Dan was able to draw on observations of how the teachers had decorated their rooms to prompt questions and conversations about citizenship, race, identity and social change. These questions, prompted by observations of the interview space, uncovered unexpected topics and concerns that would otherwise have been overlooked.

Transcribing interviews and focus groups

Transcribing interview and focus group recordings is an integral component of the research process, but one that is often overlooked or undervalued. While some would argue that you should only use the audio-/video-recordings of an interview so as to be able to draw on a broader set of cues and data, it is generally accepted that a transcript can provide a summary of an interview and is necessary in order to store, organize and analyze the data (Cook 2009; Mero-Jaffe 2011). Transcribing interviews is also a big commitment: the time-consuming nature of transcription is frequently underestimated. You can realistically expect to spend six hours or more transcribing one hour of recorded interview. Poor-quality recordings can significantly increase the time required for the transcription process, sometimes even rendering an interview unusable. Even without these added complications, transcribing interview and focus group data can be a lonely and tiring process (Roulston et al. 2003).

Due to the amount of time the transcription process can take you may be tempted to pay someone else to transcribe your interviews: this is not a cheap option. In addition to the financial cost, there are other issues to be considered if outsourcing your transcription. First, how does this fit to your agreements with informants regarding anonymity and confidentiality? Transcription providers should commit to/provide you with a confidentiality commitment at the very least, but will this meet your ethical obligations to your informants if you have said that no one else will have access to the raw data? Furthermore, having someone else transcribe your interviews can lead to problems of understanding and interpretation, as well as difficulties in understanding accents or colloquialisms.

Transcribing interviews yourself mitigates these concerns to an extent, while also allowing you to pick up on non-verbal cues in the recordings – delays, changes in emphasis or tone – which can provide important insights but which may have been lost if someone else transcribed the interview. Depending on your approach to transcription, the capturing of pauses, hesitations and intonation may not be an issue. If you are wanting a naturalized transcript, one that includes as much detail as possible, then this will need to include markers of intonation, emphasis and hesitation as well as (wherever possible) notes on non-verbal details such as body language and gestures. The other main form of transcript is a denaturalized transcript, this being one that focuses solely on the spoken words and provides a clear, coherent account of these without the inclusion of pauses and emphasis (see Mero-Jaffe 2011). While this approach limits the ability to analyze and unpack how knowledge is co-produced during an interview, it is a far faster and simpler method of transcription. Whichever approach you adopt, it is important to be thorough and accurate: simple mistakes in punctuation, spelling or grammar can result in very different meanings being presented.

If opting for a denaturalized transcript approach, your research diary takes on added importance as this is a means of recording reflections on each interview including notes on body language, demeanour, levels of interest and engagement, gestures and other non-verbal data. These observations can then be used alongside a denaturalized transcript to support your analysis of the data (see Poland 1995).

It is increasingly common – and generally seen as good practice – to involve interviewees in the transcription process by inviting them to review the transcript of their interview. Allowing the interviewee to read over the transcript of their interview is advocated as a means of empowering the research participant but also of providing them with greater control through the co-production of knowledge. This practice is also a mechanism through which the validity of interview data is enhanced, as interviewees can correct significant errors and check that the content of the transcript is an accurate reflection of the intended meaning. It is also possible to use this activity as a means to prompt further conversations through a follow-up interview where you can seek greater clarity or detail on specific topics or issues. At the same time, however, there is the potential for interviewees to edit their responses – either in an effort to appear more eloquent and considered in their responses, or by asking to delete sections or provide different responses to the ones provided in the interview (Mero-Jaffe 2011).

What to take from this chapter

You should now have a grasp of the kinds of research for which interviews and focus groups are appropriate methods and a sense of the various types of interview and question you might consider using. You should also recognize that interviews and focus groups are not an 'easy' way of doing research, but that to use these methods effectively they require careful preparation, focused application and attention to detail in recording, transcribing and analyzing the materials produced.

If you want to explore the ideas covered in this chapter in more detail, you may find these further readings useful:

Edwards, R. (1998) 'A critical examination of the use of interpreters in the qualitative research process', *Journal of Ethnic and Migration Studies*, 24: 197–208.

Ezzy, D. (2010) 'Qualitative interviewing as an embodied emotional performance', *Qualitative Inquiry*, 16: 163–70.

Foddy, W. (1993) *Constructing Questions for Interviews and Questionnaires: Theory and Practice in Social Research*. Cambridge: Cambridge University Press.

Perks, R. and Thomson, A. (eds.) (1998) *The Oral History Reader*. London: Routledge.

Roulston, K. (2010) 'Considering quality in qualitative interviewing', *Qualitative Research*, 10: 199–228.

8

ETHNOGRAPHY AND PARTICIPANT OBSERVATION

Ethnographic methods offer a potentially powerful approach to development research that can capture both the nature of everyday life and the ways in which daily life is informed by various development challenges. The depth of understanding gained through sustained engagement with daily life, noticing the unnoticed and exploring the mundane can provide vital insights into social relations, community functioning and everyday practice that are often overlooked. The tacit knowledge, trust and rapport that are developed during a period of immersion within a community can capture additional layers of detail and nuance through observations, interviews and other interactions that can be used in monitoring and evaluating development projects and designing more effective and contextually appropriate development schemes.

Ethnographic fieldwork is far more than the popular view of 'hanging out'; it is about developing a detailed and systematic account of a culture (Wood 1997) that provides insights into and understandings of everyday life. There is no single method to ethnography: ethnography is a methodological approach comprising the use of a range of methods. In other words, in conducting ethnographic research you draw on a toolkit of methods in order to develop layers of understanding through inductive reasoning. To this end, you need to conduct a prolonged period of fieldwork in order to build the relationships needed to facilitate access and engagement required for ethnographic knowledge co-production. To provide you with a sense of how such research can develop, we draw on a short anecdote from Dan's fieldwork on race, identity and social change in Cape Town. Although rather basic and mundane, this reflection captures some of the subtle shifts in relationships and access that are facilitated during ethnographic fieldwork and which provide for more nuanced understandings of daily life and development concerns. During this fieldwork, Dan adopted an ethnographic approach to his research and spent time embedded in the daily life of a number of schools in the Cape Flats (the former coloured group areas). On a daily basis, Dan would observe and participate in lessons and teaching activities, engage in the social life of the school during break times and after-school events (including coaching one of the school's rugby teams) and conduct interviews with teachers and focus groups with students focusing on the core themes of the research.

Towards the end of the period of fieldwork, Dan travelled with the school's rugby teams and supporters (comprising other staff and students) to another school in the Cape Flats to fulfil an inter-school fixture – not only the final fixture of the season but the last game involving Dan before the end of his fieldwork. The Saturday morning was bright and sunny, although a little chilly. Dan went through the warm-up and drills with his team and watched as they lost a close and hard-fought game. While this match was in progress, the first team was getting ready for its game and a consensus had emerged from the first team players and coaches that they wanted Dan to play for the final 15 minutes of this game.

A spare pair of boots and one of the team's jerseys were found and Dan was sent on to the pitch with about 15 minutes remaining, to the great amusement of the school's staff, spectators and players. While playing, Dan picked up a knee injury which prevented his conducting research at the school for a few days. On his return, hobbling with a walking stick and dosed up on painkillers, he was met with a barrage of banter and jokes from staff and senior students. This change in interaction denoted a subtle but significant shift in rapport and fieldwork relations.

The unexpected – and certainly unanticipated – outcome of this injury was an increased camaraderie and a noticeable deepening of insights as a result of increased openness and acceptance of his presence. The gradual movement along multiple insider–outsider continua and the (re)negotiation of identity and the meanings associated with markers of belonging (language, class, race, education, nationality) facilitated an evolving set of insights into the practices and meanings of everyday life, not least into the role of humour as a coping strategy in conditions of poverty. These engagements contributed to a detailed understanding of the local community and politics of identity amid a set of changing social dynamics, not only through interview and focus group discussions but also by analyzing the mundane actions, interactions, gestures, comments and practices of everyday life.

What is ethnography and why use it in development research?

As noted above, ethnography is not a single method but a research approach concerned with providing a detailed and systematic account of a culture 'that claims to take the reader into the actual world of its subjects in order to reveal the cultural knowledge that is working in a particular place, as it is actually lived through its subjects' (Yon 2003: 412). To realize this goal, ethnographers use a range of methods during an extended period of fieldwork that combine in an iterative research process, culminating in the accrual of layers of understanding through the use of inductive reasoning (Crang and Cook 2007; Hall 2009).

This depth of understanding is developed through a prolonged immersion in the social world being studied, allowing you to recognize and record the actions and inactions, presences and absences in daily life and to explore the reasons behind these and the meanings produced through them (see Gold 1997). Ethnography is not simply the description and reporting of observations but is the exploration of the meanings associated with and produced through objects, behaviours, interactions and emotions. The ability to gain this level of insight requires not only prolonged periods of time immersed with the community, organization or group you are working with, but also the careful negotiation of access and building of trust and rapport. As Sian Lazar outlines in her reflection (Box 8.1) on working with trade union movements in Latin America, these processes can vary tremendously and require careful thought and contextually sensitive practices.

Box 8.1 Sian Lazar: Ethnographic encounters with Latin American trade unions

Doing ethnographic research with an organization such as a trade union is complicated by the fact that it is difficult to 'live' with trade unionists as one might live with a particular local community. This was my experience when conducting fieldwork with two unions of state employees in Argentina. Although the unionists were often very happy to grant me an interview, it was hard to convince them of the validity of the kind of participant observation that I, as a social anthropologist, aspired to. I was often asked to outline precisely what I wanted to do at a time when I had yet to figure out what exactly I might be choosing between (which events to attend, which people to interview). This sits uneasily with a more flexible notion of ethnographic method where the researcher attempts a kind of immersion in the workings of a small, localized community. In organizations that operate on multiple sites and with multiple gatekeepers, such a holistic approach is very hard to achieve – if, indeed, it were ever possible even in the traditional sites for anthropological research.

I worked with two unions, one with a very strong vertical structure of leadership and a sense of itself as 'organic' and 'disciplined'; the other with an equally strong self-image and practice of 'horizontality' and 'openness'. I found the 'horizontal' union easy to approach and to elicit interviews with informants, who were very open and honest with me; some informants even got in touch with me themselves when they heard of my project and we could meet up to talk at length about their activities and views. I could attend public demonstrations, make friends with people there and ask them for interviews later. After a few months, I was also allowed to attend a training session for new union delegates.

With the more 'vertical' union, access took longer to negotiate. After an initial couple of scoping interviews, and the chance to attend some events to celebrate international women's day, the union's internal elections meant that the leadership had little time to consider my requests to be allowed into more intimate spaces and contact dried up. Once the elections had finished and I also developed a more concrete sense of what I wanted to do, I was able to present a 'wishlist' to the union hierarchy, and was given permission to attend training courses for new and established delegates. I could triangulate the content of these courses with that of interviews, of materials produced by the union and with my observations of people's behaviour and notes of informal conversations, in order to gain a sense of how collective mobilization is organized and strengthened in this particular context.

As I returned in subsequent years, I continued my association with the school for union delegates and was allowed to spend time with the internal commission for a ministry, to gain a sense of daily life for ordinary union activists. I noticed that the fact that I had kept going back gave me greater credibility in the eyes of my informants. They saw me as someone with a reasonably long-term commitment to the union and seemed pleased that I wanted to maintain the connection.

The framing of ethnographic knowledge has shifted over time. It has moved from early claims to detached, objective observations through to increasing recognition of its subjective nature. The development of ethnographic approaches has been informed by several theoretical approaches including social interactionism, cultural studies, critical theory, feminism, post-structuralism and post-modernism. Overall, ethnography is informed by hermeneutics: the idea that interpretation is at the core of human life and systems of meaning frame our lives. It is not, therefore, an approach suited to positivists: ethnographers are not concerned with finding out a 'truth'; neither are they proponents of scientific rationalism, but are engaged in non-repeatable, subjective research (Sherif 2001).

The initial development of ethnographic approaches and accounts of the lives of other groups and societies was entwined with the ideals of Western Enlightenment and the colonial period. Consequently, early ethnographic accounts were laden with colonial relations and focused on describing and recording the 'exotic other' from a detached and objective position. These approaches shifted with growing recognition of the influence of the researcher's presence and background on knowledge production. Increasing efforts were also made to go beyond mere description and to locate and analyze these engagements in relation to broader social questions and dynamics. Bronislaw Malinowski is recognized as a key figure in this methodological shift (see Box 8.2), challenging anthropologists to 'get off the veranda' and immerse themselves in the daily life of the culture and society being studied. This practice of immersion in daily life has come to be known as *participant observation* and remains a core tenet of ethnographic approaches today.

Box 8.2 Bronislaw Malinowski

Bronsilaw Malinowski (1884–1942) is recognized as a pioneer of participant observation. An Austro-Hungarian passport holder, Malinoswki lived and studied in the UK before travelling to New Guinea as part of a research trip in 1914. The outbreak of World War One prevented his return to the UK and, with the support of the Australian government, he remained in the Western Pacific and conducted fieldwork with island communities until the end of the war. Malinowski advocated the benefits of immersion within the culture being studied and the importance of language fluency as essential to capturing and analyzing the nuance and intricacies of discourse. Malinowski is recognized not only for his development of participant observation, but also for his contributions to theories of functionalism and of reciprocity and exchange.

Reading Malinowski's (1922) *Argonauts of the Western Pacific* provides a sense of the methods and ideologies of early anthropology and how these studies conceived of and represented the exotic 'other'. Malinowski's reflections on method and negotiation of access are particularly interesting and provide opportunities to critically reflect on how you might try to achieve access and how you would represent 'others' in your writing.

The emphasis and ethos of ethnography has continued to evolve, with increasing emphasis being placed on understanding dynamics, practices and cultures from the perspectives of those being studied and the potential for such studies to provide a voice to marginal communities and to support advocacy and development approaches (see Dodman 2003). This move towards more engaged and even activist ethnographies that were not solely focused on 'exotic others' but also with critically examining communities and cultures within the global north developed apace in the 1960s and 1970s, often spearheaded by educational anthropologists. Although these engagements were influenced by functionalism, the influences of structuralist and Marxist schools of thought were increasingly apparent as a paradigm shift in the social sciences turned scholars away from positivist, quantitative approaches.

The 1970s also saw the emergence of symbolic interactionism as a key approach to knowledge as scholars argued that the world could not be understood through simple causal relationships or the determination of society by universal theories or laws. Instead, ethnographers began to address the meanings given to and arising from symbols, objects, events and actions. Linked to this conceptual shift was an emphasis on capturing the nuance and detail of daily life through the exercise of 'thick description', a term often associated with the ethnographer Clifford Geertz (Box 8.3). 'Thick description' is a practice that recorded events, actions and behaviours *and* the meanings given to these: cultural practices were envisaged as texts to be read and analyzed on multiple levels.

Continued critical reflective engagements during the 1980s resulted in ethnographers emphasizing issues of power and voice. On one side, this has contributed to a growing role for ethnographic work in social and political advocacy for marginal groups and in relation to development concerns. On the other side, this has focused attention on your role as a researcher as someone who is simultaneously embedded in the social world being studied and your 'sending' society. As Sherif (2001: 437) notes, ethnographic 'knowledge

---

Box 8.3 Clifford Geertz

The classic example of thick description remains Clifford Geertz's (1972) *Deep Play: Notes on the Balinese Cockfight*. *Deep Play* not only describes observations, participant observation and conversations but analyzes these data to explore their meanings and importance. You may find his discussion of the experience of doing ethnography including shifting positionality, the movement along outsider–insider continua, and the role of unexpected events in building rapport, useful to reflect on.

The opening chapter of *The Interpretation of Cultures* (Geertz 1973) outlines Geertz's rationale for and method to accomplish thick description, with analysis and explanation at its core, as integral to an interpretative understanding of culture. You may find this chapter useful for prompting you to think about knowledge production practices and how these locate you in a position of power to construct representations and understandings of people and places.

is produced in a historical and social context by *individuals* . . . process and product has become political, personal and experiential . . . [there is growing] recognition that the ethnographer's self affects every aspect of the research process, from conception to final interpretation.' This has led to increased concern with understanding positionality, culture and development as dynamic and evolving: '[I]t is not enough for researchers to identify where people are (both socially and spatially) – they must also question where they/we are coming from, going to and where on this path the research encounter has occurred' (Cook and Crang 1995: 7). Developing such contextual and subjective understandings can provide powerful insights into developmental challenges and the (often unexpected) outcomes of development projects.

Recognition of the potential benefits of ethnographic approaches for developing detailed understandings of cultures and communities has gradually spread across the social sciences. Ethnography can provide detailed and nuanced understandings of societies, cultures and context-specific development challenges and opportunities. However, development scholars and practitioners have rarely engaged this research approach. This is because ethnographic methods require investments of time and money that are often at odds with the demand, within the development field, for rapid studies and results (Van Donge 2006). We would argue that despite these costs, this research approach is important for development fieldwork.

Experience of ethnographic research equips you not only with a set of research skills but also an understanding of the complexities of local contexts and their influence over development project outcomes. Ethnographic work can also play a vital role in informing development practice and policy. The nuanced understanding produced through ethnographic work can play a vital role in the design and implementation of sustainable and appropriate development policies. Ethnographic methods can also be used in monitoring and evaluation of development projects as a mechanism for identifying and understanding unexpected outcomes (cf. Van Donge 2006). Ethnography is a useful approach for understanding how and why questions in development can be addressed through building understanding from other people's actions and the functions and meanings associated with these everyday practices.

How to 'do' ethnography

Ethnography has, at times, been mythologized as something that cannot be taught. Rather, it was situated as a deeply personal experience comprising skills and knowledge that could only be developed through time in the field. While there is a grain of truth in this – doing fieldwork is the most effective way of developing research skills – you still benefit from a solid engagement with ethnographic approaches prior to fieldwork. Few of the tools used by ethnographers are particularly complicated. Indeed, most of the tools you will use are extensions of your everyday practices based on heightened awareness and a more explicit emphasis on continually questioning and reflecting on how, why, where and when things are done (or, alternatively, not done).

As an ethnographer, you can deploy a wide range of techniques for collecting data as you attempt to develop knowledge and understanding about and through a range of practices and interactions. Simply put, the practice of ethnography is 'participat[ion], overtly or covertly, in people's daily lives for an extended period of time, watching what happens, listening to what is said, asking questions; in fact, collecting whatever data are available to throw light on the issues with which he or she is concerned' (Hammersley and Atkinson 1995: 2). If you are to collect 'whatever data are available', you need a broad toolkit. Potential research tools within an ethnographic approach may include specific methods covered elsewhere in this book such as interviews and focus groups (Chapter 7), documentary analysis (Chapters 10 and 14), participant observation, solicited diaries, visual methods and field notes. Using multiple tools allows you to engage with multiple types of data derived from multiple forms of interactions, providing you with greater opportunities to capture different facets and performances of everyday life.

Over the next few pages, we provide some ideas and advice on some commonly used tools in ethnography that are not addressed elsewhere in this book. As well as the content in this chapter, we would encourage you to reflect on issues around negotiating access to the field (see Chapter 4) and questions of ethics and positionality (see Chapters 5 and 6). These discussions and reflections are particularly important for ethnographic research. The ways in which you are situated within a web of affiliations and connections, of how you live in the field and how these are interpreted and used to position and locate you in the minds of your research participants have an important bearing on your data collection and analysis (for a detailed discussion of these issues, see Bornstein 2007). In addition to the materials presented here we would encourage you to read a range of ethnographic studies (a list of some useful studies can be found in Box 8.4) and through these readings to identify common practice, experiences of mistakes and challenges and solutions to these and to think about how you can draw on these insights in thinking about your own research design and practice.

Box 8.4 Some useful examples of ethnographic research

Barnard, A. (1992) *Hunters and Herders of Southern Africa: A Comparative Ethnography of the Khoisan Peoples*. Cambridge: Cambridge University Press.

Chagnon, N. (1992) *Yąnomamö*. Fort Worth, TX: Harcourt Brace College Publishers.

Evans-Pritchard, E. (1940) *The Nuer: A Description of the Modes of Livelihood and Political Institutions of a Nilotic People*. Oxford: Clarendon Press.

Geertz, C. (1973) *The Interpretation of Cultures*. New York: Fontana Press.

Turner, V. (1967) *The Forest of Symbols: Aspects of Ndembu Ritual*. Ithaca, NY: Cornell University Press.

Weiner, A. (1988) *The Trobrianders of Papua New Guinea*. Fort Worth, TX: Harcourt Brace College Publishers.

Participant observation

Participant observation is a central tool in ethnographic research. In many ways, this method appears simple and straightforward: it could be seen as merely an extension of how you engage with the world around you on an everyday basis. Participant observation is far more than this. It involves the recording of interactions, thoughts, reflections, (un)certainties, (mis)behaviours, (in)actions and other symbolic or ephemeral phenomena to build up a picture of everyday life. To keep track of everything that is happening, and, indeed, not happening, and then seeking to unpack and question the reasons behind and the meanings generated through the observed phenomena, is challenging work. You are not only recording and analyzing the exceptional and the mundane, but also building up your understanding of daily life and tacit knowledge about a new and different context, and participating in the daily life of your research participants (see Figures 8.1 and 8.2).

This tool requires a significant investment of time (and often, therefore, money) to undertake. The pay-off for this investment comes from the detailed insights produced as a result of the prolonged period of engagement. Over this period of time you can develop detailed observations of daily life, reflect on seasonal or longitudinal changes and

Figure 8.1 Your research site might be a market or a community shop

Figure 8.2 Your research site might be a youth centre

differences, and develop understanding of overt as well as unspoken and implicit power relations and hierarchies. In the course of your fieldwork, you will also gain different insights and information sharing as your positionality changes and you move along insider–outsider continua (see Chapter 4 for further discussion of this). The indicators of this movement, and the changing narratives and insights afforded by this, are both subtle and more obvious. During Dan's fieldwork with teachers in the Cape Flats this movement was signified by shifts in the presentation of thoughts in conversation which shifted from 'Let me tell you . . .', as guidance and expert instruction from the local teacher on local culture and context, to 'Ach, you know how it is . . .', as sharing a confidence and of Dan being seen if not as an expert, then at least as a non-novice. In such situations, you can reflect on your changing positionality and the shifting ways in which information is shared to uncover hidden meanings, contradictions and other issues that would otherwise be overlooked.

It also allows the researcher to engage in informal conversations and develop a greater depth of rapport and sensitivity to the local context ,which, in turn, helps to deepen your understanding of the intricacies of local everyday life. It can also provide a way to challenge and expose power inequalities, to allow the powerless within a community a way of speaking while giving you a means through which to hear what they are saying. On the flip side of these benefits are the costs – the time, energy and money required to complete such longer periods of fieldwork.

Participation observation requires you to participate in the daily life of your research environment and is a vital part of the production of knowledge and understanding. Getting your hands dirty, literally or metaphorically depending on your research context, and engaging in everyday practice and interactions affords you a different viewpoint on events, immerses you in informal conversations and can build trust and rapport with your host community. As Laurier (2003: 135) notes, 'the best participant observation is generally done by those who have been involved in and tried to do and/or be a part of the things they are observing': so, it is down to you to get involved and immerse yourself in the culture being studied. In other words, the participation is a vital part of the process as this allows for the production of commentary on activities and interactions that would be missing from simply observing a context.

In conducting participant observation, you need to develop a good memory and a system for note taking and keeping up with recording and analyzing your observations. In some instances, you may be able to sit and take notes openly while conducting this method, but in many situations this will not be suitable or appropriate. At the very least, such a practice clearly marks you as an observer and locates you in a powerful position of surveillance: imagine how you would feel if you were going about your daily business being watched by someone making notes on your (in)actions and interactions and how this might change the way in which you act. In most encounters, you are likely to have to make mental notes of things to write up later and perhaps find ways of excusing yourself from a situation for a few moments in order to jot down a few notes to write up later.

In situations in which participant observation is not possible or suitable, detailed observational notes remain an important repository of data for ethnographers. Exploring and understanding everyday life, from the way people move around specific areas or the performativities and rituals involved in street trading and haggling, can be assisted through repeated and detailed observation. Jeff Garmany's reflections (Box 8.5) on his use of institutional ethnography while working on governance and urban development in Brazil highlight how these observational practices provided strong evidence that contradicted materials developed through interviews and indicated other issues and power relations at play. The use of institutional ethnography to understand the (non)functioning of governmental, corporate and civil society organizations can produce detailed and influential findings. Similar methods can be employed when working in less formal settings, such as marketplaces and public spaces to uncover issues of power, control, in/exclusion, informal economic activities and many other topics of interest in development research.

Solicited diaries

Solicited diaries are documents written by a research participant in response to a specific request from a researcher and intended for the researcher to read and analyze. While these diaries do not contain the private thoughts and reflections of a personal diary, they do provide the participant with a voice through which to reflect on and talk about everyday experiences related to the topic or theme requested by the researcher. These documents can be particularly effective when working with marginalized groups as not only can they promote participation and engagement with your research question but can

Box 8.5 Jeff Garmany: Observing development in northeast Brazil

In 2009 I interviewed nearly 200 people while conducting research on governance and urban development in Fortaleza, Brazil. I spent hours talking to local residents and state representatives, analyzing what they told me and thinking of new questions to ask others. These interviews, I felt, gave my research quantifiable data; quotes I could draw on and numbers to point to in justifying my results. After months of work and analysis, my findings and insights began to grow. But bombarding people with questions, as I came to learn, was not always the most effective strategy: an important breakthrough came for me one afternoon when I finally shut up and started watching what was going on around me.

As researchers, we tend to ask a lot of questions, especially when trying to understand the processes, power relations and discourses of development. It makes good sense to ask questions; after all it is the people involved with development – and those affected directly by it – who tend to reveal new insights. But all research methods have their limitations. Trite as it may be to say, there is often a gap between what people say and what people do. So how should one go about addressing this mismatch? Are there alternative methods that might help to verify and critique one's research findings?

The answer, not surprisingly, is that of course there are. Just as there are numerous ways to pose questions to research participants, there are several techniques for crosschecking those very same data. In my own case, institutional observation (sometimes called institutional ethnography) proved highly effective and dynamic. The methodology was not part of my original research design, but as I discovered while visiting a municipal building in Fortaleza, watching an institution at work can be incredibly educational.

As I waited (for hours) in the lobby, I came to observe how constituents interacted with their elected officials. And this experience enabled me to understand two important findings: 1) contrary to what I was often told during interviews, the relationships between politicians and the general public appeared painfully clientelistic; and, 2), the *real* gatekeepers – and indeed, from what I could tell, the most significant people in the entire building – were not the politicians at all. Rather, they were the administrators, the ones who managed the waiting area and decided who could enter, who were masters of protocol and navigators of bureaucratic roadblocks and who, ultimately, would carry out whatever political orders came from within. These people, the administrators, were the ones on whom everything depended. The politicians held a certain level of authority, but the administrators were the ones who wielded it.

This experience led me to conduct several more hours of institutional observation fieldwork. I continued to gather interview data, but I found it helpful to contextualize those findings using other methodologies. By taking seriously the spaces in which development unfolds, and thinking critically about the relations (social and otherwise) that characterize those spaces, we, as researchers, can gain new perspectives into a host of processes regularly taken for granted. In my own case, that moment finally occurred when, at last, there was no one left in front of me to interview.

also function as a space for participants to speak about their own priorities and concerns (Meth 2003a).

Solicited diaries also allow for a longitudinal narrative to be developed through a sequential record of reflections and recording of experiences. Regardless of whether diary entries are maintained sporadically or continuously it is possible to use these to chart and explore how ideas, perceptions and outcomes of interventions change over time. The freedom that solicited diaries provide to participants to develop personal content and discussions at their own pace, building the confidence to share information with the researcher, can be tremendously useful when dealing with sensitive topics or vulnerable groups (Meth 2003a).

Whether or not the type of information gained through solicited diaries will be of use to you will depend on your research questions and the type of data needed to address these. You can use solicited diaries to address a range of topics and to capture various forms of data. You may ask participants to maintain a record of activities and time spent on these, to describe particular challenges or opportunities (for instance in relation to health and wellbeing, employment, education) or to reflect on experiences of crime or violence. Solicited diaries provide highly individualized information and can be used to understand individual's concerns and experiences and from these to gain a general sense of issues within a community (Meth 2003a, 2004).

As well as considering whether or not solicited diaries are an appropriate tool with which to address your research question, you need to determine if this is a suitable method for the type(s) of people involved in your research. These considerations should include awareness of factors that may prevent people from completing diaries (for instance, literacy levels or disability) and reflections on whether completing a diary is a practice that targeted participants would find appropriate.

Once you have addressed these concerns, you also need to think about the location and sampling of participants. It is unlikely that you would be able to find a stratified, systematic or random sample willing to complete this activity, so it is likely that you will use this method with a strategic or purposive sample. Further concerns that may influence your sampling process may include issues of language, literacy and safety. Language may present a challenge if working with communities with a different first language from your own. In which language will you conduct the research? In which language do you ask people to complete the diary? If the diary is completed in a participant's second language, how much of a danger is there that certain meanings and nuances will be lost? If completed in the participant's first language but this is not yours, how do you organize translation of this material and how do you minimize the risk of nuance being lost in this process?

Layered on to this concern is a question of the respondent's literacy level: will someone with poor literacy be able to keep a written solicited diary? Will someone with poor eyesight or a disability that makes writing difficult or impossible be able to complete this activity? In this situation, can you offer them a means of keeping a verbal diary by using a Dictaphone? If so, how might this change the type of data recorded?

Regardless of the medium used to keep the diary, what are the safety and ethical implications? If you are asking someone to reflect on a sensitive topic – say their HIV status or experiences of domestic violence or political persecution – what are the potential implications for their safety and wellbeing if their spouse/partner finds this information? Is there a possibility of further persecution or threats towards participants if community members, criminal gangs or state security apparatus become aware of the content of diaries? Clearly the severity of these concerns will vary tremendously between different projects, but are worthy of consideration. This consideration extends to include thinking about how you will store, analyze and draw from this document so as to adhere to agreements regarding confidentiality and anonymity.

Another important consideration relates to the amount of time taken for the participant to complete the diary. Depending on the nature of the diary solicited, the frequency of entries requested and the duration for which the project will run, this time commitment can be a considerable burden on your participants. Thought needs to be given to ensuring that the purpose and nature of the diary (and individual entries) is explained clearly in order to encourage participants to make effective use of their time in completing the task and that it is not an overly onerous undertaking. In essence, if you ask someone in a very broad and ill-defined way to record their daily activities or to capture experiences and reflections of daily life, you are unlikely to gain good data. Instead, you need to think carefully about the specifics of what you want your participants to record and provide a clear steer as to the topic(s) of interest and why these are important. By providing clarity of focus and ensuring your participants understand what you are asking them to do and why you are more likely to gain detailed and useful data for your work. When

---

Box 8.6 Solicited diaries

Solicited diaries are often used with marginal groups or with people who are suffering from or vulnerable to discrimination or violence, perhaps linked to gender, race, health conditions, nationality or domestic violence. Common features to studies involving the successful use of solicited diaries are clarity of focus in the project, detailed consideration of ethics and safety, and the encouragement of participants to use their diaries as safe spaces of reflection and thought. While there are many good examples in academic literature of these practices, those noted below provide useful and accessible insights both into the method itself and the ways in which data are analyzed and presented:

Dodson, B. (2010) 'Locating xenophobia: debate, discourse and everyday experience in Cape Town, South Africa', *Africa Today*, 56: 2–22.
Langevang, T. (2007) 'Movements in time and space: using multiple methods in research with young people in Accra, Ghana', *Children's Geographies*, 5: 267–82.
Meth, P. (2004) 'Using diaries to understand women's responses to crime and violence', *Environment and Urbanization*, 16: 153–64.
Thomas, F. (2006) 'Stigma, fatigue and social breakdown: exploring the impacts of HIV/AIDS on patient and carer well-being in the Caprivi Region, Namibia',

contemplating this research tool, you also need to think carefully about how much time it will take and the extent to which this consumes resources required for subsistence or livelihood activities. You may also want to consider offering some form of payment or compensation for participants (see previous discussions on payment for participation, as well as Meth (2003a)). While there are a lot of issues to consider before using solicited diaries, a range of academics have used them within development research: the examples in Box 8.6 provide further insights into this method.

Visual methods: autophotography and digital storytelling

Visual methods can play an important role in ethnographic data collection. You, the researcher, can capture images and video of events, ephemera and behaviour that can be used in description and analysis. Capturing this imagery requires the negotiation of a number of ethical concerns, including gaining permissions from those whose images you capture and negotiating local cultural norms regarding photography. Alongside these concerns, it is important to remember that these images are captured from a particular (privileged) position and are implicitly about things that *you* think are important. Despite these concerns, such media can be useful for your research.

A set of alternative, more innovative, approaches have emerged that aim to disrupt the power relations inherent in researcher-taken imagery. These tools aim to provide participants with greater control over the data production process by allowing them to develop visual research materials. One tool in this approach is *autophotography*, a method by which you provide participants with cameras and instructions to take images that record impressions and interpretations of local places, people, events or life stories from the participants' perspectives.

If you are planning on using autophotography, there are a number of questions to consider. First, is it appropriate for the questions being asked and the data required to answer these? If so, then is it suitable for the sample you are working with? It is a method that has been used with youth or marginal populations with good effect (see, for instance, Dodman 2003; Jensen 2008; Lombard 2012), but is unlikely to work with professionals or business people. Once you have decided on the appropriateness of the method, you need to think about how you explain the activity to the participants – what do you ask them to take photos of, how do you brief them on the topic you want to find out about? How do you deal with ethical concerns regarding permissions to take photos and the use of these? How long do you give them to complete the assignment? What equipment do you provide them to take the photos – do you rely on their owning their own camera or do you provide them with one? If providing one, is it a disposable film camera? A cheap digital camera? If it is a cheap digital camera, do you make use of the video function if it has one? Are there safety considerations for the participants and if so how do you ameliorate these (this might be in terms of the spaces or times at which you want them to take photos or simply them having possession of a camera)? What are the comparative costs of the cameras, media/storage (film or digital memory card) and printing? (How) Do you recompense people for taking part? Do you provide them with copies of photos (perhaps also encouraging them to take some photos of places or people they would like photos of) or perhaps let them keep the camera at the end of the activity? Could you hold a community exhibition of the photos? There are a vast range of ethical and logistical issues to consider (see Young and Barrett 2001; Johnsen et al. 2008), but these should not be a deterrent as these methods, used appropriately, can produce invaluable research findings.

Then you must also think about how to analyze the photos – you can do this yourself, but in so doing you need to accept that this invokes your own interpretation and gaze on to the photos. You could work with the participants and ask them to explain the importance and meaning of the photos, allowing them to decode the images using their own language and explanation to unpack the subjective meanings of the images and their content. You may consider taking your own set of photos using the same brief and then work with your participants to analyze these photos and the differences in meaning, interpretation and understanding between the sets of images.

Digital storytelling or *participatory video* (see Box 8.7 for links to further information) offers another means of developing data within an ethnographic approach. This tool uses low-cost digital cameras and basic computing facilities with video-editing software to allow people to develop and tell stories that they want told through video. The challenges to using this method are largely the same as those to be addressed in regard to autophotography, but with further consideration to safety, equipment, breakdown, power availability and skills (see Mistry and Berardi 2012). Digital storytelling does offer opportunities for individuals and collectives to tell stories in different ways, to create meaning and understanding through the process of recording and editing materials to produce their narrative and can provide unexpected understandings and representations of concerns. (On a different level, you may also consider how this tool could be used in communicating your research findings or, alternatively, by looking at existing materials online how these could provide the basis for a research project.)

There are many other visual methods and tools that can be used in ethnographic work – far more than we can address here. To explore other methods, you may find the journal *Visual Anthropology* a good source of inspiration and advice.

Box 8.7 Digital storytelling

Digital storytelling is an expanding arena for researchers. Much of the pioneering work on digital storytelling has been driven by the *Center for Digital Storytelling* (http://storycenter.org/) and its website contains useful information on ethical practice, training workshops and other resources, including links to a number of YouTube channels showcasing digital stories developed through its activities (http://www.youtube.com/user/CenterOfTheStory). A useful further source of guidance on digital storytelling is Lambert, J. (2013) *Digital Storytelling. Capturing Lives, Creating Community*. London: Routledge.

Language

One of the biggest challenges in ethnographic research relates to language ability. Being fluent in the local language of your fieldsite allows you to pick up far more texture and nuance than if you are relying on a basic linguistic competency or working with a guide or translator. A consideration before undertaking ethnographic fieldwork therefore relates to your language abilities and the need (and possibilities) for language training. If you are already fluent or proficient in the required language then this puts you in a strong position. If you are not, you may need to develop your language skills prior to entering the field – perhaps through courses offered by your university or another institutional provider, or through private tuition or language-sharing/tandem language-learning scheme. You may also be able to spend some time in your fieldwork country or region conducting immersion language training, although this can be economically costly. Each of these options has its own costs and benefits and it is down to you to weigh up the best option for your research.

Even where these options are unavailable, you can benefit hugely from developing local language skills during the course of the fieldwork. If you only speak the relevant colonial language of your fieldsite, you may exclude many potential participants and restrict your respondents to elites. The ability to hold a basic conversation in the local language is an important ice-breaker, demonstrating respect for the local community and culture, and showing a commitment to your research and the host community. It is also important to 'soften' your language skills while in the field, to pick up and use appropriate slang terms and local dialect as this allows further understanding of conversations and everyday life. This was noted by Dan during his fieldwork in the Cape Flats in Cape Town, South Africa. Prior to undertaking the fieldwork, he had taken some lessons in Afrikaans and used some 'teach yourself' resources to develop some language skills. When he started working in the Cape Flats, it was quickly apparent that the more formal Afrikaans he had been learning was vastly different in structure, tone and content to that spoken in daily life in the Cape Flats. The development of his language abilities overall, and the softening of the formal language he had learnt, were integral to deepening his engagement during the fieldwork period.

Reflexivity and research diaries

On the one hand, reflexivity is vital to good ethnographic work, not least because you, as an ethnographic researcher, have a critical role in the production of data. Your positionality, (dis)comfort and practice in the field inform your relationships with your host community and frames what data are collected, how they are recorded and how they are analyzed. On the other hand, this reflexivity is important as a check on the ethnographer's version of Stockholm syndrome, of becoming overly attached to and connected with the host community. Historically, this process was referred to, rather problematically, as 'going native' – an allegation that the researcher became uncritical (or even overly critical) in their approach to specific informants or groups.

While we would not suggest that you should remain detached from your host community, it is important that you maintain a continual, detailed account of your field experiences in order to facilitate reflections on and analysis of your changing positionality as well as the content and meaning produced through interactions over time. These reflections are also vital in helping you avoid ethnocentrism in your research and analysis. You will carry the baggage of your own background, lifestyle and experience into the field with you and you need to acknowledge this and seek to put this to one side in order to develop an understanding from the perspective of the community or group you are immersed in.

Continual reflection on this process allows you to consider your evolving position and relations within the host community and subsequent influences on data collection. This might be in terms of a gradual movement along insider–outsider continua or the continual negotiation of multiple and competing positions that you claim and/or have foisted on you. Sherif (2001) provides an insightful discussion of these concerns in relation to her experiences as an American-Egyptian single woman conducting fieldwork in Cairo. To develop this continual reflection, your notes will encompass a range of components: descriptive notes of observations and interactions, methodological notes (what you have done, not done, successes, failures, complications, ideas for new approaches or further questions to ask/reflect on) and theoretical thoughts and engagements (from what you have recorded and linking through to ideas in the literature). They will also involve reflections and analysis as you locate and analyze an interaction or observation in terms of context, symbolism and meaning.

Field diaries and notes are not only important for this kind of reflexivity, but are also useful for consolidating data and knowledge. The details and observational notes made while in the field can be vital to your analysis. You may well find that the notes you made on an apparently tangential concern or peripheral event while in the field emerge as key to your analysis once your data collection is complete. In our research, we have found this to be the case on many occasions, as have many other researchers (for instance, Neimark 2012).

Ethics and ethnography

Ethnographic research requires consideration of a range of ethical dilemmas, from informed consent to maintaining ethical relations in the field and including your positioning in the field and openness about your research (see Chapter 5). As discussed in Chapter 5, it is often not feasible to gain informed consent from everyone encountered during ethnographic fieldwork, so who do you secure consent from, how do you try to make people aware of the reasons for your presence, and what data is it ethical to record and use? Although focused on ethnographic health care research, the discussion presented by Murphy and Dingwall (2007) relating to differing levels of risk, power relations and differences in working in public and private spaces, resonate with ethnographic practices in development research (useful further conceptual discussions around the complexities of integrating ethics into ethnographic practices can be found in Parker (2007) and Ryen (2009)).

As your positioning evolves, how do you respond to and acknowledge the implications of this for your work? Ethnography is not objective, but your subjectivity needs acknowledging. This becomes increasingly important as your relations in the field evolve and develop. Depending on which group(s) or community(ies) you are working with, you may find yourself drawn into activist work with marginal communities, into friendship with research participants or perhaps even romantic and sexual relationships. There are no clear-cut, definitive rules regarding your management of such relationships and you must negotiate these relationships and interactions delicately and appropriately. Remember that while you may leave the field, the community remains and will have a collective memory of you and your actions.

These negotiations are particularly important as it is likely that you will hold multiple positions during your research project, particularly if working with practitioner, activist or community groups where your role is not simply as a participant observer but also perhaps an organizer, personal assistant, event steward, sports coach or teaching assistant. In such situations, you need to reflect on and sensitively manage your relations with research participants and handle the differing power dynamics (see Bachmann 2011; Rubin 2012). In such situations and encounters, you need to think about how other people will see you. Are you at that moment a researcher or fellow activist or teacher or steward? What roles take priority and when? How do you decide what data and information it is ethical to record and reflect on? It may be that you are told things in one role that are not meant to be known in another role. How you communicate your position and role to others is therefore vital, as might be a negotiation of any limitations to how certain information can be used in the research. Throughout these processes it is also important that you reflect on your role and actions since, as Bachmann (2011: 364) notes from her ethnographic fieldwork, there is always the potential that 'the participant observer turns into a participant (without observing) and loses the researcher's approach to the (field)work and the data obtained.'

What to take from this chapter

You should now have a basic grasp of the development of ethnography as a methodological tool over time and why it can be an important and useful tool in development research. You will also have gained a sense of some of the specific tools or methods that are used by ethnographers and the importance of developing your skills in a range of these in order to build up layers of understanding. A central concern to all of these methods is a focus on the mundane and every day, of capturing the unobserved and unnoticed features of daily life, as well as the episodic and unusual. For researchers concerned with subjective understanding of daily life and development, ethnography can be a vital research approach. For those concerned with objectivism and 'truth', this method is not appropriate.

If you want to explore the ideas covered in this chapter in more detail, you may find these further readings useful:

Geertz, C. (1983) *Local Knowledge: Further Essays in Interpretative Anthropology*. New York: Basic Books.

Gottlieb, A. (2006) 'Ethnography: theory and methods', in Perecman, E. and Curran, S. (eds.) *A Handbook for Social Science Field Research: Essays and Bibliographic Sources on Research Design and Methods*. London: Sage.

Wieder, A. (2003) *Voices from Cape Town Classrooms: Oral Histories of Teachers Who Fought Apartheid*. New York: Peter Lang.

Wieder, A. (2004) 'Testimony as oral history: lessons from South Africa', *Educational Researcher*, 33: 23–8.

9

PARTICIPATORY METHODS

Participatory methods encompass both a set of diagramming and visual techniques as well as the underlying principles of grassroots or local participation. Therefore it is both method and approach. In this chapter, you will learn about how participatory approaches evolved and the key challenges to conducting empowering and reflective research. You will learn that visual exercises are important but not the central focus of participatory research and that at its core is a focus on power, co-production and shared learning. The chapter will describe the key methods and techniques that you can use and talk you through how to prepare for the field, use the methods and record the data while keeping the principles of grassroots participation at the core. To give you a sense of how these methods are used (and why they are useful) in the field, reflect on Chasca's experiences of using participatory methods while studying community environmental management practices in Nigeria and Botswana.

Some years ago, Chasca was working in a rural village in central Nigeria. She was conducting research on the management and use of tree resources and working with a small mixed group of men and women ranking different tree species according to their value to the group. The species had been identified previously on village walks with a wide range of participants. Local names for the trees were recorded and a sample of the tree (twig and leaves) had been collected. As the research was being conducted during the farming season, these longer participatory exercises were carried out in the evening, often by candlelight. On this night the group had started late and Chasca was keen not to keep people away from their homes too late into the evening. She suggested they conclude the exercise, but the group protested and said that they had not finished going through all the species. It was quickly agreed by the group that they should close as some of the women were keen to return to their houses. However, all agreed to meet at 6am the next morning to complete the ranking exercise and Chasca was then informed about this plan. Chasca felt this was a really good sign that the people with whom she was working were taking ownership of the exercise and wanted to pursue it for their own means as well as being helpful to the researcher. The group felt the exercise was helpful in reminding them of why certain species were important and provided information they could potentially share with 'projects' and others in the village.

In another research project, Chasca was working in Botswana on community-based natural resource management. She was working with the local San groups in Western Botswana and part of the research intended to map the natural resources around the village and conduct a preliminary ranking exercise to start unfolding how the biodiversity was valued by the community. The villages were extremely poor and people had very low levels of literacy. Working with two or three participants whom she knew quite well, Chasca started to introduce the idea of a map, with the option of either working with paper and pens or creating a model on the ground. It was clear quite quickly that the exercise was not going to work.

The participants were not interested in abstracting their spatial environment onto a two-dimensional surface. When discussing different 'veld products' around the village, participants kept suggesting that they go there to show Chasca so she would know where things were, rather than trying to make a map. The participants could see no point in the map and there was no interest in producing it. Chasca chose not to pursue this exercise or the ranking, focusing instead on field walks, participant observation and repeat interviews.

These two stories demonstrate the need for flexibility and sensitivity in your approach and in planning the methods for your research. They also show how you need to respect your participants and their wishes, giving them control and power over knowledge and information. These are some of the fundamental principles of participatory research and the methods and techniques that it comprises. In this chapter, you will learn about how the idea of participatory research emerged and how it has been, and, indeed, remains, a complex and contested terrain of methodological enquiry.

Short history of participation

Hickey and Mohan (2004) trace the emergence of participation back to the 1940s and participatory approaches have been strongly influenced by the work of Paulo Friere (Long 2001). However it was the 'popular participation' of the 1970s, particularly the work of Robert Chambers, that laid the foundation for the lexicon and practice of participation as we know it today (Chambers 1983). Its initial applications were in the field of applied research and action research that was common among the intersection of agricultural economics and development studies (Chambers 1997). At this time in the 1970s it was seen as increasingly important to work with the poor and marginalized, especially farmers, and this was the era of farming systems research and a greater focus on extension activities.

During the 1980s, these areas of applied research were influenced by anthropology and the realization that more sophisticated understandings of poverty and the 'human face of development' was needed. By the end of the 1980s a more coherent approach that combined the applied nature of research with farmers and diagramming techniques evolved into a flexible approach called rapid rural appraisal (RRA). Parallel to these developments in the global south were the emergence of methodologies such as soft systems analysis and cognitive mapping from management consultancy and organizational research and planning in the North (Mayoux 2001).

By the 1990s it was becoming clear that while these 'rapid' and formulaic approaches had some merit, they were also failing to capture the differential views of complex communities and key groups such as women, children or the socially excluded were sometimes marginalized from the research process. However, there was renewed interest in the methodologies of participation drawing on earlier traditions of participatory action research that had been long established as an integral part of many grassroots organizations in the South (Mayoux 2001). These approaches were based on underlying principles of rights and aimed to use the research process itself as a means of empowerment through the use of diagrams and visual techniques as a focus for discussion and activities such as role play and drawing to stimulate debate (Mayoux 2001).

From this wave emerged the terms participatory rural appraisal (PRA) and later participatory reflection and action (PRA) and participatory learning and action (PLA). Participatory approaches today are seen as a family of approaches, behaviours and methods (Chambers 2002). Today, the remit of methods or tools has expanded considerably to include photography and video by grassroots groups and the ways in which people can record their own information in diaries, photos and other accounts (Mayoux 2001; see also Chapter 8).

Emerging from this work are many attempts to provide typologies of participation. While many relate to 'development' more generally, they can be equally applied to the research context. Box 9.1 contains a useful framework by Pretty (1995) that is still one of the most comprehensive typologies around.

Participatory approaches are now viewed as mainstream and even institutionalized (Mikkelsen 2005). However, there is still anxiety about inherent power imbalances imbued within participatory research. These have been eloquently detailed in Cooke and Kothari (2001) and countered in Hickey and Mohan (2004) (see Box 9.2). These authors raise important debates about power, promise and bias within the development and research process.

Participatory approaches can be easily misused and there are things that must be kept in mind when you are planning and doing participatory research. Participation can come to mean 'a way to get people to do what we want' (Chambers 1991: 516) and at its worse a form of 'development tourism' that confirms biases, misconceptions and stereotypes (Chambers 1997). At the same time we need to be careful not to compromise academic integrity, as well as aid the disempowerment of participants (Goebel 1998). Cooke and Kothari (2001) warn that the lexicon of participation should not cover for schemes or approaches that remain top down and extractive where power relations between researchers and participants remain unequal. A participatory approach should seek to reveal and validate local knowledge, destabilize the notion of the outside expert as the only true 'knower' and include communities on an equal footing in planning and implementation. It should not adopt the language and methods of participation without acknowledging the complexity of social realities or properly absorbing or practising the intended notions of 'participation' (Goebel 1998).

When using participatory methods you, the researcher, should be a facilitator of knowledge creation, a self-conscious interpreter of complex and competing 'stories', rather than an extractor of data (Goebel 1998). As such participatory methods, if used well, emphasize the interaction between researcher and research subjects (Goebel 1998). However, participatory methods can obscure rather than reveal social complexity and can validate dominant views that become portrayed as the common view (Goebel 1998). One of your key challenges in using participatory methods is to overcome such tendencies through considering who is involved in, and excluded from, your work, and identifying

> ## Box 9.1 A typology of participation (after Pretty 1995)
>
> *Passive participation* – people participate by being told what is going to happen or what has already happened, with no ability to change it. The information being shared belongs only to professionals or the researcher.
>
> *Participation in information giving* – people participate by answering questions posed by extractive researchers and practitioners. People do not have the opportunity to influence the proceedings as the findings are neither checked nor shared.
>
> *Participation by consultation* – people participate by being consulted and external people listen to either view. External professionals or researchers define both problems and solutions and may modify these in light of people's responses.
>
> *Participation for material incentives* – people participate by providing resources such as labour and land, in return for food, cash or other material incentives. People have no stake in prolonging activities when the incentives end.
>
> *Functional participation* – people participate by forming groups or committees that are externally initiated. The groups are seen as a means to achieve predetermined goals. The groups tend to be dependent on external initiations and facilitators but may become independent.
>
> *Interactive participation* – people participate by being involved in the development and analysis of research and projects. Participation is seen as a right and not just a mechanical function. Groups take control over local decisions and so people have a stake in maintaining structures or practices.
>
> *Self-mobilization* – people participate by taking initiative independent of external institutions although the latter may help with providing an enabling framework. People retain control over how resources are used, or how time is spent. Such action may or may not challenge existing inequalities.

and finding ways to work around internal power hierarchies that determine who can speak on what topic and in front of whom. Key issues for you to consider will include composition of research sample, composition of specific groups while carrying out certain methods, finding ways of locating and engaging with marginal groups or individuals within communities and adopting and adapting methods (including the logistics of when, where and with whom methods are used) (see Figure 9.1). This will promote inclusion of different groups and ensure that their voices are heard within the activities conducted.

These efforts are vital as knowledge is embedded in power relations. There is no one local knowledge; there are simply multiple competing perspectives. Stronger and more powerful voices can dominate. Participatory activities can work to hide power relations through an emphasis on group work, on consensus in data expression and presentations, and can silence the marginal or dissident views (Goebel 1998). So, while these community

Box 9.2 The tyranny of participation?

Cooke and Kothari's book *Participation: The New Tyranny* came out to great acclaim in 2001 as it challenged the seemingly universal fashion of participation, which promised empowerment and appropriate development, on the one hand, while, on the other hand, in reality, consultants and participatory activists imposed development decisions and actions on people in ways that were antithetical to their participatory aims. The book helped bring into sharp focus the pitfalls of participation especially as it shifted from marginal to mainstream practice and as it shifted from small-scale to global policy initiatives.

In 2004 Hickey and Mohan edited a retort titled *Participation: From Tyranny to Transformation* and called for a deepening of some of the debates raised by Cooke and Kothari (2001). They put forward that 'politics mattered' and more focus should be given to understanding the ways in which participation related to power such that it provided a basis for moving forwards towards a more transformatory approach to development. Essentially, they looked to (re)constitute participation as a viable and emancipatory approach, taking on and using the critique provided by Cooke and Kothari (2001).

exercises can enable people to find common ground and consensus, there are downfalls as well (Laws et al. 2013).

In academic research, using participatory approaches means increasing the space for participants to express and control the knowledge being created. While academic research will essentially remain extractive in nature (that is there is no project or action to follow), the researcher needs to look for ways to promote positive change and empowerment through the research process, for example through training and employment for local people, by hiring local assistants, providing people with information that is difficult for them to access, validating and respecting local knowledge and realities and facilitating local dialogue on problems (Goebel 1998). It is worth remembering, however, that seemingly uncontested data can be implicated in complex ways through participatory processes in broader historical social and cultural dynamics. Information and data in some forms of participatory research are rarely always as transparent as may be desirable and the sharing of information is never apolitical. As a researcher, you should always be critically aware of your own positionality at all times in the research process (see Chapters 3 and 5).

Today, participatory research retains its dual identity of method and approach and the dichotomy of being simultaneously empowering for some and disempowering others. It is the potential for misuse alongside enlightening creativity that still divides many practitioners and academics. However, the call today, and for you, is not to underestimate the pitfalls and to be transparent in the aims of your research such that your research process is honest and open. If you adopt a participatory approach, then you will need to think carefully about why this suits your research.

Figure 9.1 Some exercises may require allowing space and time for people to work alone before talking in a group

Why use participatory methods?

As demonstrated above, participatory methods can be useful on two fronts. First, they can rebalance the power question associated with more traditional extractive methods, giving more control to participants over how information is collected, recorded, analyzed and kept. Participatory methods can also, through their focus on diagramming and visual techniques, open up debates and questioning in creative and innovative ways helping researchers and participants to view issues from different perspectives, facilitate participation of people with low literacy levels and specific disabilities and thus dig deeper into the findings.

There are two common ways of conducting participatory field research, either with large teams conducting intensive studies over a short period (perhaps through a fieldclass), or single researchers working in a more low key fashion usually over a longer period (perhaps for a dissertation). Each has its advantages and disadvantages. The former approach can allow greater community control over activities that can be empowering. Materials generated are often tailored for specific community needs such as funding proposals or management and actions plans. But these large-scale events can be intrusive, time consuming and exclusionary for those within the community. This approach, however, if done well, can be highly rewarding for both the research team and the community. The single researcher approach can be extremely effective and less intrusive. However, this approach is usually conducted over a longer time period and while it can

Box 9.3 Lorraine van Blerk: Doing participatory research with children on the African continent

For the last 15 years I have engaged in participatory research with children who are living in particularly difficult, and often dangerous, situations; usually on the streets. My ethical motivations for undertaking this research are based on a commitment to issues of justice, equity and empowerment, framed by a conceptual understanding that children are experts on their own lives, albeit constrained by the adult world in which they live. For me, using a participatory approach is therefore essential, as my conceptual and personal motivation does not always sit well with many of the more traditional approaches to, and methods used for, conducting social research.

It is vitally important then to work with young people, discover what and why research might be important to them and then invite them to work with researchers or even lead the research process in ways they see as appropriate for achieving research goals with academic, practical and policy value. A participatory process such as this is important to me as I have learned more about children's lives from being in their company, listening to them and facilitating them to take part in their own research than I think is ever possible from talking to policy makers and practitioners. To explain this I want to share two different examples with you.

Almost 15 years ago I undertook my PhD research in Kampala, Uganda, with street children. As part of the process the children used disposable cameras to represent their lives visually. Despite being advised against giving cameras to street children ('you'll never get them back') all but two were returned (one was stolen, the other broken). Every camera roll contained exceptional images and was accompanied by rich verbal insights into life on the streets: a life, in many respects, I would not have had such privileged access to without first gaining the trust of my research photographers, gained in part by trusting them. This experience taught me the value of listening to, and trusting, children.

My second, more recent, example is taken from a current project where I work as a research director on the Growing Up On The Streets project with StreetInvest. This research is on a much larger scale than my PhD but equally recognizes the importance of participatory engagement. In each of three African cities, the project is facilitating street youth to carry out their own ethnographic research: playing a central role in enabling others to really understand what it means to grow up on the streets. However, through a recent conversation with two street workers involved in the project management, I really began to understand the need to integrate research and practice: bringing everyone into participatory conversations. By stepping back and allowing the young researchers to take the lead, the street workers reflected that they had learned the value of participatory research. When one girl explained to them that she could not continue to engage in hairdressing even though she had been trained and given some equipment, because she had nowhere to keep the equipment, they realized that they had not really listened to what young people need.

Participatory ethnographic research, or active listening as our young researchers call it, can enable 'the experts' (academics, practitioners, policy makers) to see that the children are the real experts and without them we will never get it right. So, I encourage you to go and find out how to understand the needs of children, or whoever you are working with, through facilitating participation.

be transformative it is often much more subtle in its explicit ambitions. Both approaches, however, have clear potential for community engagement in knowledge production, as evident in Lorraine van Blerk's discussion (Box 9.3) of her decision-making processes when conducting participatory research with children in different parts of Africa.

Lorraine Van Blerk therefore encourages you to think about taking a participatory approach and shows how valuable this can be for your research and for those with whom you are working. The refection highlights clearly that by giving people some control over the research process, different, innovative and sometimes unexpected but valuable insights can be gained. So taking a participatory approach allows you and your research participants to think differently, to think creatively, to engage in different ways and, in many cases, to have fun while doing this.

These creative engagements can also allow insights that break down the researcher/researched assumptions. We know that knowledge is socially constructed, that all interviews and interactions are forms of performance and information sharing is done in specific and often deliberate ways (Mosse 1994; see also Chapter 7). Communities and participants may therefore construct their answers based on what they understand the researcher's interest to be: there is nothing unusual in this. However, sometimes these constructs are based around what the participant thinks is relevant or not relevant and thus they may choose to withhold or not mention information they deem to be irrelevant or private.

When Chasca was working in Nigeria some years ago, she was aware of this potential public/private view when working with a focus group of women on the Jos Plateau. They were conducting an exercise about what key problems women faced in the village and what the potential solutions might be. While Chasca facilitated the exercise, the women soon took ownership and control of the matrix diagram that was being drawn. One key issue that emerged was that there were really significant issues that affected women but these were not discussed at community meetings; neither were they discussed at a recent formal community-wide PRA event as they were deemed private issues with solutions that could only emerge from within the community and not from outsiders. The women discussed why these issues were private and where solutions lay, examining in turn the pros and cons of turning to government, community leaders, household heads etc. for help resolving these problems. The women themselves found this exercise illuminating and realized that some of the problems they faced alone could be shared in empowering ways.

While participatory methods are most commonly used to help those with less power have a voice in the research process, there is no need for these approaches to be confined to certain sections of society. Philips and Johns (2012) say that there is 'no obvious way' that participatory approaches could be used with elites as you are not trying to empower your research subjects. However, we would contest this and suggest that elites are themselves still working within often strongly hierarchical power structures. Participatory methods and approaches, used creatively in such circumstances, have the potential to give very different and unconventional views. These approaches could allow participants to reflect very differently on their roles, actions and behaviours in quite exciting ways.

So, you have chosen to use participatory methods because they fit your approach to research and provide a 'toolkit' of methods that will help you to achieve your research aims. You are aware of the potential pitfalls and complex power dynamics in the approach and you now need to work out which participatory exercises will suit your research the best.

Types of participatory method

The participatory toolkit comprises a diverse set of methods and activities, some of which are found in more conventional research, others of which may be very new to you as a researcher (see Figure 9.2). Table 9.1 sets out the different groups of methods and what sort of information they can elicit.

Figure 9.2 Participatory mapping can provide detailed data on local understandings of location, resources and mobilities

Preparation for and using participatory methods in the field

When choosing different methods and activities the key thing is to be flexible, experiment, invent and adapt. Having a go with friends and family, even about everyday topics such as which flavour of ice cream people prefer or mapping the area in which they grew up, is a valuable experience. It is surprising what you can learn about these techniques from simple experiments. Experience will help you chose which activities might work well for different types of data. You can gain this experience through reading extensively, asking colleagues or supervisors and by working with the participants themselves to decide which activities might work for different areas of the research. Engaging with these information sources can provide tremendously useful tacit knowledge about the complex realities and often unintended experiences of using participatory methods in the field. Some of these complexities are highlighted in Nicola Ansell's reflections (Box 9.4) on her experiences of researching the impact of AIDS on young people in Southern Africa.

There are many other concerns beyond those mentioned by Nicola Ansell to consider in your planning of participatory research. Deciding who you will be working with is difficult. For truly participatory research, this should be openly discussed with communities but this is not always feasible and at times not desirable. Bias can creep into sampling and selection of participants from many quarters and while local people can help ensure inclusion of different voices in the research process, they can also exclude and omit mention of some people.

When Chasca was working in Eastern Namibia a few years ago she discussed with community representatives the boundaries of the village and their grazing lands. The community was quite enlightened about its rangeland management demonstrating clear self-motivation and empowerment to manage their natural resource base. However, while on a drive around the boundary of their area, mapping and classifying rangeland types, Chasca asked about a small group of houses right on the boundary that she was being shown. It was explained that these people were not part of their community (despite living on community land for several years) as they were from an 'inferior' ethnic group and only temporary residents. These people were not included in community or committee meetings and Chasca felt it appropriate to visit them separately to elicit their side of the story. For further details of this research story, see Twyman et al. (2001).

Another issue to think through is in what order to carry out exercises. Often mapping and transect walks can be useful introductory exercises to start people talking about their community and for getting an overview of issues in villages or areas of cities that people frequent. This worked well when Chasca was working in Nigeria. However, in South Africa, there was more suspicion of researchers coming into rural communities and so Chasca and her field assistant decided to conduct semi-structured interviews for quite some time, gradually getting to know the community, before attempting collective exercises that required people to open up to the researcher, and to each other, publicly. This worked well in South Africa and repeat visits to communities at different times during the farming year again helped build rapport and trust, making the more formal participatory exercises highly successful.

When working with people for any research you need to keep in mind people's time commitments. As Chasca's story at the start of the chapter showed, people can be busy

Table 9.1 Types of participatory method

Key categories of method	Activities	Commentary
Review of secondary sources	– Documents, statistics, reports, books, files, films, aerial photos, maps etc.	– While many of these sources are available on the internet, much historical material or less public material will be logged in obscure libraries and offices and are well worth the time and effort to seek out – For example, minutes of key meetings may be made available to you illuminating how decisions have been made in the past – These materials may be subject to discourse analysis or provide useful descriptive background information. Further information on historical sources can be found in Chapter 10
Direct observation	– Looking, seeing and doing	– Akin to ethnography, direct observation and reflection in an ongoing field diary can provide much needed context and understanding to research findings – You will end up with text documents that can be coded and analyzed
Semi-structured interviews	– Key individuals – Focus groups – Chain of interviews	– Detailed discussion of interviews is covered in Chapter 7 but interviews based on written or memorized checklists of open-ended questions are often the foundation for research activities – Alternatively, chain interviews really hand power to communities with participants interviewing each other and analyzing information as it comes in – You will end up with extensive notes or transcriptions from recordings that will need to be coded and analyzed
Ranking and scoring	– Pairwise ranking – Matrix scoring and ranking – Preference ranking – Wellbeing and wealth ranking – Ranking by voting	– Good for eliciting information about difference and inequality – People place something in order and may attribute values through scoring – Often the discussions within these exercises are as illuminating as the results themselves. – You will end up with diagrams that can be photographed so you keep a record of them. – The results can be interpreted alongside other data – You can also keep notes of the discussions while the exercise is happening – These group events usually do not produce good audio recordings
Construction and analysis of maps, models and diagrams	– Social and resource maps – Topic and theme maps – Mobility maps – Census maps and models – GIS-based aerial maps – Transects	– Community maps can quickly provide overview information or maps produced by different groups can provide interesting comparisons in terms of what is included and excluded – The incorporation of remote sensing and GIS-based maps further extends the potential for these activities. Transects are particularly useful for building up layers of information – You will end up with maps or diagrams that can be photographed so you keep a record of them. The results can be interpreted alongside other data – You can also keep notes of the discussion while the exercise is happening (see Figure 9.2)

Method	Examples	Notes
Diagramming	– Causal, linkage and flow diagrams – Force field analysis – Time lines and trend analysis – Seasonal diagrams and calendars – Activity profiles – Daily routines – Venn diagrams – Pie charts	– These highly visual exercises can be useful for analyzing as well as collecting information. Each has its own specific uses and can be adapted for many different situations – Venn diagrams are particularly useful for looking at the relations between different institutions and groups – These exercises can also be carried out for different time periods e.g. historical seasonal calendars to prompt discussion about change. You will end up with diagrams that can be photographed so you keep a record of them. The results can be interpreted alongside other data – You can also keep notes of the discussion while the exercise is happening – These group events usually do not produce good audio recordings
Case stories	– Life histories – Narratives – Diaries	– Case stories provide depth and explanation to the sometimes broader brush exercises. Be careful of generalizing from such cases – If possible, as with all interviews, audio recordings allow you to produce transcripts for coding and analysis. If people keep diaries, be aware that they may want to keep them, so think of ways to record these data – Many smart phones now have apps that convert pictures to pdfs that can then be emailed to you
Drama, games and role play	– Plays – Songs – Proverbs – Games	– Often used by NGOs and professionals to bring sensitive issues into the open for discussion. If you do decide to record these events make sure you have full consent from all parties before doing so
Photography	– Diaries – Topic based	– As Van Blerk showed in his reflection, handing over the camera can give some surprising results. But as Ansell warns in her reflection, cultural considerations should always be thought through – Again, make sure participants have the option to keep some of the results. You will need to analyze this visual material carefully
Local categorization	– Soil types – Plant taxonomies – Local concepts	– Working with communities to understand their own classifications of soil, plants, landscape or whatever issue can give deep insights into practices. The focus is on local terms and ways of grouping, dividing and describing things. While these categorizations are commonly used in relation to environmental or biological terms, they can be successfully used to categorize how local people discuss illnesses or fevers within children or types of poverty and insecurity – You will end up with lists and matrixes that can be presented alongside the analysis of your notes of the discussion produced from these categorizations

Source: After Theis and Grady (1991); Mikkelsen (2005)

Box 9.4 Nicola Ansell: Participatory research with AIDS-affected young people in Lesotho and Malawi

When conducting research about the impacts of AIDS on the lives of young people in Southern Africa, I have used participatory methods for three purposes: 1) generating information about young people's experiences, 2) producing data about how young people (and adults) think about the impacts of AIDS and 3) getting feedback from participants, practitioners and policy makers on preliminary research findings. The methods have varied but in almost all cases, participants have generated a product (drawing, diagram, poster, photograph, drama) that has then formed a focus for discussion.

The wide variety of possible participatory research methods means each new activity is an experiment. Some methods prove more successful than others. Sometimes the outcomes are unexpected. One important lesson is that participatory group methods are seldom very effective as a means of uncovering individual experience, particularly where the subject is a sensitive one such as illness and death associated with AIDS. People are seldom willing to share and reflect on matters that are usually private in the presence of their peers (see Ansell et al. 2012). In my experience, when asked to do so, children become reluctant to speak, talk of less-sensitive matters or copy one another's ideas. In research undertaken with colleagues in rural Lesotho, we used an 'emotional storyboard' exercise to provoke discussion of situations that provoked happiness, sadness, fear and hope. Two children drew pictures of snakes: one said their saddest moment had been being bitten by a snake, the next child said theirs was seeing a snake. Given that both had experienced the death of their parents, this suggested they were deliberately evading a difficult topic for conversation. By contrast, given the opportunity to spend half an hour drawing a series of pictures representing their 'migration stories' for one-to-one discussion with a researcher, young people were able to offer detailed accounts of their personal histories, including the emotions they evoked (see Young and Ansell 2003).

Most participatory research is undertaken in a group setting and involves participants in discussion about a particular issue. While people doubtless draw to some extent on their own experience, such discussion is deeply embedded in local discursive constructions of the issue. This means that we should be somewhat cautious concerning the data produced in such settings and not assume that it reflects empirical 'reality'. For example, in collective workshops, participants repeatedly talked of children whose parents had died being inevitably 'doomed', but in individual interviews most young people insisted that the death of a parent had relatively little material impact on their lives. Nonetheless, group participatory activities are a very effective way of learning about how people think and talk about sensitive issues.

Participatory research not only reflects popular discourse, it is also a way of constructing new knowledge. This can be very productive with young people when they are interested in a controversial issue, particularly if a series of repeat discussions is possible. I organized repeated group discussions among high school students in Zimbabwe, who relished the opportunity to exchange ideas on gender relations (see Ansell 2001). Through the sequence of discussions, they repeatedly reworked their personal views through debate with one another. Although we began with separate groups of girls and boys, both groups asked to come together and challenge one another's ideas in a mixed sex setting. The participatory research process undoubtedly encouraged young people to think differently about gender.

Finally, a cautionary tale. Participatory research not only offers opportunities to construct new discourse, it can have unanticipated impacts. In Malawi, we loaned a digital camera to young people to photograph significant aspects of their lives and allowed them to keep some of the photographs. One young woman selected photographs she had taken of her baby. A year later, when we returned for follow-up research, she told us the baby had died shortly after our first visit. Culturally, some people are suspicious of the power of photography and the young woman's husband insisted that she destroy these mementos of her child's life.

at different times of the year and days of the week and some exercises may impinge inappropriately on daily lives. Be aware that in rural areas the timing of your research might coincide with migrant workers being home, or away, specific farming activities, children away at boarding school, cultural events, all of which will affect who you can talk to and how much time they may be able to give you. In urban settings, simply communicating with your groups may be more complex and require gatekeepers to facilitate access to certain organizations and sections of the community. Meeting people in the evenings after work is not always ideal as again, many have family commitments or there may be safety issues about being out at night.

When using participatory methods in development research, you may be working in cultures other than your own and in communities in which you do not speak the local language. In these circumstances, you need to work closely with translators or research facilitators. Discussion about working with translators is covered in Chapter 4 but it is worth mentioning here that they can be pivotal in whether your participatory approach works. A translator who feels superior to the people with whom you are working will struggle to engage with the empowering nature of the research. Chasca was worried about this with research she did in Namibia several years ago and she discussed at length, with her translator, the issue of accepting a chair when visiting different households for interview.

Something as simple as accepting, or not, a chair to sit on for an interview can be a tricky thing in very poor communities. Do you allow people to show you respect and borrow their neighbour's chair for you to save them the embarrassment of having their visitors sit on the ground or do you reassure people that you are very comfortable sitting with them, as they sit, round their fire while you talk. Luckily, Chasca's translator was very well tuned to the sensitivities and complexities of their collective positionality and when appropriate they would even discuss these dilemmas with their research participants. It is in these circumstances that your translator is crucial.

You may also need to provide some training for your translator especially when using some of the diagramming and visual techniques. Explaining that the *process* is just as valuable as the final product helps ease the pressure the translators may feel to make sure people 'do the right thing'. Again, if you keep the principles of participatory research at the heart of your planning, then these issues are less likely to arise.

The expanding availability of digital and other technologies are providing new ways of recording data while using participatory methods. However, the role of technology in the field is not unproblematic: it can be both liberating and highly divisive. Some of these comments are common to many different methods, but are worthy of mention in this context. Recording devices such as audio and video are much easier to access today with smart phones but be careful that they do not become intrusive. As mentioned in Table 9.1, apps exist that convert pictures of documents to pdf files that can be emailed to you, or others, providing a quick record of visual outcomes from different exercises. Similarly, photos can be shared more easily with communities with new technologies but be sure to ensure full consent for use of pictures and videos especially if you are planning to post any online at any stage.

When in the midst of planning research, even when taking a participatory approach, it is easy to forget at times that ownership of the research process and the outputs should be embedded within the community and with the participants themselves. While academic research, as discussed, will essentially remain extractive in nature, you need to look for ways to keep the principles of participatory research embedded within your practice. Think carefully about how the information gathered can be left within communities and think through and discuss the politics of who you leave it with.

What to take from this chapter

From this chapter, you should have developed a sense of the complexity and potential creativity of participatory methods. In this chapter, you have also learned that participatory methods encompass both a set of diagramming and visual techniques as well as the underlying principles of grassroots or local participation: it is both method and approach. The complex history of participation still divides academics today, but the potential of participatory research can be transformative and done well is worth pursuing. You have learned that visual exercises are important but are not the central focus of participatory research and that at its core is a focus on power, co-production and shared learning. We have outlined the key methods and techniques that you can use and raised issues for you to think through how you might prepare for the field, use the methods, and record the data while keeping the principles of grassroots participation at the core. The challenge is to remember that research and learning are intertwined (Laws et al. 2013) and truly participatory research will manage expectations, co-produce learning, foster knowledge exchange and keep people themselves at the heart of the research.

If you want to explore the ideas covered in this chapter in more detail, you may find these further readings useful:

Mayoux, L. and Chambers, R. (2005), 'Reversing the paradigm: quantification, participatory methods and pro-poor impact assessment', *Journal of International Development*, 17: 271–98.

Narayanasamy, N. (2009) *Participatory Rural Appraisal: Principles, Methods and Application*. London: Sage.

Theis, J. and Grady, H. (1991) *Participatory Rapid Appraisal for Community Development: A Training Manual Based on Experiences in the Middle East and North Africa*. London: IIED and Save the Children.

10

ARCHIVES, DOCUMENTARY AND VISUAL DATA

Archival data underpin a vast array of development work and understanding, but are often overlooked in development research. This chapter conceives of the archive as comprising written documents as well as climatic records, photographs, broadcast media, government documents, postcards, political cartoons and images. We outline a range of methods that can be deployed to explore development concerns including livelihoods, democratization, othering, tourism and economic development and urbanization. Critical reflections on the challenges of accessing archives and the use of these data are engaged in relation to power dynamics, ethics and accessibility.

This chapter starts with a reflection from Lawrence Dritsas (Box 10.1), who eloquently highlights the challenges of archival work alongside the revelations and rewards that it can generate. Lawrence reflects not only on the content of the archives and what they can reveal, but gives an insight into some of the practicalities that you will have to face when undertaking this type of research.

What is an archive?

Archives are simply collections. They are published and unpublished assemblages of materials. Archives lie at the centre of national, regional and personal histories and may be housed in a variety of different places. Formal archives may be held in museums, national archives or libraries, while informal and more ad hoc archives may lie in the dusty corners of offices or in people's attics (see Table 10.1). Whatever the collection, archives are a valuable and insightful research tool that can open up, animate and make history come alive (Mills 2013).

Montello and Sutton (2013: 68) define archival research as a 'type of data collection in which existing records that have been collected by others, primarily for non-research purposes, are analyzed, often after coding'. The way in which you use an archive in your research may be varied. Archive materials may provide context to your study, they may provide contrast to present-day experiences or they may be the central focus of your study. As with other methods they can be combined with other methodologies (see Chapter 14).

Archives comprise sets of documents or texts (reports, newspapers, letters, manuscripts etc.), photographs, films or sound recordings and may be in material or digital form (see Table 10.1). These assemblages are rarely complete but are nevertheless excellent sources

Box 10.1 Lawrence Dritsas: The challenges and revelations of the archive in Zambia

The reading room in the National Archives of Zambia is probably the coldest place in the city of Lusaka. The air conditioners hum powerfully from their perches up near the ceiling and the cold air spills out directly onto the readers' backs. 'Wear warm clothes', is the first piece of advice I have for someone unaccustomed to research in this particular part of tropical Africa. But it is not all bad. The cold can be a relief from the heat outside and gives one assurance that this is an institution that takes climate control seriously; the records of the past are safe here.

I have spent a great deal of time in archives such as this one and wherever I have gone, there are striking points of common experience: each archive has its own idiosyncratic catalogue that will take time to learn, fetching files takes around a half an hour or so – possibly much longer, only pencils to be used under threat of expulsion, there may not be an electrical supply for your laptop, don't even consider using a camera without asking and deciphering the content of the files you have found is never straightforward, if not downright baffling. With these challenges in mind, why enter the archives at all? Why spend all day squinting at illegible handwriting searching for a fact that may only be a footnote in your next article? The answer is because the rewards can be rich.

In my research on the history of exploration and the practices of scientific research in colonial Africa, archives have provided data unavailable anywhere else. Explorers wrote hundreds, if not thousands, of letters from the field, most of them never published. This correspondence provides a wealth of insight into expeditionary practices, including (very important for me) details about interactions with local informants and assistants that never made it into popular narratives or scientific reports. Colonial records are equally rich. The files were created to ease the work of the office: bringing memoranda, internal reports and correspondence together, usually chronologically, in one place for ready reference. These bits of paper tell a complex story of the work of colonial government as it happened, on its own terms. Here you find first drafts with edits in red pencil scrawled across the page alongside intra-office correspondence that reveals the undiplomatic opinions of officials who later produced more 'politically correct' prose for wider consumption. Expressions of doubt and concern are revealed in internal memoranda as decisions are being made. You also find here the wider concerns that all workers have in their careers but that never surface elsewhere: job security, pension arrangements and personal issues.

For those of us who pursue research into how knowledge about the developing world was made, how it was socially constructed in specific spaces and places, then the archive is an invaluable source for insight. Archival work allows you behind the façade of an institution to see its inner workings and you never know what you will find.

Table 10.1 *Types of archive*

College or university archives, e.g. University of Sheffield archives	https://www.shef.ac.uk/library/special
Corporate archives, e.g. archives from the Unilever corporation	http://www.unilever.com/aboutus/ourhistory/unilever_archives/
Government archives, e.g. Republic of Botswana or India	http://www.mysc.gov.bw/?q=dept_bnars# http://nationalarchives.nic.in/
Archive hubs – specific, often online, hubs collating resources/links for archival research	http://archiveshub.ac.uk/
Historical societies, e.g. Jamaica Plain Historical Society	http://www.jphs.org/sources-archive/
Museums, e.g. the Eliot Elisofon Photographic Archives at the National Museum of African Art, in the US	http://africa.si.edu/research/archives.html
Religious archives – the Mundus Gateway is a web-based guide to more than 400 collections of overseas missionary materials held in the United Kingdom	http://www.mundus.ac.uk/
Special collections, e.g. an online archive of personal materials linked to Vietnam hosted by a US university.	http://www.vietnam.ttu.edu/virtualarchive/
Private family collections – rarely online but this example is a collection of letters linked to novelist Olive Schreiner	http://www.oliveschreiner.org/
Small local institutions – rarely online	Ask around in country

Source: after Schmidt 2011 (examples added)

of information. Archives contain unique, specialized and sometimes rare collections. For archives to be useful in your research, you will need to think about this method in a similar way to many of the other methods described in this book. You will need to ask yourself what data (for example, documents) are available for your study area or topic. Can you negotiate access to the material and how will you sample them? You will need to work out how will you sift through the vast array of materials you will find (or not) and record the information you need. You will need to think carefully about how will you analyze your material and what meaning will you be able to infer from your data and how will this enable you to answer your research questions. In all, despite the inconsistencies and ambiguities of many of the archive materials, you will need to be as rigorous and careful in your planning and execution of your research as with any other methodological approach.

What types of data do archives contain?

Archives contain many different types of material collected or assembled for many different purposes. The purposes and ways in which the collection was put together, however formal or ad hoc, affect its interpretation and this is followed up in the next section. Here, in Table 10.2, we simply describe the wealth and variety of materials that you may find and use from an archive.

Table 10.2 *Typology of archival materials*

Form	Examples	Examples of research using these materials
Texts	• Directories, yearbooks, almanacs, registers • Government, company and organization reports • Official correspondence (government, commercial, civic) • Logbooks (coastguard, army, naval etc.) • Memos and telegrams • Published accounts (books, newspapers, magazines) • Letters • Diaries • Telegrams	Ashmore et al. (2012); Craggs 2008; Craib (2010); De Leeuw 2012; Dritsas and Haig (2013); Endfield (2010); Hazareesingh 2012; Hoffman and Rhode (2011); Jacobs 2002; Kreike 2003
Photographs and pictures	• Newspapers • Postcards • Exhibition materials • Cartoons • Ephemera (posters etc.) • Stamps (philatelic archives) • Maps • Currency	Brooks (2005); Campbell (2007); Child (2005); Cornelissen (2005); Edensor (1998); Fuller (2008); Fürsich and Robins (2004); Gruley and Duvall (2012); Hammett (2010, 2011a, 2011b, 2012, 2014); Hoyo (2012); Kevane (2008); Mawdsely (2008); Mwangi (2002); Penrose (2011); Wallach (2011); Willems (2008)
Film and sound	• Films made for public broadcast • Radio broadcasts • Recordings of special events • Research collections (e.g. British Library oral history collection)	Beinart and McKoewn (2009); Duncan et al. (2006); Hongslo et al. (2009); Lekgoathi (2009); Mwesige (2009); Paleker (2010); Sullivan and Rhode (2002)

The examples cited above provide useful starting points demonstrating the diversity and variation of archival research. People's written and pictorial records of their experiences travelling in foreign lands provide informal sources of information about people and places visited (Montello and Sutton 2013). Travel journals and diaries have been systematically mined as an archival source of data to study the motivations and experiences of long-distance international migration (Endfield 2010). Comments about rainfall in journals, logbooks of explorers, expedition leaders and missionary letters (Nash and Endfield 2002) have been analyzed as proxy measures of historical weather patterns. Photographs can be analyzed in themselves (Hartman et al. 1998), and sequences of images through the decades have shown how landscapes have changed over time (Leach and Fairhead 2000). Postcards have been explored (both sides) to reveal different views of history (Adedze 2010; Vokes 2010) and brochures, leaflets and stamp collections give insights into the social and political histories of different cultures (Hammett 2012).

In the reflection in Box 10.2, Chasca discusses an incidental finding when working in the archives in Botswana. This reminds us to sometimes follow our curiosity, and seek out the unexpected.

Several years later Chasca supervised a PhD student who worked with the descendants of those original settlers. The student worked closely with key families to document environmental changes over the 100 years since settling the area. He used family archives that included farm records (of stock numbers, rainfall etc.), photos, letters and diaries as well as aerial photos and remote sensing to build up a picture of environmental and social change on these farms over the century. Building trust and rapport with the 'keepers' of these private archives and records was essential for the success of his research.

You will find archive material online and in UK institutions as well as capital cities around the world. There is a big move to digitize many archives but only a fraction of material will be available online so think carefully about what sources you are thinking

---

Box 10.2 Chasca Twyman: Capturing the feeling of a moment in history in Botswana

When I was doing my PhD research in Botswana I spent some time in the National Archives looking at planning documents from the 1960s that were held there. As this can be slightly tedious work, I also called up some historical documents from the 1890s, the time when white settlers were moving into the area of Botswana in which I was working. The historical documents made fascinating reading and unfolded a real and raw side of history I had not expected to experience.

The context of this time was that a route across Western Botswana to Lake Ngami had become important and numerous explorers and prospectors travelled through the area. Boers from the Cape were encouraged by the colonial administration to move up to the area to settle it for cattle production, irrespective of the territories of indigenous groups. The politics around this period are fascinating (see Russell 1976; Biesele et al. 1989; Wilmsen 1989) and still have resonance today (Hitchcock 2002; Hitchcock et al. 2011) as a band of freehold ranches stretch across the Ghanzi ridge and their presence has transformed the livelihoods and landscape of the region. Therefore, when I came across the following original handwritten telegram my mind was immediately transported to a different era and a different view of this history (see Figure 10.1).

This telegram captures a moment in history when someone asks some very poignant questions about what is happening around them. It expresses the anxiety of the chief at the arrival of the Boers who went on to settle in the area and whose descendants still farm much of the freehold land today. Although this telegram was not central to my PhD research at the time, I found it so emotive that I kept a copy for many years. Other documents that I found from this moment include the trekkers' correspondence working out how many provisions they needed to take, adverts promoting the 'free land' that was available for settling, telegrams requesting information on water points for cattle along the route. These documents brought to life an important historical moment, highlighting both the significance of that moment as well as its ordinariness.

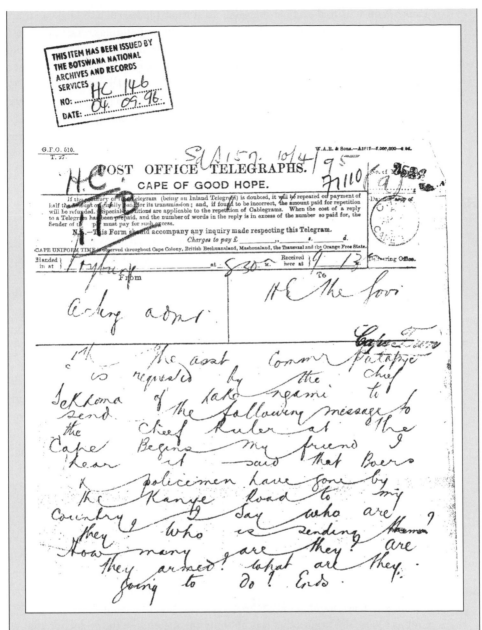

Figure 10.1 Telegram from Botswana National Archives, which reads '
The assistant commissioner Palapye is requested by the Chief Sekoma of
Lake Ngami to send the following message to the Chief Ruler of Cape.
Begins. My friend I hear it said that Boers and policemen have gone by the
Kanye Road to my country. I say who they are. Who is sending them? How
many are they? Are they armed? What are they going to do? Ends'

Source: Botswana National Archives: The research in Botswana was conducted under the Global Environmental
Change Research Project research permit (No. OP 46/1 XLII (105) PAO I), as required under the
Anthropological Research Act 1967

of using. You will find archive materials in formal places such as dedicated archives, museums and libraries. Not all of these are old dusty buildings. As Laurence Dritsas commented in the introduction, many archives are located in extremely modern buildings with climate control and air conditioning to preserve documents. There is even academic debate about climate change and keeping the archives cool (Hong et al. 2012). Mahajan (2011) encourages us to research beyond the confines of the written archives, away from official records, to the villages and homes in which people gather. In moving outside the formal archive, be courteous and respectful and if borrowing materials, always return them when you said you would and in the same condition in which they were lent to you.

How are archives produced?

The materials that you will find in an archive are secondary sources, i.e. they do not comprise primary data that you have created (it is not your interview transcript). However, you will be doing original analysis of the material so in many ways you can treat the texts and photographs in similar ways to other texts you may be more familiar with analyzing. What is crucially important in the archival context is how the material has been produced. When you conduct an interview, you are fully aware of the context of the interview, the surroundings, and the relationships with the interviewee. All this is recorded in your field diary to be analyzed alongside your transcripts. In other words, you have documented carefully the *production* of the interview transcripts, down to whether you mark up pauses, laughter etc. on your transcript. With archival material, you have not been present when the document or photo was produced or when it was catalogued or archived and this is something of which you must be aware.

Archives are not neutral storehouses for the consignment and transmission of the memorandum of history (Archives of Exile 2010). They are sites organized around specific habits, routines and programmes for the keeping of knowledge, even if only partial and selective. They may have been created and kept for idiosyncratic or personal reasons (Montello and Sutton 2013). Archives can suffer from inherent bias as material may have been collected for a range of non-scientific and objective purposes. Many have been collected for various administrative, journalistic or record-keeping purposes (Montello and Sutton 2013), but some, as we have seen, may be in private collections. The key is to see the analysis of the archive production, not as a limitation to your study, but as an opportunity to learn more about the materials that you are using.

Craggs (2008: 48) argues that it is not only the material in the collection that we need to question, but also the practices and technologies of assembling, classifying, cataloguing and displaying what goes into the archive. Archives, whether online or in a building, will have a cataloguing system. You need to be aware of how the cataloguing is being presented, what key words are used to tag entries, how documents or pictures are classified and whether this affects how you will come across certain materials. In some ways, the cataloguing is someone else's first level of coding. All the sifting and sorting that has been done before you get to the materials has been done with a purpose and you need to be aware of what influences and biases this has on your study. Thus the work inside the archive is as important as the wider networks of stories that the archive tells (from colonial administrators, collectors, travel diaries, maps and cameras) and which produce 'knowledge' about the world (Craggs 2008).

Reading along or against the grain?

Archival research generates considerable academic debate. Dwyer and Davies (2010) call for researchers to animate the archive by bringing the material and documentary properties of archives into play. This call is echoed by Mahajan (2011) and her work tries to bring history alive through combining oral histories with archival documentary sources. Dwyer and Davies (2010), however, caution against a simplistic notion of archival research. They point out the contradictory processes of archives that give form to the identities and capacities of past communities, space and landscapes, while simultaneously erasing that which cannot so easily be captured (Dwyer and Davies 2010). In a similar fashion, Lorimer (2003) calls for us to hear the small stories of history. Therefore, identifying what is missing or unseen in the archives is as important as what is presented and documented. For example, what is not in the photo is as interesting as what is there and how it is framed and who is behind the camera all tell different stories. Furthermore, hidden or small stories within archives often connect big things and events and can be highly illuminating.

So, while some call for us to seek out what is missing and to read against the grain when investigating archives, Craggs (2008) and De Leeuw (2012) caution this approach. They advocate reading against the grain to rescue the alternative memories or other lives from the colonial archives, but, they argue, we need to read *with the grain* first to make sense of the archives and how and why they were produced. Therefore, while looking for alternative or hidden stories can be exciting, you will need to understand the archive and its compilation before you can make assumptions and inferences about absences and unorthodox entries.

Practicalities of using archives

Once you have familiarized yourself with the academic debates around archives, and identified what materials you need for your study, you need to work out how to arrange access to the materials. You may be able to request materials remotely (via inter-library loans or via request for scans or photocopies), although these will often incur costs. If you are planning to visit the archive in person, you will need to make prior arrangements with the archive. It is very rare that you can arrive at an archive without an appointment and sometimes you will need a research permit too. If the archives are overseas you may not be able to set up appointments before you leave, but always make sure you find out as much as you can before you arrive.

When planning your visit to an archive you will need to correspond with the archive staff to determine that you will be able to access the material that you want. You should read their guidelines about use of the archive and find out the archive opening hours and entrance fees. Dan and Chasca have both been caught out when planning archival research to find unexpected public holidays that close all government buildings, including archives.

So be specific about the dates you would like to be there. You will also need to ask about their policies on how many materials can be requested at any one time and you will need to find out their restrictions on reproducing materials, i.e. what can be photocopied, scanned or photographed and how many you may copy at any one time.

Most archives will ask you to register on arrival, so you will need identification (passport) and you may need a research permit to access archive material. You should check this prior to travel as some research permits can take months to apply for and can be costly. You may also need a letter of introduction from your home university or a local host university. Each country will have different and specific requirements so make sure you find out what you need to do before you go. At the archive, you will often have to leave your bags and coats etc. in a locker (a precaution to reduce theft of materials) and it is usual to restrict any food or drink in archives (to mitigate against spillages and thus damage). You may be allowed to take paper and pencil (not pens) in with you and archives will also have policies about electronic equipment. Some materials may not be permitted to be reproduced so restrictions on laptops, tablet computers and mobile/cell phones may be in operation.

In most archives, you cannot simply walk along shelves of interesting files, as you might with books in a library. You will have to engage with the cataloguing system and make specific requests on request forms. Some digital systems may allow you to put in requests before your arrival, others may not. Allow plenty of time for archival research as it takes time to search for, and retrieve, individual documents and items. Keep in mind through this process, how the cataloguing system has been produced and record your thoughts about this as you work. Phillips and Johns (2012) provide a useful reminder suggesting that you ask for help both with existing collections you are looking at, but also where you might find other materials. For example, if you are working in museums, curators may allow special access to collections not on display, so talk with staff and seek their advice.

As well as ensuring you have permission to conduct research in a country and its archive, you should also be aware of copyright, restrictions and unprocessed collections. Copyright legislation is complex and should be checked in detail with the archive and for the country in which you are working, even if you access the collection online. For example, although the archive may own a document or image, the copyright of that item may remain with the creator or publisher of it and you may have to trace the source before you can publish or cite from materials (Schmidt 2011). However complex, breach of copyright is a serious issue and you should ensure you work with the archive staff to make sure you know exactly what you may and may not do with materials. In many cases, materials may be reproduced for educational purposes, but remember that at a later stage you might want to publish an extract from a document. In this case, sorting copyright issues when you are at the archive can save a lot of time and effort. Some materials in collections may have specific restrictions imposed by donors or governments, perhaps to protect identities or personal information, often linked to a date at which this information can be released. Unprocessed collections refer to materials that are yet to be classified in archives and these are often unable to be viewed (Schmidt 2011).

You should also familiarize yourself with *how* to reference archival materials. This is important as with all references, other researchers should be able to identify the precise location of the cited document. For archival material, you will need to include the location of the document (which archive, what reference in that archive) as well as the authorship of the material and its date. Jennings (2006) provides a useful outline of how to reference materials in detail.

Schmidt (2011) provides a further extremely comprehensive and user-friendly guide to using archives and can be downloaded free (http://www2.archivists.org/usingarchives). It is based on conducting archival research in the US, but it is general and many of the lessons can be applied to the wide range of contexts in which you may be conducting archival research.

What to take from this chapter

So, when you visit an archive prioritize your requests and know the procedures so that you can be efficient with your time and that of the archive staff. Ask for help when you need it. Make sure you take thorough notes and full citations as in many cases you cannot go back if you are missing something later down the line. Lastly, as Lawrence Dritsas reminded us, take a sweater as it may be cold.

If you want to explore the ideas covered in this chapter in more detail, you may find these further readings useful:

Derrida, J. (1996) *Archive Fever: A Freudian Impression*. Chicago: University of Chicago Press.

Mills, S. (2013) 'Cultural-historical geographies of the archive: fragments, objects and ghosts', *Geography Compass*, 7/10: 701–13.

Schwartz, J. and Cook, T. (2002) 'Archives, records, and power: the making of modern memory', *Archival Science*, 2: 1–19.

Steedman, C. (2001) *Dust*. Manchester: Manchester University Press.

Stoler, A. (2009) *Along the Archival Grain: Epistemic Anxieties and Colonial Common Sense*. Princeton: Princeton University Press.

11

QUANTITATIVE DATA AND SURVEYS

Quantitative approaches to research provide development scholars and practitioners with an important set of tools for exploring the context, causes and outcomes of development concerns. Quantitative methods are used to develop an understanding of the 'big picture' based on rigorous objective measurement. Within development research, you can use quantitative approaches to engage with a vast array of topics and concerns, from health to education, from governance to aid flows, and from perceptions of corruption to understanding migration. Using quantitative methods, you may base your research entirely on existing datasets or you may develop your own survey through which to generate new data. In either case, you are not seeking to develop a deep understanding of an individual's feelings, perceptions or identity. Instead, you are concerned with generating broader findings based on large samples from which you can draw generalized findings and recommendations.

To illustrate this approach, and some of the complications that can emerge in conducting survey research, we can consider a project addressing the health and wellbeing outcomes of a housing upgrade scheme in an informal settlement in South Africa. This project applied a socio-ecological perspective to an in situ upgrading of housing within an informal settlement to determine whether self-reported physical and mental health outcomes and sense of community belonging differed between informal and formalized housing. A survey questionnaire and cover letter explaining the research were developed and administered to over 200 households in the community. Due to the size of the sample, both closed and open questions were included in the questionnaire (see later discussion on the pros and cons of this approach).

In order to complete the survey, a number of questions and challenges needed to be addressed. First, the researchers needed to negotiate access to the community. This was done through the local 'street committee' and required Dan to attend a number of meetings and allay suspicions that he (a white, English-speaking male) was working for the Democratic Alliance (the main opposition party in South African politics, historically associated with white voters and at the time controlling the municipal government: Imizamo Yethu was a staunchly ANC neighbourhood).

Second, English was not the dominant language in the informal settlement. The home language of the majority of residents was isiXhosa, a language that neither researcher spoke. This factor, coupled with concerns over safety issues, contributed to the hiring of local residents to conduct the survey. These individuals had previously been trained in and had experience of using survey techniques through a local NGO, the Development Action Group. The assistants were provided with a refresher training session and supplied with the necessary materials to conduct the survey. The plain language statement/cover letter for the questionnaire was translated into isiXhosa, but not the questionnaire itself.

No sampling frame (the set of units from which the sample is drawn – this might be an entire population or a subset thereof) existed for the informal settlement so a two-tiered sampling process was utilized, with assistants surveying upgraded housing sampling every second property and those covering informal dwellings every fifth house. The majority of surveys were completed and returned, although it was clear that a number of these contained misunderstandings and mistranslations of questions, missed questions and evidence of guided responses in response to open questions, raising issues about potential shortcomings in the training of assistants and highlighting the lack of careful piloting of questions that may have prevented some of these issues (see later discussion). Data were cleaned and inputted for analysis using SPSS software. Even in this relatively small survey, the inclusion of open questions resulted in a disproportionate time burden for data capture and analysis. The findings of the research illustrated the complex relationships between housing and health and the need for a holistic approach to in situ upgrade projects (Shortt and Hammett 2013).

The experience recounted above of conducting a small-scale research survey demonstrates a range of factors to consider in designing and implementing quantitative research. The next part of this chapter outlines why quantitative methods are of use in development research. We then address the use of existing datasets in research before considering how and why you might develop your own dataset through a tailor-made survey.

Why use quantitative methods?

Quantitative methods are a useful tool for development researchers, allowing you to develop information about experiences, attitudes, behaviours and characteristics across a population. Whether you use existing datasets, develop your own data or integrate new and existing datasets, quantitative methods provide powerful analytic tools at a range of scales. These methods can allow you to monitor and evaluate changes over time or to produce comparative analysis between case studies and places. You can present profoundly powerful stories based on quantitative data: think about the compelling evidence you have seen presented on topics such as HIV/AIDS prevalence levels and deaths at a global scale, between countries and before and after policy interventions for instance.

Survey-based research includes a broad range of survey types, from working with national census data through to small-scale project-based surveys. Surveys may include mapping of land use or conflict, locating resources and facilities, identifying zones of risk or safety, as well as developing understandings of demographics, voting patterns, attitudes and behaviours. In essence, questionnaires and surveys are powerful tools for addressing readily observable phenomena or experiences, but they cannot get at the processes behind opinions or the context in which those opinions are formed (Galasinski and Kozlowska 2010).

However, surveys and quantitative data analysis do not need to be used as lone methods. They can also be effectively used in conjunction with other methods such as focus groups, interviews and participatory methods. In these situations, you may use the survey to provide background and contextual information to a more in-depth qualitative study, and to indicate key questions and topics or specific geographical areas for follow-up research. Or you may do the inverse and use the qualitative data to inform your survey work by identifying common themes of concern or important issues of which to gain a broader understanding.

Regardless of whether you use existing datasets or conduct your own survey, and whether you combine quantitative and qualitative methods, there are a number of key decisions to be made. First, does this methodological approach fit to your epistemological view? If not, then the type of data and analysis produced will not be suitable for your work. If there is a match to your approach to knowledge and understanding, you must then consider if/which quantitative tools will be suitable for your research questions and, derived from these, the specific objectives you want to address. At this stage, you can begin to think about the data you will need and how you will then generate these and the logistical concerns associated with this process.

Using existing datasets

The range of existing datasets suitable for use in development research has grown rapidly in recent years (see Findlay 2006; White 2010). This expansion of resources is in part due to the expansion of census and other large-scale surveys conducted by government and non-government agencies in the developing world. The rapid growth of the internet and computing technology has also facilitated increases in the breadth and speed of data sharing and analysis. However, the scale and complexity of data handling and processing from national-scale surveys such as the census can result in data being released several years after the census was completed. This can be a frustration for development and other researchers who want to work with these data to explore the human, social and economic indicators of development, but feel that this work is out of date before it is begun due to the time lag in accessing such data. Further complications in accessing data can arise due to bureaucratic issues and inefficiencies and myriad other causes associated with governance structures and challenges in many states. Gijsbert Hoogendoorn's reflections (Box 11.1) on accessing and analyzing quantitative datasets in South Africa provide an insightful view on the realities of such endeavours.

As you can see from the sample list below, you can now access datasets on political conditions, censuses, child nutrition, labour, demographic and household surveys, international aid flows and many other topics from around the world. In many cases, access to these datasets is free but in some cases you may have to pay the organization or government department in order to access the data (and you may be subject to certain restrictions in how you make use of the data). Such payments are usually modest, especially when you consider the costs incurred in collecting large-scale survey data. While the list in Box 11.2 includes a range of sources for such data, your own searches of government department websites and those of international donors, civil society

Box 11.1 Gijsbert Hoogendoorn: Quantitative research methods in South African human geography: making things up?

The use of quantitative methods in South African human geography research faces a range of challenges, from accessing reliable databases hosted by governmental institutions to complications resulting from researcher positionality in the current political climate (Hoogendoorn and Visser 2012). These challenges affect both national and international researchers in South Africa and are echoed in many countries in the global south (see Fraser 2012; Rubin 2012; Schuermans and Newton 2012). The researchers' positionality and ability to access raw data from governmental institutions play a major role in developing quantitative databases that are reliable and of good quality. My own experiences are of mixed successes, from instant service with access to high-quality data, to being sent from one office to another for days on end with no access to any kind of data. Therefore, when and if these hurdles have been overcome and raw data acquired, it is vitally important to consider the validity, relevance and accuracy of these databases generated from governmental organizations. Often data gathered from governmental organizations are outdated, incorrect or inaccessible and procedures taken to verify the validity of the data can be prohibitively labour intensive. As a result, you – the researcher – need to consider the available, and often innovative, data-mining and data-cleaning techniques that could be used to improve the validity of quantitative data and the overlaying of different datasets to improve accuracy in database construction is often essential (see Pienaar and Visser 2009).

The frequent strikes and industrial action affecting many government departments and parastatals, such as the post office, further complicates data collection practices, resulting in unreliable responses and increasing the costs of research. In such situations, you should consider whether quantitative methods are necessarily the ideal approach to your research. Qualitative research may be a more appropriate means of conducting research, where the researcher can consider the responses of the researched more critically than could be achieved with quantitative data using statistics that are incorrect. However, with the increasing development of information technologies in Southern Africa, especially in remote rural areas, the quantitative researcher may still be able to find avenues to do research. As Hall (2011) notes, web-based resources are in the majority of cases not qualitative but quantitative. Web-based resources can be especially valuable as municipalities are incorporating technologies in previously unlikely locations, but reliability should still be questioned. The use of online survey programs similarly could allow the researcher to target particular groups that were difficult to contact in the past. Therefore, the researcher needs to consider the accuracy and validity of their data very critically in terms of what innovative quantitative methods can be used.

Box 11.2 Existing datasets

There are numerous sources of existing datasets of relevance to development researchers. The list below is not exhaustive but should provide some inspiration for potential sources of data:

AfroBarometer – presents country-specific datasets from across the African continent based on research surveys addressing social, economic and political indicators.

AidData.org – a website hosting various datasets capturing aid transfers and spending.

Asian Development Bank – carries a range of datasets accessible through its website covering macro-economic and social data, key indicators, aid flows, environmental data as well as trade and economic growth indicators. Many of their data are free to access but some datasets require payment for access.

Bureau for Research in Economic Analysis for Development (BREAD) – acts as a portal for accessing datasets from around the world, including longitudinal, health, living standards, nutrition, poverty and education-related datasets.

Devecondata.blogspot.co.uk – a blog run for development economists that contains links to multiple relevant datasets including demographic and household surveys from around the world, climate data and road network spatial data.

Eldis.org – provides access to a range of development-related datasets through its open access Knowledge Services API.

Geohive – offers access to a variety of population and economic datasets. Also provides a comprehensive listing of national statistical offices/bureaux through which you can access a wealth of national-level statistical data from around the world.

Open Data Impacts – another blog-type site that provides reflections and commentaries on quantitative datasets as well as hosting a spreadsheet of development-related datasets (opendataimpacts.net/2011/06/a-whole-lot-of-development-datasets).

Population Reference Bureau – provides a range of population-related datasets.

United Nations – holds 34 development-related datasets, comprising 60 million records (http://data.un.org/).

World Bank – hosts an extensive array of development-related datasets covering economic, social, political and other development indicators. Its open data resource provides a range of search and analysis tools and APIs.

Box 11.3 Adam Whitworth: The politics of poverty analysis: challenges and opportunities in building a small area multidimensional poverty index for the South African government

The South African Index of Multiple Deprivation (SAIMD) was a project I worked on with colleagues at the University of Oxford in the late 2000s for the Department of Social Development, South Africa. Since the end of apartheid the South African government has spent considerable resources trying to reduce poverty through employment and a still evolving welfare system of cash transfers and services. Although the South African Census 2001 was not collected primarily for the purpose of poverty analysis, its detailed data offered policy makers the chance to measure progress on living standards via SAIMD (Noble et al. 2009).

SAIMD conceives of poverty as a multidimensional concept measured using 13 indicators across five equally weighted domains (income and material deprivation, health, education, employment and housing). Unlike survey data, census microdata (i.e. records for each individual) enabled us to build the measure at an unprecedented level of geographical detail in the African context – just a couple of thousand people in each area analyzed. This spatial detail allows the measure to identify small 'pockets' of deprivation that might otherwise be hidden. Altogether, SAIMD results in a measure of multidimensional poverty at small area level for the whole of South Africa that government and researchers can download in Excel or look at using maps (see Figure 11.1). It gave national, provincial and municipal policy makers unprecedented detail on small area multidimensional deprivation so that they could better target interventions or allocate resources.

They wanted it. We wanted to do it. So far so good. But when politics and policy makers engage academics in commissioned research, it can often be difficult to reach shared understandings and expectations about the research. Policy makers tend to want quick answers that cannot easily be questioned by their opponents. But research often takes time, involves difficult and sometimes uncertain decision making and may be constrained by data, time or financial limitations. And policy makers tend to have little interest in these technicalities. One early challenge we encountered was that for a measure of deprivation such as this to be 'fair' it needs to compare areas of a similar population size. But no such statistical geography existed in South Africa and while we could (and did) build it (a new 'Datazone' geography consisting of around 22,000 small areas across South Africa with around 2000 residents in each) that took time (Avenell et al. 2009). So while we were making the Datazone geography we were asked to quickly build a poverty index for geographies of uneven sizes. We all knew it would be imperfect but it was still progress for the policy makers and helped to develop policies and debates in South Africa. The politics of the apartheid era made the building of the Datazone geography itself difficult. In some areas, the existing small areas that we were using as the building blocks for our new Datazones had been deliberately made so that

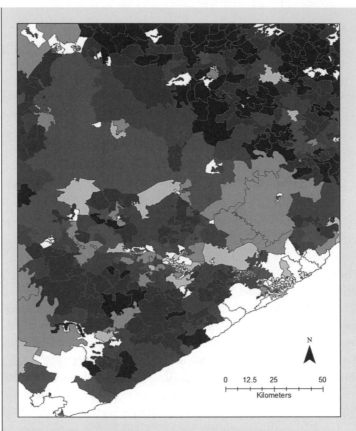

Figure 11.1 SAIMD in practice: Small area multidimensional
poverty in East London, Eastern Cape province.
Small areas are shaded on a gradual scale from
darkest (most deprived decile nationally) through to
lightest (least deprived decile nationally)

they made no real sense geographically or administratively, particularly in the rural areas and former Bantustans. So we just had to do as best we could and explain to the (hopefully patient) policy makers why this seemingly dry technicality mattered and why their poverty research was not yet on their desks. Once this was (finally) made, the actual calculations and building of the poverty measure were easy enough, although one challenge we faced was that Statistics South Africa wouldn't actually give us the data and one of our team had to fly to South Africa and work non-stop in its offices for the week before he was allowed in to do it. He had a shock on arrival when they initially stated no knowledge of his trip and refused entry and I don't think he got much sleep that week, but all was well in the end and he got it made.

It can be challenging working with policy makers and engaging with the inevitable demands and politics that come with that kind of research. But it also opens new doors and opportunities and means that you get to be involved in research that gets directly into policy and that makes a different to the real world.

It is rewarding to use existing data such as the South African census and put it to good use to tackle poverty even though that was not particularly what the census was collected for – there are plenty of data out there (and increasingly so with growing interest in administrative data, social media data, crowdsourced data and so on) and we can put that to good use in our varied research without necessarily having to spend time and money collecting new data specifically for our task.

organizations, research funding bodies (such as the ESRC's data repository) and libraries will uncover many other datasets that may be of relevance to your research.

The vast repository of data collected by other researchers provides a rich resource for development researchers. However, you cannot simply take these data at face value and assume their validity and rigour (see White 2010 for a detailed and accessible summary of how to use secondary data in research). Whenever you are using a dataset generated by someone else, whether by a government, NGO or academic, you need to question it. Who produced it? For what purpose? What biases might it contain (due to the (wording of) questions asked or under-/over-reporting depending on the focus and topic of the survey)? Who collected the data (and how well were they trained)? Where and how were the data collected (and what potential biases exist therein)? What was the sampling frame and procedure? What was the response rate?

These questions are all important as they help you to understand the quality and potential weaknesses of the dataset (see Park 2006; White 2010). A weak dataset will undermine the quality of your analysis and findings, so it is important to be sure of the quality of the data before you begin working with it. Assuming that you can access good-quality data, this provides you with the evidence to be able to assess the outcomes of development policies, to analyze changes in development indicators over time, to compare development trends within and between countries and so on.

Analysis of multiple datasets can provide you with a means of exploring and understanding various interactions between variables as well as developing comparative or longitudinal analysis. For instance, analysis of a series of census returns can provide evidence of longitudinal change. Examples of this technique can be seen in Smiley's (2012) use of Tanzanian census data to explore longitudinal changes in housing provision or in Barro and Lee's (2013) comparative analysis of census data from 146 countries to explore changes in educational attainment, disaggregated by gender and age. Booysen's (2007) combined analysis of existing datasets from the South African Independent Electoral Commission with a project-specific primary survey allowed detailed analysis of issues relating to political participant, voting and service delivery. These examples are indicative of a few of the ways in which survey data can be utilized within development-related research. To give you a more detailed example, Adam Whitworth (Box 11.3) reflects on his experiences of working with colleagues at the University of Oxford to produce an index of multiple deprivation for the South African government and the various technical and practical challenges encountered in this work.

Developing your own dataset and conducting survey research

You may decide to conduct your own survey research and to develop your own dataset. This may be because no existing dataset addresses the topic or questions you are interested in, because you are unable to access existing data or because you do not trust the data within an existing dataset. Before you decide to conduct a survey, you need to consider if this is an appropriate method for your research questions and approach. Remember: surveys are useful for the rapid generation of large-scale data relating to opinions, attitudes, social relations, personal perspectives, behaviours and experiences. They are used to identify causal relations, to develop generalized findings and can be used for comparative work, such as in Sawada and Loskhin's (2009) work in Pakistan. In this project, Sawada and Loskhin conducted a field survey across 25 villages in Pakistan to generate descriptive statistics and econometric data from which to identify a range of factors constraining school progression.

Once you have decided that a survey is appropriate and required for your research, you will need to address a series of questions relating to how you will conduct the survey. You will need to think about the questions to ask in the survey and how to format the survey document. You will also need to consider how you will sample for the survey, how you will collect the data (and the implications they will have for response rates and sampling strategy), how you will clean and code the data, how you will analyze the data, as well as various practical considerations for the administration of the survey.

This list of considerations can seem rather overwhelming, but a bit of careful thought can usually address these considerations relatively easily. We will start by thinking through the construction of a survey document, and then continue on to consider how you might conduct a survey – from sampling to coding. Our main focus will be on surveys that generate statistical data about a sample population. However, at the end of the section, we do provide some brief notes on other forms of surveys you may consider using in development research.

Designing a survey document

Designing and constructing a good survey questionnaire takes time and patience. A poorly worded and conceived survey will not produce robust and reliable data. If you think for a moment about the process of a survey interview, you can see how easy it is for misunderstandings and misinterpretations to occur and for this to compromise the data being collected. First, the researcher *encodes* her request for information into a question. The respondent then *decodes* this question and *encodes* an answer to it, which the researcher must then *decode*. In each step of this process, differences between intended meaning and understanding can develop, particularly when working across language differences or in different cultural contexts. In order to ensure the veracity and comparability of data collected, it is vital that you spend time developing questions and response options that minimize the potential for misunderstandings to develop and bias your data.

The language used in the survey document will influence how people engage with and respond to your questions (Galasinski and Kozlowska 2010). It is important not to assume that respondents will have an (informed) opinion or view on every topic and to recognize that responses will vary depending on how you present a topic and question to them. To this end, the first lesson is to keep your questions simple and straightforward so they are easy to understand, use clear terminology, be culturally sensitive and are specific in their instruction and request. In this sense, your questions should be *reliable* in that they mean the same thing and are interpreted in the same way by all respondents and the researchers. They should also provide a *meaningful and accurate measure* of a key concept or concern (although an older text, see Nichols (1991) for a useful background to these concerns).

Your questions should be appropriate to the local context in which the survey is being conducted (this can include working around cultural practices that result in the avoidance of disagreement or negative responses or working with different units of measurement (such as different calendars)). They must also be relevant to both the main research question(s) being addressed and the population being sampled. If you start pursuing other lines of questions, you are likely to confuse the respondent and run the risk of boredom and withdrawal of their participation.

You must avoid leading questions (see Box 11.4 for further guidance on developing questions). A leading question will bias the responses given as this will influence how the respondent answers you. This bias can be introduced in a number of ways, either through the poor use of prompts (for example, giving a list of possible/examples to responses to a question such as 'What are the biggest threats to your safety in this community – for instance Q, Y, T' will increase the chances of respondents identifying these topics as threats while overlooking other issues) or through the phrasing of a question that suggests a preferred answer (for example, 'Do you agree that local education provision could be improved?' or 'Would you like to benefit from greater service provision?').

You should avoid jargon and overly technical terms as these will confuse and alienate respondents. It is likely that they will result in respondents either making a guess at an answer to the question or the assistants having to try and explain terms or ideas in other terms. Through this process, the danger for divergent understandings and readings to

---

Box 11.4 Developing questions

You may find this detailed but accessible guide to developing good survey and questionnaire questions useful:

Foddy, W. (1993) *Constructing Questions for Interviews and Questionnaires: Theory and Practice in Research*. Cambridge: Cambridge University Press

For a useful critical reflection on the complexities of using survey questionnaire, you may want to read:

Galasinski, D. and Kozlowska, O. (2010) 'Questionnaires and lived experience: Strategies of coping with the quantitative frame', *Qualitative Inquiry*, 16(4): 271–84

develop is heightened, which can reduce the veracity and comparability of your data. You should also avoid double negatives and double or multiple questions. Both of these practices can be highly confusing to respondents and can lead to misunderstandings and incorrect responses (while an older text, Foddy (1993) provides accessible but detailed guidance on these concerns).

Whenever possible, you should pilot the survey to ensure your questions are clear and meaningful and culturally sensitive. You can pilot the survey using a qualitative approach (using focus groups to test understanding of and the meanings given to questions) and quantitative approach (complete a small number of survey interviews using drafts of the questionnaire and identifying difficulties and misunderstandings). Piloting questions can also allow you to identify potential and likely responses, so that if you include lists of possible answers to be selected, prioritized or ranked you can ensure that you are including all of the most likely responses. The time spent piloting a survey in the design stages can save a lot of time and trouble later in the research process by pre-empting problems with questions that do not work or need revision or rewording.

Not only do you need to think carefully about the wording of each question, but also the responses offered to respondents as these affect both the accuracy and reliability of data collected as well as the ease with which data can be cleaned and coded. For some questions, you may include binary responses ('yes' or 'no'), for others it may be a range of categories covering different ages, time periods, topics, facilities and so on. Other questions may seek a response along a continuum or sliding scale. In some instance, this may involve marking a position along a line or in others selecting a score along a scale (usually a Likert scale of 1 to 5).

Linked to this concern is consideration of the inclusion of open questions in your survey. Open questions are advocated as a means of developing deeper insights and understanding and a way of allowing the respondent space for free thinking and through which to draw out new issues. While these questions do provide the potential for detailed, reflexive responses, there are a number of significant drawbacks to them. Often the data provided in the responses are irrelevant or tangential to the research being conducted. Of greater concern is the time and resources required to read, code and analyze the responses provided. With small-scale surveys, this process is often feasible and can provide useful quantitative and qualitative data. (See, for instance, Shortt and Hammett (2013), a paper based on the survey outlined at the start of the chapter in which we use responses to open and closed questions. Even for this relatively small survey (201 households), the resources required to work through the open question responses was significant.)

For larger scale surveys this becomes incredibly difficult if not impossible due to the time and other resources required and are incredibly difficult to standardise. These challenges are compounded by the response bias these questions often generate through eliciting more responses from confident and articulate participants, even if these opinions are not representative of popular belief. Further challenges arise from the lack of control over the interpretation of questions and framing of responses given. We are not saying that you should never use open questions in surveys, but do think very carefully about this. You may find that using open questions during a pilot phase of the work helps you to fine tune your closed questions and response categories in ways that can supersede open questions as a means of data collection.

A pilot or trial run of your questionnaire would also allow you to ensure that you are asking the correct questions and that the questions are phrased and framed in an understandable and accessible manner. Do try to build this into your research design, but it may be that you cannot do this due to time, money or logistics (for instance, if it would require a separate trip to the field, the timeframe for completing the survey is too tight to allow this iterative process or you would not be able to print and produce copies of the questionnaire once in the field). If you are able to complete a trial survey then spend the time to continue through the entire process, including data cleaning and entry so as to identify any weaknesses or problems with these steps in the process. Once you have completed the trial run, you can draw on these experiences to modify and amend your survey to make it as good as possible.

A pilot study can also help you hone the structure and flow of your questionnaire. An initial consideration should be the length of the questionnaire – it needs to be long enough to generate the detail needed to answer your research questions but should not be so long as to be an undue burden on respondents. On a practical note, longer survey interviews also mean you can achieve fewer interviews in a day, increasing the time and financial cost of completing a survey.

Other structural considerations include the progression of questions. You should try to position the most important questions early in the survey, so that if a respondent finishes the interview early, you still have responses to core questions. This said, the initial questions should be easy and straightforward to answer as this allows the respondent to settle into the survey interview and build their confidence and comfort. This means that as you introduce more complex and/or potentially embarrassing or political or personal questions later a degree of trust and rapport has developed, increasing the probability of an open and honest response to such questions.

When considering the flow of your questions, remember that a person's response to one question in the survey will inform his response to answers later in the survey. Respondents, generally, do not want to be seen to be contradicting themselves and will often maintain a stance or position later in the survey based on an earlier response. However, if you need to include questions that may influence one another then try to keep them apart in the survey. An alternative approach would be to test this concern by running a randomized experiment within the survey by using different orders of questions for groups of respondents or locating the questions consecutively for half of the survey respondents and separated by other questions for the other half. Keeping a careful record of who answers which version of the survey (through a variable placed in the dataset) can allow you to begin to assess if the response to one question does influence another answer.

You can also use filters and signposting through the questionnaire to improve the speed and flow of the interview. It is likely that there will be certain follow-up questions or entire sections of questions that are only applicable to some respondents. In this situation, you can use filters and directions to move people on to the appropriate questions while missing out those that are irrelevant to them. You should also leave plenty of space on your survey document if you are using a printed questionnaire: do not cram things on to the page, leave space so it is an easy document to navigate and so that there are spaces for notes and comments.

Including coding details in the questionnaire can increase the efficiency of data capture and input. In this practice, rather than having a separate code sheet for data input, the codes are included on the questionnaire itself. This allows the person completing the data inputting to read the code straight off the main document without the need to skim back and forth between documents. To give you an example of this practice, see the pages extracted from a survey conducted by Dan and his colleague Niamh Shortt, which show how useful having the codes on the questionnaire itself is for the data input process (Box 11.5). The first page of the survey included codes from the outset, the second page did not but as you can see from the revised version of this page, adding the codes to the document (in this case, as superscript numbers linked to each response) would make the coding process quicker and easier.

Conducting the survey – sampling

At the heart of a good survey is the question of how representative are your data, which depends to a great extent on your sampling strategy. The need to sample comes about because of the costs incurred in surveying an entire population. The most common understanding of this full population survey is a national Census, a survey carried out by most states in the world on a ten-year cycle. While the resultant dataset is tremendously extensive and detailed, it requires huge resources to conduct and process. You could also conduct a census at a sub-national scale, for instance of a village or neighbourhood, but you will still face issues of cost, logistics and handling non-responses. In general, you do not need to survey everyone within a population (everyone in a target area/ community/country) to gain an accurate picture of the population. Instead, you can get good sense of this from a sample (a smaller subset of individuals within the population).

To overcome these challenges, you can use a sampling strategy in order to develop insights into a subsection of the population from which it is then possible to generalize up to the full population with a reasonable degree of accuracy. The population is not necessarily everyone in a country, but is the community or group you are studying – be this residents of a particular village, those working in a particular economic sector, or a particular socio-demographic group. Ideally, you want to work with a *representative* sample of the population you are studying as this provides a more accurate reflection of the population to which you are intending to generalize your findings. Sampling provides a more efficient approach to research, saving you time and money that can be used for other research activities without compromising the quality and veracity of your findings.

To identify your sample, you must first clearly identify the population from which this is selected. The population of interest depends on your survey objectives, so you must first be clear about your research focus, objectives and questions. From this, you can then identify the population and from this then identify the sample based on your research focus and the characteristics of those who will be included or excluded from this. For instance, if the intention is to study the outcomes of microcredit loans as a poverty alleviation strategy for widows in an urban city then those eligible for inclusion in the study are fairly self-explanatory – you are not going to be recruiting married men in a rural village!

Box 11.5 Example of coding details within survey document

First page of the survey:

Q1		Person Number				
		01	02	03	04	05
A	Male or Female 1 = Male 2 = Female					
B	How old is ...? Less than 1 year = 00					
C	What population group does ... belong to? 1 = African/Black 2 = Coloured 3 = Indian/Asian 4 = White 5 = Other (please specify) 6 = Information refused					
D	Has ... always lived in Imizamo Yethu 1 = yes 2 = No If no, state Province of previous residence.					
E	What is the relationship of ... to the head of the household 1 = mark the head of household 2 = Husband/Wife/Partner 3 = Child (blood, adopted or step) 4 = Brother/sister or step sibling 5 = Father/Mother or step parent 6 = Grandparent or Great Grandparent 7 = Grandchild/Great Grandchild 8 = Other relatives (e.g. inlaws, aunt/uncle) 9 = Non-related persons					
F	What is ...'s marital status 1 = Married 2 = Living together like husband and wife 3 = Widow/Widower 4 = Divorced/Separated 5 = Never married					

Original version of the second page:

Section 2: I would now like to ask you several questions about your property.

Q2 A. Indicate the type of main dwelling and other dwelling that the household occupies?
Dwelling/house or brick structure .... O
Traditional dwelling/hut/structure made of traditional materials......................... O
Informal dwelling/shack in backyard.. O
Informal dwelling/shack not in backyard .... O
Other, specify (hostel, caravan/tent, etc)_____

Q2 B. Thinking back to your previous home what type of dwelling/dwellings did this household occupy?
Dwelling/house or brick structure .... O
Traditional dwelling/hut/structure made of traditional materials......................... O
Informal dwelling/shack in backyard.. O
Informal dwelling/shack not in backyard .... O
Other, specify (hostel, caravan/tent, etc)_____

Q2 C. Is your current dwelling the only dwelling on the site?   Yes ......O   No......O
If no then please state other property type present _____

Q2 D. How long have you been living in your current home? (years and months) ___   ___

Q2 E. What is the tenure status of this household?
Owned and fully paid off...O
Owned but not yet paid off.. O
Rented..O
Occupied rent free..O
Other (please specify) ....O

Q2 F. How many rooms, including kitchens, are there in this dwelling (rooms must be separated by a solid wall and not a curtain)? (exclude bathrooms) _____

Q2 G. How many rooms in your dwelling have no form of heating? _____

Q2 H. Thinking back to 2001 how many rooms in your dwelling had no form of heating? ____

Q2 I. How many rooms in your home suffer from condensation, mould or damp?

Q2 J. Thinking back to 2001 is the level of CMD:
    Worse   O
    About the same   O
    Better   O

Revised version of the second page, with codes added:

Section 2: I would now like to ask you several questions about your property.

Q2 A. Indicate the type of main dwelling and other dwelling that the household occupies?
Dwelling/house or brick structure .... O^1
Traditional dwelling/hut/structure made of traditional materials.......................... O^2
Informal dwelling/shack in backyard.. O^3
Informal dwelling/shack not in backyard .... O^4
Other, specify (hostel, caravan/tent, etc)_____ 5

Q2 B. Thinking back to your previous home what type of dwelling/dwellings did this household occupy?
Dwelling/house or brick structure .... O^1
Traditional dwelling/hut/structure made of traditional materials.......................... O^2
Informal dwelling/shack in backyard.. O^3
Informal dwelling/shack not in backyard .... O^4
Other, specify (hostel, caravan/tent, etc)_____ 5

Q2 C. Is your current dwelling the only dwelling on the site? Yes ......O^1 No......O^2
If no then please state other property type present _____

Q2 D. How long have you been living in your current home? (years and months) ___   ___

Q2 E. What is the tenure status of this household?
Owned and fully paid off..O^1
Owned but not yet paid off.. O^2
Rented...O^3
Occupied rent free...O^4
Other (please specify) ....O^5

Q2 F. How many rooms, including kitchens, are there in this dwelling (rooms must be separated by a solid wall and not a curtain)? (exclude bathrooms) _____

Q2 G. How many rooms in your dwelling have no form of heating? _____

Q2 H. Thinking back to 2001 how many rooms in your dwelling had no form of heating? ____

Q2 I. How many rooms in your home suffer from condensation, mould or damp? _____

Q2 J. Thinking back to 2001 is the level of CMD:
Worse O^1
About the same O^2
Better O^3

Once you have identified a target population, you need to develop a sampling frame and determine how you will identify your sample from this. In making these decisions, you need to think about the relationship between the population, your sample frame and the sample itself. Your priority here is to ensure representativeness – that your sample provides a good representation of the sample frame, which, in turn, is an accurate representation of the population – and to recognize the limits to the generalizations you can make from the sample identified.

You also need to ascertain the sample size needed. This may relate to the minimum size needed to ensure statistical validity in your analysis and to ensure you can generalize from your sample to the whole population. Logistical factors of time and money may also be influential here, but it is important to ensure you have a large enough sample to make valid generalizations: it is a balancing act between getting the most accurate results you can (i.e. increasing the sample size and providing for the testing of more variables) and minimizing the costs of the survey (i.e. decreasing the size of the survey, reducing the accuracy of the survey while increasing the sampling error). There are various ways in which you can identify or estimate the required survey size for your research based on having enough cases to ensure the generation of statistically significant results dependent on the number of variables and relationships being analyzed and the confidence level and interval you need. Detailed discussion of these techniques is beyond the scope of this chapter, but you may find the following sources of use: Ardilly and Tillé (2006) or Czaja and Blair (2005). A number of online tools exist that can provide estimations of required sample sizes, including: http://www.surveysystem.com/sscalc.htm and http://www.raosoft.com/samplesize.html.

There are two main ways you can sample, *probability* and *non-probability*. Probability sampling provides a statistically representative cohort of the population in which each member has an equal chance of being selected. Non-probability sampling depends on judgement or convenience and is less statistically robust as not all members have an equal chance of inclusion. There are a number of specific methods of sampling that you will need to select from. We provide a brief overview of some of these methods below and more detail can be found in the books noted below.

Common approaches to probability sampling include *simple random sampling*, which uses a random number generator or table to identify members of a sampling frame for inclusion in the survey. To do this, you need a full list of the sample population – for example, this may be a list of every household in a village, a list of pupils attending a school or a list of all recipients of food aid in a particular area – and to assign each individual a unique identifier (usually a number). You then select individuals for inclusion in the sample based on the random numbers generated. This is a straightforward way of securing an unbiased sample but may not provide you with big enough samples of different sub-groups within the sampling frame for robust analysis. To overcome this, you may use *stratified random sampling*, by which you employ the same random sampling approach but to specific sub-groups within the sampling frame. This is a more complex approach and needs careful management to identify an appropriate number of sub-groups without the survey becoming overly complex.

Another approach would be to use *systematic sampling* (as in the study Imizamo Yethu outlined at the start of the chapter) whereby you sample every nth item encountered in the sample. This is another simple approach but can be problematic if the sampling frame has a recurring order (for instance, sampling the presence of street traders every seven days would not give a random sample), when trying to gain a representative sample becomes unlikely with this method. In these situations, you may opt for a *stratified systematic sampling* approach.

In some instances, you may feel that non-probability sampling is needed for your research. This may be in order to access marginal or hard-to-reach groups or because

Box 11.6 Sampling practices

For more detail on sampling procedures and practices, the following texts are useful and accessible sources:

Bulmer, M. and Warkwick, D. (eds.) (1993) *Social Research in Developing Countries*. London: UCL Press.
De Vaus, D. (2002) *Surveys in Social Research*. London: Routledge.
Fink, A. (1995) *How to Sample in Surveys*. London: Sage.
Fowler, F. (2002) *Survey Research Methods*. London: Sage.

your work focuses on particular (sub)groups. Perhaps the most common mechanism in this situation is to use *snowballing sampling* where you rely on previously identified members of the population to facilitate introductions to other members who would be willing to participate. While non-probability sampling can be useful in specific situations, there are much bigger concerns over bias and challenges to generalising from these data. You can find further details on sampling practices referenced in Box 11.6.

Conducting the survey – data collection

There are various ways in which you can carry out a survey, each with its own strengths and weaknesses. The first way of completing a survey is to conduct each survey interview in person. This will ensure consistency in practice and maintain an understanding of the research questions and commonality of language used by the surveyor across all the survey interviews (Simon 2006). However, this approach is vastly time consuming and will rapidly use up your fieldwork time. You may also encounter problems of access, safety, transport/travel and language differences in trying to administer a survey yourself. In reality, this approach is only suitable if you are conducting a very small-scale survey. In most situations, you will need to utilize a different means of administering the survey.

To overcome some of these challenges, you could opt to conduct a survey by telephone. This approach removes difficulties associated with travel, safety and access. However, the time-consuming nature of conducting each survey interview remains. In addition, further challenges arise as your sample frame will alter: you can only speak to respondents with a telephone, either landline or mobile, and even then how do you identify and access a list of relevant people with a telephone? Further complicating this are considerations regarding inherent biases in such samples due to economic factors (the cost of a phone, air time, charging (for a mobile phone), etc.) as well as in relation to technological literacy, disability and so on) and your response rate is likely to decline (just think about how often you ignore or hang up on unsolicited telephone calls). In many development research scenarios, this approach is not a viable option. A very similar set of limitations – but more pronounced – would relate to efforts to use internet- or email-based surveys. This is not to say such approaches are not possible and indeed various organizations do use SMS-based surveys and crowdsourced data to compile information and complete surveys. If an internet-based or facilitated approach is possible for your work – this may be the case if you are working on/with elites, university students or NGOs or the press – then these tools can provide advantages in terms of speed and cost both in terms of collecting data but also then in downloading data directly into analysis software (for more details on such methods, see Madge and O'Connor, 2004).

Similarly, the distribution of survey questionnaires for self-completion, either by post or hand delivery, is often inappropriate. Challenges here can include the lack of a (reliable) postal service, safety and access considerations for hand delivery and collection, language differences and literacy levels, as well as different understandings and interpretations of meanings and phrases used in the survey. This mechanism for collecting survey data will be practical only in certain contexts.

It is likely, therefore, that in conducting any large-scale survey data collection within a development research project, you will need to employ field assistants. The use of assistants can facilitate the collection of a much greater volume of data within the same timeframe. They may also provide a mechanism through which to mitigate access, safety and language barriers. On the flip side, there are certain costs involved – financial costs being but one. If you intend to employ field assistants this will, obviously, be a drain on your finances and you may also need to budget for the time taken to train assistants in how to administer the survey and for any data input or cleaning you want them to do. You will need to think about how you will recruit assistants. Do they have the necessary literacy for the work? Appropriate language skills? Previous research experience? Are they appropriate for the research in terms of socio-demographics and local cultural and religious norms? For instance, if you are conducting a survey relating to Muslim women, it is likely that you will need to employ female assistants due to sociocultural norms of interaction. If you were conducting research on health and sexuality in rural Kenya, you may need a mixture of assistants in order to access different groups within a community due to cultural norms around age and gender.

Further consideration must also be given to the safety and access of any assistants working for you – what are the potential dangers to their working in a particular area (if they are from outside the community or area is this a concern, particularly if there are ethnic differences and tensions)? What are the potential implications for the accuracy and honesty of responses – will respondents answer questions honestly and fully to the assistant if they are from the community/from outside the community? The references in Box 11.7 provide further guidance on these, and other, concerns.

You will also need to ensure that the assistants are trained in relation to ethics and confidentiality as well as in terms of the precise conducting of a survey. Absolutely critical to this training is to ensure that they *always* ask each question in exactly the same ways using exactly the same words (those printed on the survey document) and in exactly the same order in each and every interview. Ideally, they should use the same neutral tone of voice and body language throughout. Even the slightest change in the wording of a question can compromise the comparability of the data by changing the meaning or emphasis of a question and the response given. Assistants should also understand that

---

Box 11.7 Research assistants in survey research

For further discussions of the use of research assistants in conducting survey work, see:

Nichols, P. (1991) *Social Survey Methods: A Fieldguide for Development Workers*. Oxford: Oxfam.

Scheyvens, R. (2014) *Development Fieldwork: A Practical Guide* (2nd ed.). London: Sage.

Simon, D. (2006) 'Your questions answered? Conducting questionnaire surveys', in Desai, V. and Potter, R. (eds.) *Doing Development Research*. London: Sage.

Smith, F. (1996) 'Problematising language: limitations and possibilities in "foreign language" research', *Area*, 28(2): 160–66

responses, including non-responses, must be recorded accurately, that they must not lead responses through prompts or put words into the respondents' mouths, and so on. The time spent training your assistants is time well spent: poorly trained assistants can render a survey unusable if it is conducted badly.

You may also want to consider mechanisms through which to check on responses and completed surveys at specific points during the survey period to ensure the accuracy and validity of data capture and comparability between different assistants. It may be useful to give clear guidance to your assistants as to how long a survey should take and how many should be completed per day and to reiterate the importance of following the sampling protocol for the project (including how to deal with non-responses).

Your handling of non-responses is important as these affect the representativeness of your sample, particularly if the characteristics of non-respondents are different from those of your respondents. For instance, if you receive responses only from men and non-responses only from women to a survey of community experiences of domestic violence your findings will clearly not be representative of the community as a whole. Such practices can lead to non-participation bias and a skewing of your findings. It is important, therefore, to record the number and characteristics of those who decline to participate in your research and to compare these details with those of your participants in order to identify and attempt to compensate for any bias.

You should also consider how you want responses to be recorded. Traditionally, survey interviews have been recorded using pen and paper through the completion of a survey questionnaire. However, technological advances have provided options for the computer-assisted capture of information. Depending on the type of survey being conducted, the location(s) of the survey and your budget you may consider how new technology can assist in survey work. For instance, you may feel that having GPS readings of household locations or the location of resources or land use will be important data to capture, in which case mobile computing and GPS handsets can allow you to do this (Barker 2006). More generally, technology such as touch screen tablet computers may allow direct inputting of data into programmes, the use of maps and other visual aids in surveys, allow the collection of metadata, and increase the speed and accuracy of data collection (Caviglia-Harris et al. 2012).

If you are thinking of using computer-assisted data capture then you should also think about the practicalities of this (see the sources in Box 11.8 for further details and discussions). Are there any added safety and security considerations – will carrying and using this technology put the surveyors at greater personal risk due to mugging etc.? What safeguards are in place to protect confidentiality of data on a computer? Are the research assistants trained and comfortable in using the equipment? Will respondents accept and be comfortable with the use of this technology by the surveyor? Is the computer rugged and robust enough to cope with being used in the field? Will the surveyor be able to see the screen in bright sunlight? Will the equipment cope with getting rained on? Are both the hardware and software reliable and affordable? Is it practical to use the equipment in the field – in terms of weight and portability, as well as battery life?

Depending on how you conduct the survey, you will need to develop a prompt or instruction sheet for the interviewer to explain the survey questions and process to complete the document. It is also worth including prompts and directions between

Box 11.8 Technology and surveys

For examples of how different forms of technology have been incorporated into development survey research, the following articles provide some useful insights:

Barker, D. (2006) 'Field surveys and inventories', in Desai, V. and Potter, R. (eds.) *Doing Development Research*. London: Sage.

Barry, M. and Rüther, H. (2005) 'Data collection techniques for informal settlement upgrades in Cape Town, South Africa', *URISA Journal*, 17(1): 43–52.

Caviglia-Harris, J., Hall, S., Mullna, K., MacIntyre, C., Bauch, S., Harris, D. et al. (2012) 'Improving household surveys through computer-assisted data collection: use of touch-screen laptops in challenging environments', *Field Methods*, 24(1): 74–94.

Smith, H. (2005) 'The relationship between settlement density and informal settlement fires: case study of Imizamo Yethu, Hout Bay and Joe Slovo, Cape Town Metropolis', in Oosterom, P., Zlatanova, S. and Fendel, E. (eds.) *Geo-information for Disaster Management*. Berlin: Springer.

questions at appropriate points within the survey document to reduce the potential for them to be overlooked during an interview. If you are conducting the survey yourself, then this may simply be a few prompts and reminders. If you are employing field assistants or researchers to administer the survey for you, then you will need to provide them with various directions through the survey. These may include directions regarding sampling and dealing with non-responses (see above) or guidance as to the noting of socio-demographic details that may not need to be asked but still need to be recorded (e.g. gender of respondent) or metadata. These instructions may also relate to filters and flows through the survey questions (for instance, if response X is given to question 3 then proceed to 3a, if response Y is given then move on to question 4). These types of instruction will also be required if you are using a self-completion survey, along with directions as to how to return the completed survey to you.

Whichever mechanism you use to administer the survey, you will need to provide a plain language explanation of your research and ensure that your respondents are providing informed consent to participate in the survey. In this explanation, you should ensure that potential participants are made aware of the time it will take to complete the survey and whether or not there is any definite or potential direct benefit for participation (see Chapter 5 for further discussion on this topic).

You may also find it useful to look at articles based on survey data and quantitative approaches to development research. The articles listed in Box 11.9 provide examples of a range of techniques and engagements that are useful in understanding how these tools can be effectively used in development research:

Box 11.9 Examples of survey-based development research

The articles listed here provide a few examples of development-related research that has utilized quantitative methods and data collection and analysis in various ways. The examples may provide some inspiration as to topics for research as well as the practice of quantitative research:

Akbar, D., van Horen, B., Minnery, J. and Smith, P. (2007) 'Assessing the performance of urban water supply systems in providing potable water for the urban poor', *International Development Planning Review*, 29(3): 299–318.

Akresh, R., Lucchetti, L. and Thirumurthy, H. (2012) 'Wars and child health: evidence from the Eritrean-Ethiopian conflict', *Journal of Development Economics*, 99(2): 330–40.

Baruah, B. (2010) 'Women and globalization: challenges and opportunities facing construction workers in contemporary India', *Development in Practice*, 20(1): 31–44.

Batista, C., Lacuesta, A. and Vincente, P. (2012) 'Testing the 'brain gain' hypothesis: micro evidence from Cape Verde', *Journal of Development Economics*, 97(1): 32–45.

Chowdhury, M., Ghosh, D. and Wright, R. (2005) 'The impact of micro-credit on poverty: evidence from Bangladesh' *Progress in Human Geography*, 5(4): 298–309.

Frayne, B. (2010) 'Pathways of food: mobility and food transfer in Southern African cities', *International Development Planning Review*, 32(3): 291–310.

Heald, S. (2008) 'Embracing marginality: place-making versus development in Gardenton, Manitoba', *Development in Practice*, 18(1): 17–29.

Linke, A. (2013) The aftermath of an election crisis: Kenyan attitudes and the influence of individual-level and locality violence, *Political Geography*, 37: 5–17.

Robles, M. and Keefe, M. (2011) 'The effects of changing food prices on welfare and poverty in Guatemala', *Development in Practice*, 21 (4–5): 578–89

Other forms of survey

In addition to the questionnaire-based surveys discussed above, development researchers make use of a range of other survey methods. These may include the use of a field inventory survey. This approach involves the development of a checklist of items, attributes or phenomena you expect to see in the field relating, for instance, to types of flora and fauna through to livelihood activities to infrastructural provision. You would then mark the presence or absence of these during your fieldwork. Such checklists are generally developed prior to commencing fieldwork, based either on previous surveys and existing literature or on previous personal field experience, but can be updated and modified while in the field. This approach can also be modified to allow you to record the frequency or intensity, or to give a ranking or relative weighting (see Barker 2006).

Field mapping provides another form of surveying that can be of use to development researchers. Mapping projects may seek, for instance, to explore agricultural land use, to map zones at risk of flooding, to record urban land use changes, to record perceptions of safe and dangerous spaces or to identify patterns of (im)mobility. In essence, this approach depicts in spatial terms the phenomena being surveyed and can be carried out in a number of ways. You may opt to use the pen and paper approach and either draw a new map or to modify or annotate an existing map of the area being studied. With recent advances in technology, you may consider using GPS positioning and mobile computing to record and map the phenomena being addressed. This approach has the advantage of increased accuracy in plotting and recording data and can allow the quick crosslinking of photos and visual imagery to the map product being produced. However, practical and safety considerations should be borne in mind (see discussion earlier), particularly if you do take photos when these may cause problems with local sensibilities or have legal implications (in many countries, it is not permissible to take photos of government buildings for instance). If you do use field mapping as a tool, do bear in mind questions of logistics and safety, as well as thinking through how you will record and code information – for instance, how do you record multiple land use activities in the same space, categorize different forms of economic activity or capture both formal and informal economic practices within a space? For further discussion of such practices, see Barker (2006).

What to take from this chapter

This chapter has provided you with an overview of a number of quantitative approaches that can be used in development research. You should be able to identify the suitability of such methods to your research questions and interests and determine whether these tools are appropriate for your work or not. You should also have a sense of the possibilities offered by existing datasets for you to conduct quantitative analysis and research without having to gather new data through extensive survey work. If you do decide to carry out your own survey, you should have an idea of some of the preparatory steps needed to do such work and how to address various challenges that may arise.

If you do use these methods, we would strongly encourage you to consult more detailed and advanced texts on statistical analysis techniques. A comprehensive outline of statistical analysis techniques lies beyond the scope of this book but useful guidance can be found in Czaja and Blair (2005) and Montello and Sutton (2013).

If you want to explore the ideas covered in this chapter in more detail, you may find these further readings useful:

Beegle, K., Carletto, C. and Himelein, K. (2012) 'Reliability of recall in agricultural data', *Journal of Development Economics*, 98(1): 34–41.

Glewwe, P. and Kremer, M. (2006) 'Schools, teachers, and education outcomes in developing countries', in Hanushek, E. and Welch, F. (eds.) *Handbook of the Economics of Education, Volume 2*. Amsterdam: North Holland.

McLafferty, S. (2010) 'Conducting questionnaire surveys', in Clifford, N., French, S. and Valentein, G. (eds.) *Key Methods in Geography*. London: Sage.

Olsen, W. (1993) 'Random sampling and repeat surveys in south India', in Devereux, S. and Hoddinot, J. (eds.) *Fieldwork in Developing Countries*. Boulder, CO: Lynne Reinner.

Rigg, J. (2006) 'Data from international agencies', in Desai, V. and Potter, R. (eds.) *Doing Development Research*. London: Sage.

12

BIG DATA AND SOCIAL MEDIA

'Big data' have transformed research practices in the physical sciences, social sciences and humanities. There are now enormous data shadows that reflect a range of important trends, patterns and processes. This chapter outlines some of the possibilities of big data research, with a particular focus on data obtained through crowdsourcing and social media. It then concludes by demonstrating the need for particular caution, but also flagging up opportunities, in the context of development data.

To illustrate the emerging potential offered to development researchers by these advances in technology and datasets, we can look at a UN project exploring the potential links between social media postings and development challenges. The project began in 2011, when a UN report found that messages on the microblogging service Twitter that were discussing the price of rice in Indonesia followed a similar function to official food inflation statistics (UN Global Pulse 2012) (see also http://www.unglobalpulse.org/projects/twitter-and-perceptions-crisis-related-stress). The team began a project in order to broadly understand whether large datasets from social media sites could be used to address populations' vulnerabilities (e.g. affordability and availability of food, fuel, and housing) (see Figure 12.1).

Because it is often extremely difficult to get reliable and timely data about those topics, the researchers turned to data from Twitter. At the time of research, there were over 100 million users publishing 200 million tweets a day. The team wrote some computer code to collect and store data from Twitter that was in Javanese or Bahasa Indonesian. Because those languages are almost exclusively spoken in Indonesia, the research team made the assumption that content in those languages would most probably originate in Indonesia. The dataset was then further filtered to retain only tweets that mentioned buying food (using many synonyms for both 'buying' and 'food').

What the researchers found was a clear relationship between the volume of conversations on Twitter and measurements in food price inflation in official statistics. This could be useful in cases where it is important to understand economic fluctuations quickly. The geography of messages could also provide a useful indicator of where events or processes are happening.

The potentials of this sort of research for development researchers who are often operating in traditionally information-scarce environments are obvious. However, there are also many important caveats and concerns about employing 'big data' in research, all of which are discussed in the remainder of this chapter. Despite these concerns, many

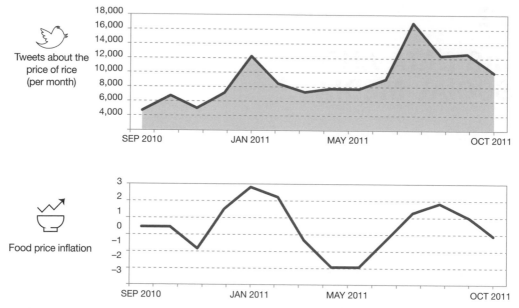

Figure 12.1 Tracking food inflation and tweets about the price of rice

people have argued that 'big data' have transformed the ways in which research is conducted. In many cases, researchers no longer have to settle for a subset or a sample of the data that they are interested in. Instead they can either use the entire population of the trend, pattern or process that they are interested in measuring or use a dataset that serves as a useful proxy for that trend, pattern or process. This chapter guides you through some of the possibilities, potentials and pitfalls of 'big data' for development research. It then offers some starting points for researchers interested in doing big data: reviewing some of the key tools and strategies that can be employed by novice big data researchers. To give you a sense of the scale and range of issues that you need to consider in developing big data-based development research and an inkling of the range of topics you could address using this approach, Charles Brigham (Box 12.1) reflects on his experience of using crowdsourced and big data in his research.

First, however, it might be useful to clarify what it is we mean by 'big data' and social media. The term 'big data' does not have any agreed definition, but generally is used to refer to datasets that have a *volume, variety, velocity* and *veracity* at a different scale from that researchers are used to dealing with.

The *volume* of many datasets has reached a point that they are now so large that they are difficult to handle with traditional research tools (such as Excel). The *variety* of data refers to the fact that data can take a range of forms (e.g. structured, unstructured, text, non-textual data). 'Big data' can also refer to both directly captured data (i.e. data that exist because the goal behind their creation was to perform some sort of direct measurement) and 'exhaust data' (data that are produced as a by-product of a process or system rather than being the main purpose of a process or system, frequently also referred to as 'trace data'). *Velocity* indicates that many data are now captured in motion. That is,

Box 12.1 Charles Brigham: Using crowdsourced data and big data in development research

Outsourcing tasks to a distributed undefined public, rather than a specific body, is becoming more commonplace in development work. Processes of sourcing data through crowdsourcing have evolved beyond mere collection of data at a generic level involving people at large, to becoming a strategy for gathering collective development intelligence to complete specific tasks. This, coupled with 'big data' systems, enables the capture and processing of massive amounts of near real-time information, making it accessible and providing an opportunity to significantly change the way we create, execute and communicate development. It indicates a movement away from an information age, with lots of data, to the idea of a knowledge age, where information and data are converted into knowledge through a collective crowdsourced voice. Consuming information and collaborating to become educated brings us closer towards the coming age of correlation of information, in parallel with stronger visualization and storytelling.

Big data tools, such as Hadoop (http://hadoop.apache.org/), process data quickly and make it useful, allowing us to ask questions and get the answer back before we've forgotten why we've asked the question in the first place. This is what is fundamentally new in using crowdsourced and big data in development and is only recently becoming widely understood. Past development work has focused on utilizing personal, siloed computing for timely processing of data, information and common social data channels to filter out noise and understand crowd voice in development projects, often resulting in insufficient gain. Today innovations in cloud computing, SAS and data science are providing the backbone for tools to allow big data for better informed development decisions. A key challenge will be in realizing potential and making these trained systems easier to build and maintain.

Business and marketing sectors are utilizing big data and crowd voice to tackle technological complexity, data accuracy and rights of use, business and technological alignment and need for data specialists. It is hoped that the development field will follow suit to strengthen collective intelligence, crowd voice and big data to utilize this wisdom within development institutions. We see this emerging through initiatives such as open data, greater access to information and improved technology tools and uptake but there is still much further to go in order to utilize new methods in order to impact the poor and marginalized. As an example, Indonesia's National Program for Community Empowerment (PNPM), is the largest community-driven development project globally. It has continually utilized social tools (networks and platforms) to enhance collaboration. While a massive number of users are linked to these networks, they in no way constitute or foster an active crowdsourcing for development environment: less than one-quarter of the thousands of locally driven social networks and groups at the community level are annually active. Over recent years fewer people are utilizing these common social networks to collaborate. *This is being addressed in real time.*

Newer beneficiary engagement platforms are emerging to utilize crowd voice and big data. Currently the Indonesian community-driven development project mentioned above is utilizing a multi-mode approach to collecting feedback from citizens (i.e. mapping, SMS, mobile and web) for implementing and development partners around project-specific development outcomes. The aim of the platform is to improve the current mechanisms of citizen input and feedback loops. The streamlining and regular sharing of large volumes of data at small area geographies will effectively engage citizens to provide actionable feedback and provide tools for multiple stakeholders (project implementing units, government, service providers and relevant staff) to view, act on, monitor and analyze crowdsourced and big data as needed.

data are not in neatly stored databases, but rather captured as traces from ongoing events and processes. *Veracity* is a way of describing the uncertainty and complexity in much 'big data' because of imprecise and unreliable data and data types, uncertainty in unstructured data from conversions and probabilities used in machine learning and uncertainty in how good a proxy the 'big data' that we have are for the more general phenomena that we wish to measure.

The reason that many of these data exist is because ever more facets of everyday life are both digital and digitized. Those everyday activities, transactions and behaviours together are encoded into datasets of sizes that most researchers have never had to deal with before. 'Big data' are also not limited to activities in rich countries. The high penetration of mobile phones around the world has meant that billions of people around the world have turned into digital sensors: emitting and sharing data. Telephone metadata can reveal movements and socio-spatial interactions (e.g. the number of people in one town who communicate with people in another town).

Many of the 'big data' available to researchers come from social media platforms. Social media can be particularly useful because they offer a platform for people to share information about their lives on. This information, in the aggregate, can potentially offer social, economic and political insights that would otherwise be challenging to gain. 'Big data' are not only volunteered, but are sometimes unknowingly emitted. People riding transit systems, applying for permits, paying tolls, triggering sensors, agreeing to terms of service (without reading those terms) and performing all sorts of other acts can all contribute to repositories of 'big data'. The question is what can we do with those data and how can they be useful to development researchers?

Why use 'big data'

The overarching reason that we might want to use 'big data' is because of an ability to look at broad-scale development-related patterns and processes that either are not captured by traditional sources (such as the World Bank) or are time sensitive. It is often the case that datasets of interest to development researchers do not exist, are of poor quality or are rendered inaccessible by governments or companies. Other times, high-quality and accessible data exist and are available, but are hopelessly out of date because of quickly changing contexts. A growing number of researchers are therefore looking for alternate types of direct and exhaust data that allow them to still ask broad-scale questions about patterns and processes related to development.

Much of the power of 'big data' also comes from the ability to combine a diverse range of sources in order to look for connections and associations. Those connections, in turn, can reveal patterns, processes, and relationships – that might otherwise remain undiscovered. Said differently, much 'big data' work is about exploration and obtaining insights that might be explored in more depth with other methods (albeit within the framework of an informed understanding of the field of study to be able to formulate good hypotheses and questions).

Some examples of contexts in which 'big data' might be used in the context of development research include:

Public health – an analysis of mobile phone traces (and thus, the flows of human mobility) could reveal how diseases and illnesses spread.

Transport planning – measuring flows of minibuses to understand the supply and demand of public transport.

Disaster response – user-generated reports could be used to create almost real-time understandings of human needs in information-scarce environments.

Demographic data – mobile phone usage combined with GNI or other metrics of human development could help to create up-to-date understandings of population in non-census years.

Understanding society – by knowing who communicates with whom, researchers can gain detailed insights into the micro-contours of different social groups: thereby understanding who occupies central and who occupies marginal positions.

Preventing famine – monitoring sudden price spikes in markets may lead to identification of early warning signs. In future cases, these early warning signs could prompt intervention before prices raised too far.

Economic justice – databases of market prices can also be used to allow farmers and other producers of goods to better understand the worth of their products and therefore avoid exploitation by intermediaries.

Many other examples exist, but already you can see that a core value of 'big data', especially in the context of development, can be illuminating needs and patterns in the context of previously information-poor environments.

How to use 'big data'

How then do you begin using and analyzing 'big data'? Below we offer a few strategies and tools. Some of this guidance is directed towards the novice, while other suggestions will be more useful to researchers with at least some experience writing computer code with a computer programming language. We address where you might obtain data, how

Box 12.2 Emmanuel Letouzé: Big data for development: how can we make sure big data are not the 'new oil'?

Over the past few years the concomitance of the availability of passively emitted data about human actions, generated and collected digitally as a by-product of the analogue-to-digital transition and of advances in computing and analytics capacities – dubbed 'big data', has given rise to questions about the phenomenon's potential to foster human progress. Publications, discussions and initiatives around 'big data (or data science) for development' (or 'social good') have flourished, culminating in the recent call for and numerous references to a 'data revolution' (with or without the qualifier 'big' in front). By and large, this call reflects and nurtures a salient recognition of the need for more agile and accountable policies and a widespread dissatisfaction with current systems and tools in the age of digital data and technology. So for some, big data are thought to be set to become the 'new oil' that provides, for human development, an immense source of value if only we are able of to refine it. There is some well-founded scepticism about the hype around some of these initiatives. If anything, the fact that the 'old oil' hasn't done much to foster human development among the population sitting on it (which has been coined the 'resource curse') may serve as a cautionary tale of some of the risks involved.

But what exactly are big data and what are these risks? In recent papers and interventions, I first distinguished big data and 'Big Data', referring to the former as data and to the latter as a field. Then, I argued big data's novelty is not 'about' size but about the availability of what we (with co-authors Patrick Meier and Patrick Vinck) have described as 'digital traces of human actions picked up by digital devices' (2013). Professor Sandy Pentland (2012) focuses on what he calls 'digital breadcrumbs' or data particles that individuals leave behind. The main message is that big data as data are essentially a *qualitative* shift. Subsequently, citing Gary King, I contend that *big data are not about the data*, but about the analytics, which I further interpret, following Kentaro Toyama, as being about *intent and capacities*. As such, big data are or should be interpreted and approached as a deeply political phenomenon.

Big data's potential to enhance 'our' understanding of and ability to predict mobility patterns, poverty dynamics, and more, is evident, as documented in a fast growing body of research. There is a detrimental dearth of data in many developing countries and big data may help fill part of a gap which, in the case of Africa, has been referred to as a 'statistical tragedy'. But the downside, and the main risk, is the belief that lack of data is the main issue impending human progress and the temptation to 'leapfrog' into a data-driven model of governance designed by and for an enlightened minority. The corollary risks are the creation of a new digital divide, dehumanization or de-democratization of decision making, and/or infringement on fundamental human rights – with respect to privacy and confidentiality most evidently. Three main sets of related factors create or increase these risks. One is a normative and empirical technological bias with insufficient consideration for the phenomenon's humanistic (including ethical) implications. Two is poor institutional connectivity between traditional policy actors, on the one hand, and data, web and computer scientists and ethicists, on the other hand. Three is limited structural channels and technical capacities for local actors to be fully engaged and access these resources.

So the fundamental question my work is trying to tackle is: what would a humanistic big data revolution – one informed by the principles and attempting to contribute to the objectives of human development – look like and take? How could I better identify and contribute to avoiding the traps that may impede its realization? This is a huge but exciting undertaking.

you might analyze them and how you might gain relevant skills or assistance. Indeed, as Emmanuel Letouzé outlines in his reflection in Box 12.2, the issue of analytics is vital in approaching work with big data and that care is needed to ensure the field is developed in a humanistic way – in others words, one that is both ethical and contributes to human development.

Where do your data come from?

'Big data' could be thought of as one (or more) of three varieties: machine, social and transactional data. Machine data include data gathered from machines or sensors. It could also refer to a variety of exhaust data such as web logs or mobile phone traces. Social data can refer to mining the traces that people leave behind on large social media platforms (e.g. Facebook likes or Twitter mentions) or data that they actively send and contribute (such as emails, free-form text, images, audio, videos). Transactional data include records of things like logs of processes, emails, stores of documents and a range of other transactions. Following the link at http://www3.weforum.org/docs/WEF_TC_MFS_BigDataBigImpact_ Briefing_2012.pdf provides another way of thinking about the three primary types of 'big data' source: individual (e.g. data exhausts and crowdsourced information), public (e.g. tax, facility or census data) and private (e.g. transaction data or spending data). These data will tend to be in one of three forms: *Precompiled datasets, structured data, unstructured data*:

Precompiled datasets: Measurements are already defined and data have already been collected and compiled by other researchers or organizations.

Structured data: Here there tends to be a need to compile the dataset, but the measure of interest is usually already defined. Often accessible through an application programming interface (API).

Unstructured data: Here raw data are gathered via an API or other sources, but then must be transformed to create the measure of interest (e.g. primary language of a speaker or tweets about price of rice). In other words, a proxy often has to be defined in such cases.

In some cases, you may be able to obtain such data simply by asking. Many government agencies, NGOs or firms might welcome the analytical expertise that a researcher could bring to their organization. However, because 'big data' often tend to contain sensitive, private and personal information, organizations can be reluctant to allow access to their information. If you do gain access to 'big data' in this way, you will usually be asked to sign some type of data-sharing agreement. Here it will be important to ensure that any agreement you sign provides you with the intellectual freedom to ask and answer your core research questions, and publish and share your results. There is no point in spending time carrying out research if it is only going to be embargoed by an organization that does not like your findings.

When data are not easily available from the organization that collects them, you may have to turn to programming or scripting to obtain the information that you need. Many repositories of data will have an application programming interface (API), which is a set of procedures, instructions and standards used by computer programs to request services and data. In other words, it is a set of building blocks that you can use to customize the way in which you extract the specific types, forms and quantities of data that you need (meaning that you do not have to reinvent the software wheel for common uses).

Many of us would not think twice about learning a local language before doing fieldwork and yet relatively few development researchers in the social sciences and humanities know how to 'speak' a scripting language. (Scripting in this context refers to the short computer programs that can be written to perform relatively narrowly defined tasks.) Luckily, learning the basics of a scripting language (such as Python, Perl, Javascript or Ruby) is now a relatively accessible task. Universities and colleges around the world offer short courses that allow programming basics to be understood in a few weeks. Many such courses are designed for people who have full-time jobs and so tend to either be evening and weekend courses or short, intensive programmes. Other learning resources include:

codeacademy.com – an interactive platform that offers free courses on a variety of programming and markup languages. The site is highly popular is has a very broad user base (five million users).

Coursera.org – another platform offering introductory programming courses designed in/by universities such as Stanford and Toronto. The courses themselves are free, but the organization charges fees for certificates to prove that you have done the work.

Khan Academy (khanacademy.org/cs) – features a series of popular video-based programming 'micro-lessons' in a range of languages.

Open Courseware Consortium (ocwconsortium.org/courses) – also offers dozens of free courses on programming. These full-length courses are usually designed by large educational institutions such as MIT and the University of Cape Town.

Finally, we want to mention two alternate ways of collecting data, both of which involve outsourcing tasks to distributed workers. First, there are now a variety of online job marketplaces (such as odesk.com and elance.com) that mediate interactions between people who need customized code and people who can design customized code. Although using such platforms can often be a quick and cheap way of obtaining a customized software tool to gather data, it is also important to recognize their limitations. The most fundamental one being that it is important to always understand how your datasets are put together. If you are unable to decipher the code that you are using then it is quite possible that you won't ultimately understand many of the nuances of your own data.

Distributed workers could also potentially be employed in a fundamentally different way as 'paid crowdsourcers', 'microtaskers' or 'clickworkers.' These terms imply that workers (hired through sites such as Amazon's 'Mechanical Turk' or crowdflower.com) can be paid to perform large numbers of small tasks. For instance, those paid crowdsourcers could be paid to manually put together databases of all health clinics in East Africa or they could be paid to standardize addresses in a large database into a format that allows them to be mapped. Here, again, it is important to proceed with caution. Many microworkers are based in the global south and are often paid exploitatively low wages for their labour. While microwork platforms allow, and perhaps even encourage, cheaply sourced work, researchers should make sure that they have considered what a fair salary might be.

Where does your analysis power come from?

'Big data' analytics can take a variety of forms depending on the specific data, frameworks, and questions that you have. Querying, mining, visualizing, modelling, simulating, natural language processing, geospatial analysis and voice/video analytics are some of the most common tasks that 'big data' are enrolled into. Many (but not all) of these methods employed in 'big data' analysis are existing methods that are applied at a larger scale. Below we describe some of the tools, infrastructures and methods useful to 'big data' research.

Computing cluster

In some cases, the speed and size of your own computer may be sufficient to process the data that you want to analyze. However, for many other 'big data' tasks, you need to turn to more powerful resources. Computer clusters are groups of machines that work together to solve tasks that involve large amounts of processing power or disk space. Open-source software frameworks such as *Hadoop*, which can manage tasks in clusters, have made it possible to analyze massive datasets without the availability of expensive computing resources. Solutions like Hadoop allow large datasets to be processed in a distributed manner: meaning that you could set up a 'Hadoop cluster' on a handful of computers (or, of course, many more). None of those computers needs to be high end or particularly powerful, but their combined resources (linked through the Hadoop framework) means that powerful computing resources can be made available for processing huge datasets.

Hosted cloud computing

Hosted cloud computing can be employed in a similar way to the computing clusters described above, the main difference being that you do not have to build the system yourself (however, this is not to say that the setting-up process is always straightforward). A variety of companies offer researchers the ability to outsource such tasks and pay (usually relatively small amounts) for the resources used. Amazon Web Services is one of the most used services in this category (Google's App Engine, and Microsoft's Azure offer similar services).

Cloud labour

A very different way to analyze large datasets is to rely on what has been dubbed 'cloud labour'. The term refers to a globally distributed labour pool of people willing to do small defined tasks for very small payments. This means that you could take a very large task that would be hard for a computer to do automatically (for instance, classifying tens of thousands of images) and break the task up for people around the world to help you with. Crowdflower and Amazon's Mechanical Turk are two of the largest marketplaces for cloud labour.

Positionality and ethics

Much has been said about the transformative power of big data in the contexts of development. However, it is also important to recognize some of the potential problems and pitfalls of 'big data'.

As the size, speed and scale of your data increase, your ability to adequately contextualize data often decreases. This can lead to spurious correlations and misinterpretations of results. Not only can the signals in our data be noisy (and false), but those signals could serve to misdirect us from more fruitful lines of inquiry. The saying that 'if your tool is a hammer, then every problem looks like a nail' is especially apt in the context of 'big data'-driven research. Many of the most important questions in development might not be able to be answered by large datasets and instead will require in-depth, contextual knowledge. In other words, 'big data' themselves can rarely substitute for other rigorous quantitative and qualitative research methods, but rather can be a useful complement.

'Big data' also raise a range of important privacy and ethical issues. Data are sometimes released in (or about) low-income countries that would not be released in (or about) rich countries. The EU, for instance, has relatively strict laws about the ownership, use and dissemination of personal data. But such laws do not exist in most of the rest of the world. It is therefore important to reflect on not just whether data are available but whether you actually want the data that are available (they may for instance contain sensitive information), or whether their use could be considered ethical (see Chapter 5 for a more thorough treatment of ethics in research). The Association of Internet Researchers has an up-to-date and evolving ethics guide (http://aoir.org/documents/ethics-guide/) that keeps track of developments related to technology and ethics – that could serve as a useful starting point for questions about 'big data' and ethics.

Finally, there are also concerns, alluded to earlier, about working with distributed microworkers. Many of these workers are based in Bangladesh, the Philippines, India and other low-income countries and work under poor conditions. It is your role and duty to ensure that you are constantly carefully reflecting on the wider impacts of your footprint in the world as a researcher: a task that becomes more difficult when interacting with people through relatively opaque internet platforms.

What to take from this chapter

'Big data' could serve as a way of measuring the pulse of society, tracking trends that were previously hard to measure and categorising and analyzing messy and large datasets. Some have gone as far as to claim that 'big data' represent the end of theory; that 'big data' can potentially generate more accurate or true results than specialists who traditionally craft carefully targeted hypotheses and research strategies. However, caution is essential. Understanding the social, economic, and political contexts under which data are produced will always matter (Graham and Shelton 2013). 'Big data' will rarely be able to offer a pure and unbiased reflection of any pattern, place or process, either because so much 'big data' comes in the form of closed, restricted or black-boxed datasets or because the 'big data' being used are not actually an appropriate proxy of the pattern, place or process of interest.

As Escobar (2008) has noted, the very processes of measuring and having data has in many ways created practices of development. As ever more facets of development are quantified and datafied, it will then remain crucial to consider not just what 'big data' measure and what questions they allow us to answer, but also what they omit and how to keep a focus on some of the most crucial and critical questions in development.

In sum, the ready availability of massive datasets and the tools and methods to analyze them is beginning to have (and will undoubtedly continue to have) transformative effects on research in the contexts of development. This chapter has offered a starting point for researchers who are interested in reflecting on how 'big data' might be best embedded into their own work.

If you want to explore the ideas covered in this chapter in more detail, you may find these further readings useful:

Escobar, A. (2008) 'The problematization of poverty', in Chari, S. and Corbridge, S. (eds.) *The Development Reader*. Oxford: Routledge.

Graham, M., Hale, S. and Gaffney, D. (in press) 'Where in the world are you? Geolocation and language identification in Twitter', *Professional Geographer*.

Graham, M. and Shelton, T. (2013) 'Geography and the future of big data; big data and the future of geography', *Dialogues in Human Geography*, 3: 255–61.

Hilbert, M. (2013) 'Big data for development: from information to knowledge societies'. SSRN Scholarly Paper No. ID 2205145. Rochester, NY: Social Science Research Network; http://papers.ssrn.com/abstract=2205145.

Kitchin, R. (2014) *The Data Revolution: Big Data, Open Data, Data Infrastructures and Their Consequences*. London: Sage.

UN Global Pulse (2012) 'Big data for development: challenges and opportunities'. http://www.unglobalpulse.org/sites/default/files/BigDataforDevelopment-GlobalPulseMay2012.pdf.

World Economic Forum (2012) 'Big data, big impact: new possibilities for international development'. http://www.weforum.org/reports/big-data-big-impact-new-possibilities-international-development.

13
LOCATIONAL AND SPATIAL DATA

'*Where?*' is often one of the most powerful questions that you can ask in any research project. Asking this question allows you to ground and contextualize your work into discrete geographic units. Location is often used as a proxy for many other variables, patterns, processes and meanings. If you read about a study in Rhode Island, USA, or the Rift Valley in Kenya, you immediately recognize those places as not only fundamentally different and distant from one another, but you are also likely to contextualize those places within pre-existing understandings about culture, economics, politics and environment.

It is for that reason that spatial data are also inherently powerful. Despite much work that has been carried out by critical cartographers to deconstruct the power of maps, most people still treat maps as representations of the truth. As such, there are important considerations to internalize when working with spatial data. The goal of this chapter is to guide you through some of these issues.

While other chapters have focused on a range of qualitative and quantitative data, techniques and methods, this chapter explicitly concerns itself with spatial or locational data. Too often location has been overlooked and omitted from the research process in part because it was relatively difficult to obtain. For instance, until very recently, most spatial data have been locked away in the private sector or behind impenetrable government bureaucracies. This is changing. Researchers now have access to unprecedented amounts of locational information and the goal of this chapter is to help you understand what you might do with those data. To illustrate the potential power of these methods, consider the improvement in online maps available of Haiti in the immediate aftermath of the 2010 earthquake.

In January 2010, when a magnitude 7.0 earthquake struck Haiti, there was an immediate need for spatial data. Emergency responders needed to know how to allocate scarce resources. Specifically, they needed to know where the people most in need were, and how best to get assistance and relief to them. In Haiti, streets and buildings were mostly absent from online mapping services such as Google or Apple maps that people in rich countries take for granted. As one of the world's poorest countries, there was little incentive for large corporations to create detailed and publicly available spatial databases about the country. However, after the earthquake, the demand for spatial data became immediately apparent given the urgency of relief operations.

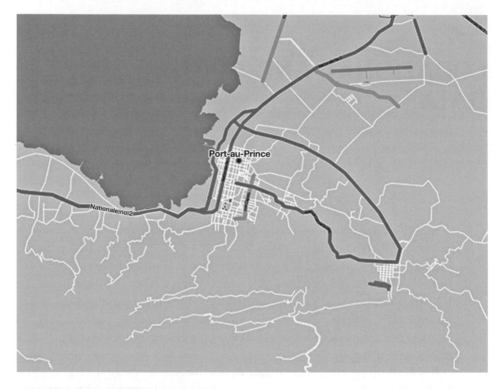

Figure 13.1 OpenStreetMap coverage in Port-au-Prince before (11 January 2010) and after (14 January 2010) the earthquake

This mismatch between the availability of, and need for, spatial data resulted in a remarkable crowdsourcing effort. Hundreds of OpenStreetMap (OSM) volunteers from around the world used satellite imagery to trace the outlines of streets and buildings and upload that content into the OSM platform. Ten thousand edits were created to the Port-au-Prince region alone in the weeks following the earthquake (see Figure 13.1)

This crowdsourced mapping was undoubtedly useful for people involved in rescue and relief efforts. However, it was not the only platform onto which people were contributing spatial data. Google's 'Map Maker' was another tool that allowed for user-generated submissions of crowdsourced data and was also used extensively after Haiti's earthquake. However, unlike OSM, all data submitted to Map Maker is not subject to an open and sharable licence. What ended up happening was a spatial division of labour, whereby some parts of Haiti were better covered in Map Maker and others better covered in OSM.

This example highlights not only some of the power and potential of spatial data (i.e. that it can be seen as baseline information essential to asking further questions and providing further services), but also some of the core considerations that need to be taken into account when producing spatial data. Because of the ways that spatial data are often used as a platform for other work, it is important to be clear about not only the coverage and limitations of those data (e.g. do we have blank spots on the map because there are no villages there or because we have no data about that place), but also the ways in which that data can be reused and shared for other purposes. The chapter begins by asking what spatial data are. It then moves to discussions of how spatial data can be collected, how they can be analyzed and, ultimately, how they can be presented and visualized.

Spatial data and spatial questions

Spatial data are bits of information (or symbols) that define a location and allow us to answer geographic questions. Spatial data typically include locational information (where?) and non-spatial information (what?). Think, for instance, of a database of rural health centres. The location of those centres, stored as latitude/longitude coordinates, would be the explicitly spatial data, while other attributes such as the number of beds, nurses and medicines would be the non-spatial data. However, it is by linking these data together that we can address interesting and important spatial questions. Steve Cinderby's work on environmental sustainability and livelihoods provides one example of the ways in which locational and spatial data can be used to develop powerful understandings of development needs and priorities. His reflection on using participatory GIS mapping also emphasizes the how important it is to develop and use tools that are appropriate for the question you are addressing and not to become dominated by a particular tool.

Spatial data tend to be stored as either raster or vector data. Raster data are stored as a grid in which the cell values of the grid are the attribute of interest (e.g. land-cover codes). Here your data could be considered as either thematic (whereby each spatial unit represents a specific feature type: for instance, land use) or continuous (for instance, temperature or precipitation). With vector data, multiple attributes can be linked to points, lines or polygons that contain the relevant information.

Box 13.1 Steve Cinderby: Participatory GIS – working across development scales

Managing environments sustainably and developing livelihoods to alleviate poverty requires understanding how systems operate across different spatial and governance scales and how resource quality and quantity interact with ecosystems management and livelihoods. My research on water usage for smallholder agriculture in Tanzania illustrates this. Different methods could have been applied, including questionnaires or field surveys, however, participatory GIS allowed the problem to be looked at in a spatially explicit framing including the interactions between different livelihoods across the landscape.

With an unlimited budget, PGIS mapping could have been undertaken in all the villages across the watershed to gain detailed local contextual understandings, which were then mosaiced in the digital GIS to produce a basin-wide overview. However, problems can then occur with data overload and deciding what is important or generalizable. A more pragmatic and rapid approach was utilized to work at nested scales with different groups with relevant knowledge and expertise.

The first phase of the nested PGIS work involved identifying with local government development officers which villages to work with. Four villages were identified as representative of different livelihood approaches and varying access to water, while being relatively accessible by vehicle and open to the research.

Workshops with local farmers who represented different types of agricultural system (rainfed, traditional irrigation, formal irrigation and pastoralists) were organized and used maps to identify what resources farmers were using, where they were producing crops and what yields they experienced. Separate mapping was undertaken with women-only groups to ensure that any gender-specific differences were identified. The mapping process proved easy for the villagers to engage with and helped enable everyone to contribute and discuss issues.

The village PGIS results allowed the researchers to understand who utilized different resources across the seasons, what benefits they gained, as well as what problems were experienced. This information was synthesized across the four villages to create typologies (generic descriptions) of what it was like to be a 'pastoralist' or a 'rainfed farmer' in the basin, what resources were critical, what was holding back production and how the problems affected poverty levels. The challenge was to move up a scale – to the whole watershed, a level useful for policy making, governance and allocating development resources.

The researcher's local contacts identified experts with an understanding of different aspects of the environment and livelihood system (foresters, agricultural extension officers etc.) across a larger spatial extent, but without the detail of the farmer's maps. A mapping workshop with these experts identified where livelihood systems existed (including systems we missed at the village scale) across the whole watershed, where there were differences in yields, access to water or other key factors in our generic typologies.

Using this nested approach and PGIS, we could quite rapidly identify and map where systems were, how they worked and interacted, and what resources were critical for their success at a scale that was useful for managing both the environmental and governance systems.

The critical success in this approach was identifying the gatekeeper organization – the local government officers who could make introductions to communities and suitable experts, a connection facilitated by local academic partners. Another key factor is to remember the purpose of what you are doing and not get bogged down in a rigid process: if a particular question does not work then think of a different way of getting to that information and introduce new questions to address key topics emerging from the work. Remember the *purpose* and be *flexible* to ensure you understand the issue you are investigating.

To find out more, visit www.sei-internation.org or follow Steve Cinderby (@s_cinderby_SEI) to see what I am working on now.

Box 13.2 Examples of spatial questions

Spatially based questions crop up in many development research projects. Think about the list of questions below (taken from Kitchin and Tate 2000: 165) and think about how each of these could relate to a development research project on: health care needs and delivery, rural and urban livelihood adaptation, or political activism and participation:

Location: Where is it . . .?

Condition: What is at . . .?

Trend: What has changed . . .?

Routing: What is the best way . . .?

Pattern: What is the pattern . . .?

Modelling: What if . . .?

Your vector data will be stored as coordinates and attributes. Coordinates with latitude, longitude and elevation allow us to determine location with two sets of numbers. Usually two of the numbers represent position on a plane (*latitude* and *longitude*) and one represents vertical position (*elevation*). Latitude refers to the distance of a place north or south of the equator and is measured in degrees. Longitude is also measured in degree, and measures how far east or west of the Prime Meridian (a north–south line that goes through the Greenwich Observatory in London) any place is. Elevation, in turn, measures how elevated any point is above sea level.

Sets of latitude and longitude coordinates can be used to define specific points, lines and polygons. Points are discrete locations used to represent an entity that exists in a single location. Lines are constructed by connecting two or more points. Lines represent features with length, but not width or area. Polygons are made up of three or more line segments. Unlike points and lines, they are two dimensional and can be used to represent features that have area. By combining all three of these features, we are able to use a vector element to describe any social or natural feature.

Our ability to locate a place is also determined by how geographically accurate and precise our data are. *Accuracy* refers to how close the measurement of a location is to the actual place that it represents. *Precision*, in contrast, refers to the level of measurement on a map or in a database. That means that you could, for instance, say that collecting data at the village level is more precise than collecting data at the state level. It is important to remember that data with high precision are not necessarily accurate and data with high accuracy are not necessarily precise.

Collecting spatial data

Spatial data can be collected in three primary ways: through direct measurement, through secondary or mediated measurements and from existing spatial data repositories.

Direct measurement

Only a few years ago, collecting accurate and precise spatial information was a highly specialized task and usually required an expensive global positioning system (GPS) device. GPS works by trilaterating the time it takes for radio waves to travel from three or more satellites orbiting the Earth to a GPS receiver on the ground. Most GPS receivers have an accuracy of around 20 meters and work anywhere on Earth.

It is now relatively easy to find a device that will allow you take relatively accurate and precise measurements. Many cameras and voice recorders now have in-built GPS functionality that allows you to record where exactly images or sounds were recorded (the usefulness of this functionality during fieldwork cannot be underestimated). Most contemporary phones now also have the ability to function as a GPS receiver.

This means that with the right application these technologies will record points or paths, add short notes or other data to recorded coordinates and export that spatial information to a computer. Using a GPS phone is a relatively inexpensive way to collect spatial data during fieldwork. Dedicated GPS devices also exist for a range of specialist purposes and may be more useful in the context of research that requires large-scale primary spatial data collection.

Secondary and mediated measurement

There might also be cases in which you need to collect and use spatial data, but are unable to conduct any direct or primary measurements yourself. In such cases, you could manually extract spatial data from repositories such as OpenStreetMap or Wikipedia.

OpenStreetMap, in particular, has relatively thorough coverage of many populated parts of the world and allows data to be easily (and freely) repurposed for research purposes.

In addition, many automated tools exist to convert geographic information. Think of a situation in which you have the addresses of hundreds of schools, but want to obtain geographic coordinates for those buildings so that you can plot them all on a map. This process of assigning coordinates to an address stored in another format is called geocoding. This could be done using tools such as Google's 'Fusion Tables' that offers an easy-to-use geocoder for data placed into a spreadsheet. Geocoders such as Yahoo!'s PlaceFinder or EDINA's Unlock can also be used for similar tasks.

Spatial data repositories

Repositories and existing databases offer a rich source of spatial data. International organizations such as the World Bank and the United Nations offer useful and standardized datasets (usually at the national scale). National statistical bodies (such as census bureaux) in many countries can often have useable and detailed data at the sub-national scale. These details can be joined by a key attribute to file with spatial coordinates.

The last few years have seen a host of 'open data' repositories that tend to be run by metropolitan, regional and national governments and the third sector. Kenya's open data initiative, for instance, shares information about poverty rates, urbanization rates, school enrolment, health facilities and many other things.

How spatial data can be analyzed

Spatial data are typically analyzed through a geographic information system (GIS). A GIS is a system that allows for the capture, storage, analysis, updating, manipulation, sorting, comparing and visualization of geographic data and can take a variety of forms. Here we outline a few categories and examples of GIS in order to give you a sense of some of the tools that may be available to you. Our goal here is to provide an indicative rather than comprehensive list.

Basic tools

Many data repositories offer tools for simple and descriptive spatial data analysis. However, these tools are often enough to answer some quite powerful questions. Tools include the following.

Platforms run by multinational organizations

The most useful tools in this category include those created by the World Bank. The World Bank's eAtlases (data.worldbank.org) allow users to compare national statistics on hundreds of indicators such as violence against women, financial inclusion and life expectancy. Its Mapping For Results platform (maps.worldbank.org) allows the variable amounts of funding that different projects receive to be compared at local scales. Its Aidflows website (aidflows.org) also offers useful functionality to measure and compare the global flows of development aid between recipients and donors. The World Bank isn't the only international organization to offer simple web-based analysis tools (the UN, for instance, also offers some based comparative tools at data.un.org), but its website does offer the most comprehensive collection of data.

Third-party platforms

Google's *Public Data Explorer* (google.com/publicdata) and *Gapminder* (gapminder.org) are two of the best examples of tools in this category. They both pull together massive amounts of third-party data into easy-to-use interfaces that allow users to compare national-level metrics. Statistics as diverse as gender balance in parliaments, internet speeds, and labour productivity can be compared and exported.

Development agency platforms

Unfortunately, there are few development agencies or departments of development that offer useful analysis tools to explore their data. Two exceptions to this statement are the United Kingdom's Department for International Development (DFID) and the United States' USAID. DFID has created a *Development Tracker* (devtracker.dfid.gov.uk) that offers a useful tool that allows all medium and large-sized expenditure by the organization to be traced and mapped. The USAID mapping website (aiddata.org/maps) similarly offers a range of geospatial analysis tools that allow data on topics such as governance, health and security to be mapped.

In all of these cases, tools are extremely useful to use, require very little technical expertise and can quickly and freely provide powerful results. However, because they tend to be designed for the general public rather than researchers, the tools tend not to be malleable and lock researchers into particular questions and answers that come from organizations with particular agendas. For these reasons, it can be useful to explore spatial analysis tools that offer more flexibility and user control.

Interactive tools

Anyone with web access is now able to make use of a range of free and interactive spatial analysis tools. These tools generally allow users to upload their own data and perform descriptive spatial analyses on those data. Many tools also come with in-built automatic geocoders: meaning that you could enter in text-based address data that look something like what we find in Table 13.1.

The automatic geocoders would then be able to process the variable amounts of information submitted in the 'address' column, match those data to a gazetteer and then assign those locations to coordinates on a map. This can be a very useful process because it saves the researcher from having to find structured latitude and longitude coordinates for every entry of unstructured data in their database.

Table 13.1 *Tools that come with in-built automatic geocoders*

Name	Address	Employees
Imperial Teas (EPZ) Ltd	Kingorani EPZ Complex, Msa-Nrb Highway, PO Box 17091–80100, Mombasa, Kenya	9
Al-Nakhil Enterprises Ltd	Refinery Rd, Mombasa	23
Arroket Factory	Kericho	10
Githambo Tea Factory	79–10201 Kahuro, Nyeri	48

Once geocoded, most of the tools below allow users to ask a variety of geographic questions to their data. For instance, you could compute distances, shade areas (or points or lines) according to the value of particular variables, create buffers around points and measure the differences between variables in difference places.

The following lists a few tools that allow basic mapping and spatial analysis to be performed (note that this is an illustrative rather than a comprehensive list and the services offered in this category are likely to quickly evolve in the next few years):

GeodesiX (calvert.ch/geodesix) is open-source freeware that can be downloaded as a plug-in to Microsoft Excel.

MangoMap (mangomap.com) is a website that allows you to upload spatial data and turn those data into sharable interactive maps.

ArcGIS Online (arcgis.com/home) offers similar functionality and makes it especially easy to create collaborative maps (e.g. between different members of a team of researchers doing fieldwork).

batchgeo (batchgeo.com) is a very simple online tool that is useful if all you want to do is map unstructured data.

Google Fusion Tables (tables.googlelabs.com) is a service that allows users to create a range of graphic outputs and manage fairly large datasets.

Advanced GIS tools

For those who are more comfortable with spatial analysis, or wish to ask more advanced questions of their data, there are a number of very powerful GIS tools that are available. These tools offer a full range of spatial analysis functionality and can be especially useful for researchers who wish to have close control over their spatial files or perform advanced inferential spatial analysis.

Because of the complexity of the subject, this book is not the right venue for a detailed description of how advanced GIS software works. We do, however, recommend some further reference works at the end of this chapter.

Two of the most common advanced GIS tools are now described in detail in order to offer a sense of the range of analysis afforded by each.

QGIS

This software is open source, free, and cross-platform (meaning that it can run on Windows, Linux, Mac OS X, BSD and Android). It allows the use of both vector and raster files (see Chapter 15) and makes it straightforward to perform the most common GIS operations (such as importing, reprojecting, spatial or tabular joining, spatial queries and a custom symbolization of data). It has a relatively small user base, but there are a variety of free tutorials available online, such as Harvard University's 'Quantum GIS Workshop' (maps.cga.harvard.edu/qgis).

ArcGIS

This software offers a full range of spatial data manipulation, editing, and analysis capabilities. It also offers a high degree of flexibility in terms of data building and map display. Because of this flexibility, the software also comes with a steep learning curve. However, the wide adoption of the software by GIS professionals means that there is a wealth of tutorials and learning resources available online. Most universities (and even many small community colleges) around the world also offer some type of instruction for ArcGIS. Potential drawbacks include the fact that the software is expensive (for researchers based in institutions or companies that don't already have a licence) and that it only runs on Windows.

How they can be presented

While much of Chapter 15 (visualizing data) is devoted to explaining strategies and practices of data visualization, the remainder of this chapter focuses on considerations that are particularly important within the context of mapping.

Projection

So far we have talked about location in a fairly unproblematic way: assuming that if we have a set of coordinates that we know – in an absolute sense – where exactly something is. However, determining location is somewhat more complicated than that. In most cases, we are attempting to transform a location that is somewhere on a three-dimensional sphere into a location that is on a two-dimensional plane. This process of transformation inevitably means that flat maps inevitably distort some combination of shape, size, direction or distance. No flat map can hold all of those attributes constant. When mapping (or even using maps), you should therefore pay attention to the specific projections, and thus the specific distortions, that are induced.

One common mistake that researchers make is to use Google Maps and other web-based maps (that use the Mercator projection) for small-scale maps. Because the Mercator projection exaggerates the size of places close to the poles, it shows Greenland as being about as big as Africa (when in reality, Africa is about 14 times the size of Greenland).

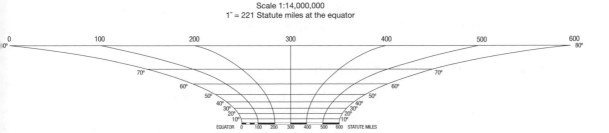

Figure 13.2 Scale bar from a large world map showing how scale can change with distance

Source: James Woodward (2010), Wikimedia Commons: http://commons.wikimedia.org/wiki/File:World_Scale_from_DMA_Series_1150_map.png

Map scale

Map scale refers to the size of an object's representation on a map compared to the object's actual size. Scale is typically shown through either a scale bar or a ratio. Map ratios tend to be written as a representative fraction in which map distance is reduced to 1 and the corresponding distance on the ground is expressed as its corresponding distance. For instance, a fraction of 1:500,000 written on a map means that a line that is 1 unit long on the map would be 500,000 units long on the ground. Using these fractions, people tend to use the terms 'large-scale' and 'small-scale' maps, with 'large scale' generally referring to maps about a local area and 'small scale' referring to maps about a much larger area. This might seem a little counter-intuitive, but it is because 'large-scale' maps (e.g. a 1:25,000 map of a village) have fractions that are larger than 'small-scale' maps (e.g. a 1:10,000,000 map of a continent).

A scale bar is typically a line that is annotated at regular intervals to show the distance on the earth's surface that a distance on the map's scale represents. It is important to note that because many maps use projections in which scale varies with latitude, scale bars are not necessarily simple or linear (see Figure 13.2).

Locational privacy

Location-based services and the proliferation of GIS have meant that spatial data about many facets of everyday life are now often readily available. For some people, location is very sensitive, for others not at all. However, in all cases, it is crucial for researchers to be aware of how their decisions may be intrusive or breach confidence and uncover positions, locations movements, routings, proximity, personal metadata.

Some questions to ask of your data include: who has access to your locational data? What facets of the data can, and should, be made available to which audiences? With what other sources can your data be linked? Are your data securely stored and encrypted and do you have a policy in place to dispose or transfer them at the end of your project?

Although ethical decisions in the context of development research are reviewed in more detail in our chapter on research ethics (Chapter 5), this chapter hopes to make you particularly aware of the need for locational privacy.

Representation and participation

Spatial data are powerful. They tell you who and what is 'there' and allow you to infer what isn't. This power that they exert has meant that many people tend to treat data on a map as *the* truth rather than simply *a* truth. But because the world that we seek to represent is infinitely complex, and maps are inherently partial and distorted documents, mapmakers therefore have a responsibility to clearly explain to their audiences not just the data on the map, but the people, places, processes and perspectives that have been omitted. Every mapmaker engages in specific choices about what to count and represent and what to omit and these decisions should be made as transparent as possible. Mapping is therefore not merely a technical, it is also a social practice. Maps are never simply neutral or objective representations of the world. They don't just mirror social and spatial realities, they also work to produce them.

This is especially true when working with data generated through any sort of participatory mapping process. Despite the recent 'democratization' of possibilities to create spatial data, we always need to ask what the variable possibilities for participation in process of data generation are, and who those participatory processes still leave out. Maps are contested documents that shape how we understand and produce the world and much responsibility thus rests with the mapmaker to use their power wisely.

What to take from this chapter

This chapter should act as a starting point for researchers interested in further exploring spatial questions. It covered some of the reasons why you might want to use spatial data, core ways that spatial data are organized, most important spatial data platforms and repositories in the context of development research, primary tools with which research can analyze and map spatial data and core considerations to take into account when mapping.

Mapping and the use of spatial data can be a time-consuming and complex part of research. However, the availability of relatively new data sources and analysis tools means that it now has the potential to enter the research practices of many development researchers. We hope that this chapter has offered a useful starting point to both the possibilities and unique potential problems associated with using spatial data.

If you want to explore the ideas covered in this chapter in more detail, you may find these further readings useful:

Longley, P., Goodchild, M., Maguire, D. and Rhind, D. (2010) *Geographical Information Systems and Science* (3rd ed.). Chichester: John Wiley.
O'Sullivan, D. and Unwin, D. (2010) *Geographic Information Analysis* (2nd ed.). London: John Wiley.

14

INTEGRATING METHODS AND ANALYSIS

One of the major concerns you will face as a researcher is ensuring that your data and findings are reliable and robust. To this end, it is important that you use appropriate methods to collect the data required for the questions you are asking. However, even when using an appropriate method, a range of biases and errors can creep into your research. For this reason, among others – including the emergence of new research questions or avenues – you may want to consider integrating multiple methods of data collection and analysis as a means to reduce these risks and ensuring the validity of your findings. These practices are often referred to as a 'triangulation' or mixed methods approach. Alternatively, you might use multiple methods due to their complementarity in order to increase the breadth and depth of your findings. In this chapter, we outline the arguments for using such approaches, as well as their weaknesses, and then address a range of analytic tools that you can use to accurately analyze your data.

To give you a sense of how useful such an approach can be, consider Paula Meth's reflections (Box 14.1) on her use of mixed methods research to explore experiences of violence in South African townships. As is evident in her reflection, the use of mixed methods not only allowed the researchers to compare findings across the types of data produced but also provided participants with greater freedom to express themselves and recount their experiences. As a result, the data from this work could not only be subject to more robust analysis but also were also richer in the content produced, allowing for more detailed insights to be developed.

Why integrate research methods?

Supporters of mixed methods research argue that this approach allows you to use different tools to approach the same question or concern from multiple directions. In so doing, you are able to check whether the data you gather from one source/approach match the findings generated from a different approach. In essence, the aim here is to enhance the trustworthiness of your research and to ensure that your findings are reliable. In doing this, you may use multiple qualitative methods or a mixture of quantitative and qualitative methods. For instance, if you are concerned with understanding the benefits of transport infrastructural upgrades to a community, you may find it useful to conduct interviews and/or focus groups with local stakeholders including village elders/chief/chairperson,

Box 14.1 Paula Meth: Using mixed methods to uncover everyday life

Using a mixed methods approach has always appealed to me because of the flexibility and breadth it offers. I work within a qualitative approach, so when I mix methods, I tend to use in combination methods in which the emphasis is on words, meanings and their interpretations. My work often mixes life history and regular semi-structured interviews with focus groups, autophotography, drawing and solicited diaries.

My work over the past decade has focused on gender in relation to violence and the living environments of relatively poor residents within South Africa, with an emphasis on residents living in informal settlements. This work is rather sensitive, particularly when dealing with questions of forced eviction, fire, domestic violence or very personal accounts of sexual violence or other distressing experiences. My work has explored both men's and women's experiences, often asking questions of men (who have perpetrated but also experienced some of the violence) about issues such as gender and power. I explain what my work focuses on, so that you can understand how mixing methods can be crucial to gaining insights into sensitive but also complex issues.

In a project working with men on violence, we found that different methods provided men with multiple opportunities to express their thoughts and experiences. Some men felt more comfortable telling difficult stories through methods that were more private (such as life history interviews, drawing and diaries) than in more public forums such as focus group discussions. In one example, a particular man – let's call him Sakhile – eventually revealed his HIV positive status as well as his anger over his wife's alleged infidelity through his diary and life history interview: both issues were deeply humiliating for him. We knew these methods provided him with a sense of safety, trust and privacy, unlike the focus groups, but the range of methods also helped to corroborate and piece together his life story in a way that a single method would have failed to achieve. Sakhile also used a combination of photography and his diary to explain how the settlement he lived in was a 'nest of criminals', his photographs providing explicit references to dangerous sites (such as paths and bridges). Through the medium of drawing, he then graphically illustrated observations that he had made in interviews and in his diary about his ideas of place, gender and sexuality. His drawings helped him to articulate his concerns about informal settlements as places where adults lacked privacy for sexual relations and hence were immoral spaces. In these last two examples, the mixture of methods deepened and specified the findings allowing for a more complex understanding of gender and place.

In this sense, I'd describe our approach to mixing methods as iterative not linear. Although we started with a coherent research proposal, we made and negotiated decisions about what to mix, what to try out and what to give up on as the project progressed. This flexibility is crucial in ensuring that the choice and timing of methods is appropriate and that they build on each other. In this sense, mixing methods provides a project with some degree of longitudinality, as you tend to meet participants on more than one occasion through the use of different methods.

Mixing then, is more than the sum of its parts. Different methods add value not simply in terms of the amount of data you gather, but through their impact they have on each other: throw-away lines and unexplained comments in written diaries become key questions in interviews, and drawings illustrate and articulate observations made in focus groups and diaries. Mixing is not without its problems, however. Your choice of mix must be justified in relation to your overall methodological approach, research questions and the practicalities of your project. Don't mix methods because you are not sure what you are doing: mix them when you are confident about your focus and you are keen to maximize your understanding of a complex issue.

traders, transport providers, service providers and the general population. You may then find it beneficial to compare these findings with the results of surveys of traffic volume and type, of product availability and cost, both at local markets but also in relation to medical, educational and other essential supplies, before and after upgrade to see if there is correlation or disjuncture between people's perceptions and measurable phenomena. In this example, the use of qualitative and quantitative methods complement each other, providing richness of data and insights that would not be possible using a single method. The benefits of using multiple methods to build up layers of understanding are evident in Katherine Gough's reflection (Box 14.2) on research with youth entrepreneurs in Kampala, Uganda. In this example, you can see how different methods were used to generate different types of data in relation to the project's core focus.

The possibilities for mixed methods are increasing with the emergence of new data sources and both data collection and analysis tools. We have identified how you can use big data (Chapter 12) and spatial data (Chapter 13) in development research and these are forms of data that can be used in imaginative ways with other methods. For instance, discussions of changing livelihoods and adaptation to climate change could be linked to participatory GIS mapping. Alternatively, if you were interested in researching questions relating to human rights and citizenship, you could collect and analyze data from interviews with international or national civil society organizations promoting idea(l)s of citizenship or human rights relating to their advocacy, awareness raising and policy work and compare these to big data analysis of their social media postings.

If you revisit the example presented at the start of Chapter 12 relating to Twitter trends and food vulnerability, you can begin to think about how you could incorporate such data into a project involving survey data collection and/or interview and focus group data collection. In relation to any number of development concerns, you could explore the possibilities of using the basic, free version of a tool such as DiscoverText (discovertext. com) and a Twitter API to collect and then analyze data from Twitter. For instance, had you been researching urban protest or citizenship claims making in Uganda in 2011 you could have used this tool to harvest and analyze social media postings in relation to the

Box 14.2 Katherine Gough: Using multiple methods and peer researchers to investigate youth entrepreneurship

In a research project to investigate entrepreneurship among young people in African economies, I was part of a team that employed a range of methods including questionnaires, focus group discussions, life story interviews and participatory videos. We entered the low-income settlement of Bwaise, Kampala, through a local youth organization, Uganda Youth Welfare Service (UYWS). UYWS selected 12 young people living in Bwaise to work together with me, with my colleagues Thilde Langevang (Copenhagen Business School) and Rebecca Namatovu (Makere University Business School) as peer researchers. These young people spoke local languages and English and they participated in all stages of the fieldwork.

Questionnaire

The peer researchers conducted a questionnaire survey with 400 young people living in Bwaise to enable us to obtain an overview of the level and types of entrepreneurship. There were numerous advantages to using peer researchers. They were not influenced by the stigma outsiders attach to Bwaise, were used to navigating the narrow pathways and were able to create a good rapport with their contemporaries resulting in a very high response rate. The peer researchers in turn, as well as earning money, benefited from gaining new skills, increasing in confidence and expanding their networks. Despite having to spend considerable time training and supporting the peer researchers, we believe employing them rather than more experienced outsiders generated data of a higher quality.

Participatory focus group discussions

Groups of five to seven young people who were relatively homogenous (i.e. of the same sex, similar education level and had businesses of a similar size) were selected by UYWS to participate in focus group discussions. These were led by the Ugandan researcher together with two peer researchers. The groups were encouraged to discuss what they understand as entrepreneurship, local perceptions of young entrepreneurs and the opportunities and challenges entrepreneurs face. During a discussion of the key factors that influence entrepreneurship, a ranking and scoring system was introduced using available items, such as coins for finance, a notebook for education, a mobile phone for social networks and a bottle of water for flooding. The participants allocated ten peas between the differing items indicating those they found to be the most important factors. Photographs of different types of business owner were also used to stimulate discussions of who is a successful entrepreneur. These exercises and the ensuing discussions proved to be highly revealing regarding perceptions of, and factors affecting, youth entrepreneurship.

Life story interviews

In-depth interviews were carried out with 20 young business owners who were asked to narrate their life story highlighting key turning points in their lives and explain why they had decided to set up a business. The interviews were conducted by one of the main researchers working together with a peer researcher who acted as a translator when necessary and asked additional questions. This form of interviewing generated detailed insight into young people's employment paths and how these are affected by changes in the socio-economic environment, institutional factors, social networks and family relations.

Participatory videos

Two groups of ten young people each made short films on a topic of their choice about young entrepreneurs in Bwaise. The young people were taught by Slum Cinema, a youth organization that trains young people to make films, how to use a video camera, make a script, use images to tell a story, film, edit and screen the finished product. Not only did this allow the peer researchers and research participants to produce a visual representation of entrepreneurship from their own perspective, they also gained new skills and in the process grew in confidence. We, meanwhile, obtained a more nuanced view of youth entrepreneurship through the many discussions that took place during the making and screening of the films as well as from the videos produced (see http://www.vimeo.com/72811329).

Combining multiple methods has generated a wealth of knowledge on youth entrepreneurship in African economies. Pivotal to the success of the fieldwork was engaging young people as peer researchers who not only improved the quality of the data collected but also made the research process more enjoyable.

For more details of the methodology, see Gough, K.V., Langevang, T. and Namatovu, R. (2014) 'Researching entrepreneurship in low-income settlements: the strengths and challenges of participatory methods', *Environment and Urbanization*, 26: 297–311.

---

walk-to-work campaign and the government's response to this. These data and analysis could then have been considered alongside discourse analysis of media coverage and both government and opposition statements relating to these protests, as well as alongside interviews with participants and others in Kampala.

Similarly, the development of crowd-mapping/crowd-reporting platforms such as ushahidi (www.ushahidi.com/) or ureport (www.ureport.ug) could allow you to monitor and analyze real-time tracking and reporting of electoral violence or service delivery challenges and responses. These could then be utilized alongside documentary analysis of official reports or press coverage, as well as interview data from affected communities. With the emergence of these new data and analytical tools, there is tremendous scope for the integration of these methods with 'older' forms of data collection in order to approach and address research questions in different ways over different scales of space and time.

It is not only multiple methods that can be used to increase the reliability of findings. Advocates of a broader approach of *triangulation* argue that as well as using multiple methods, you can utilize multiple investigators or multiple sources/informants to investigate an issue and gain more detailed and robust findings (Laws et al. 2013). The concept of triangulation is drawn from navigational practices where a position can be verified through measured relationships with other known points. In research, this idea is translated to meaning that if you are able to correlate, for example, the content of an interview with a pastoralist describing their responses to environmental variability with archive material in the form of a tree ring sample or climatological records you can make a more powerful and reliable argument than if you were reliant on the interviewee's testimony or the tree rings alone.

This ability to verify one set of data with another is often useful. However, it is not always necessary or desirable. In Chapter 7, we mentioned oral history interviews as an increasingly common method used in development research. In the case of oral histories, the perceptions and partiality of recall recounted in these interviews are often of core concern to the research – the subjectivity is of far greater interest than objectivity. Indeed in many cases, certainly for those working from a social constructionist approach, the divergences and differences in perspectives are as much of interest as the similarities. In other words, mixed methods and triangulation may help you to smooth out research findings and identify common issues; it is also desirable to identify and analyze the disjunctures and differences that emerge. A critical analysis of the different perspectives on an event or issue can uncover important findings and debates. Thus, for some projects, and to address some questions, triangulation or mixed methods approaches are certainly desirable, but this is not a one-size-fits-all approach.

On a practical note, triangulation usually requires additional investments of time and energy, both in developing the skills required to collect and handle different data, as well as in locating, accessing, collating, recording and analyzing different data. For a fieldclass project or undergraduate dissertation, these demands are often too great but such an approach may be feasible with a larger scale project. If you have developed a mixed methods approach then you will undoubtedly have multiple forms of data requiring analysis and you will need to develop an integrated approach to your data analysis.

Analyzing data

Analyzing your data is an essential but often challenging task, but is also one that can be incredibly interesting and rewarding. The main function of data analysis is to work out the links between data and to critically reflect on these links and patterns and to make or find meaning in these. Depending on the type of data you have collected, there are various tools you can use to do your analysis. Before introducing and explaining a few of these tools, it is important to note a few universal 'rules' in analysis: never make things up and do not make claims that your data do not support. You need to consistently question your data and ensure that your preconceived expectations do not distort your analysis.

Ultimately, if you make poorly or unfounded claims from your data, you will lose credibility and contribute to ill-founded development projects. Finally, although we often talk about phases of the research process as being the designing of the project and then the data collection period, which is followed by analysis and writing up, this is an oversimplified and misleading way of thinking about your research. Research is an iterative process and you should be conducting data analysis from the start of your project – be this in your analysis of existing literature, media coverage of relevance to your topic or archival materials. While you are embedded in the data collection phase, you should be thinking about and analyzing your data – even if only at a superficial level – in order to identify any areas of emergent interest or importance.

Qualitative data analysis

When analyzing qualitative data, your goal is to draw out the meaning within the data, to identify trends, patterns, correlations, causality or other factors of interest. In seeking to achieve this, you need to ensure that your methods of analysis are transparent, disciplined and systematic. You should also document how you have gone about conducting your analysis and the key decisions made during this process. Overall, this process will adhere to the core flow outlined below while recognizing that, because this is an iterative process, you will repeatedly cycle through the various stages of this process:

1   Organize and collate all of your data and metadata.
2   Conduct an initial read-through/overview of your data so that you are familiar with the material you have.
3   From this initial overview, develop a first list of key themes and topic categories that emerge from the data. (It is important that these initial categories are ones that come from the data and are NOT categories you impose on the data – if you impose categories on to the data you will limit and potentially bias your analysis and findings. It is often useful to introduce categories that you think are important (i.e. 'impose' on the data) during a second read-through, allowing you to draw out a broader range of topics for analysis.)
4   Read through your data, adding notes to your transcripts etc. of key issues or reflections while carrying out the process of coding. Coding is the mechanism by which you locate material within your dataset that relates to your key themes/ categories. You may find in this second read-through that additional categories and themes emerge and you should include these in your coding scheme (do not forget to go back to the start of your dataset and recode for any new topics).
5   Collate all the material you have for each theme/category and analyze the data.

As well as being an iterative process, qualitative data analysis is an inductive process – you are developing concepts and ideas from the data, which are then theorized (and, in due course, these theories tested against data in a deductive process).

At the heart of this process is the practice of coding. Coding is the process that allows you to organize and evaluate your data and thus develop an understanding of meaning.

In other words, it is the process through which you identify important categories and patterns of content. You might approach your data with a predetermined set of codes, you might develop a set of codes from an initial processing and analysis of the data or you might adopt an iterative process whereby you start with an initial set of predetermined codes that are then adapted and altered as you find new themes or codes emerging from the data. Whichever approach you take, it is important to have a clear and detailed code list that provides a clear definition or understanding of what is included in each code – including any sub-categories or concerns. This level of detail will help you more accurately code your data as well as providing an efficient means of identifying any sub-codes or sub-themes that you want to elevate to being main codes and where to find previously coded data.

Your coding might focus on identifying patterns, but you can also code instances relating specific types of theme. There are four key types of theme that you might consider developing codes for: conditions (this may be housing, employment, climate, etc.), consequences (outcomes because of event, condition, interaction, etc.), strategies and tactics (what people do or how they respond to situations), and interaction among actors (how people engage with others, their views of others and how they are viewed by others) (Cope 2010: 442). By paying attention to these specific areas, you are then able to draw out further themes for analysis.

There are a number of ways in which you can approach your data analysis. You may find it easiest or preferable to cut and paste the coded excerpts from your data into your thematic categories – either in paper or digital form. When doing this, be sure to annotate the materials you are moving around with details of its original source (interviewee, date, etc.). It is also important that you are careful not to simply reshuffle your interview or other data through this process: this is not analysis. You need to go further and think about what the data mean, what stories the data are telling you about the code or topic. Once you have done this you can organize and sort the data within each category and try to identify further patterns, themes or commonalities. This process of evolving and adapting codes driven by the data is often referred to as 'open coding' (Cope 2010). You may also think about using 'axial coding', a process whereby you focus on a particular theme or category and explore specific nuances to this through more precise and focused codes. Another coding approach is 'selective coding', whereby a dominant theme is identified and coding is coalesced around this theme, with any other themes being viewed through the lens of the dominant theme.

Overall, you are likely to develop an increasingly complex set of codes, starting from more general and descriptive themes and gradually becoming more specific and analytic, potentially with greater relevance or links to theoretical literature. As you build up your codes and coded data, you can begin to identify correlations and/or disjunctures between the appearance of codes and topics. From these, you can begin to identify and develop themes working across your research materials and that will form the basis of core arguments or findings.

Alternatively, you may decide to use data analysis software to assist in the analysis process. At a very basic level, you can use a program such as Microsoft Word or Excel to record key observation, reflections and excerpts in relation to each analytic theme and use different font, colour or style options to distinguish different themes or issues. While these programs are useful for data entry and storage (and, in the case of Excel, for basic statistical analysis), you may want to invest the time and effort to learn how to use more advanced, specialist qualitative data analysis software such as NVivo. These programs allow you to conduct various forms of analysis, including discourse analysis (analysis of language use to address relations between text/discourse and power, context, interaction or memory) and conversation analysis (analysis of social interaction), on written, visual and social media data. These types of program do take time to learn and to get the most out of but this can be time well invested when conducting large-scale projects.

However you approach your analysis, you need to ensure rigour and validity. It is important, therefore, to reflect on how your positionality influences your analytical engagement and it is often useful to provide a few biographical details in your project write-up so that the reader is aware of these. You should also keep a clear record of the decisions you make in developing your codes and analysis. This allows you to answer any questions that may be raised about your analysis but would also allow another researcher to repeat your work if needed. You should also ensure that you are not making oversimplified or over-reaching conclusions, that you have explored alternative explanations to your findings, that the examples and excerpts used in your analysis are representative and appropriately selected and that your analysis is grounded in existing literature and knowledge.

The availability of new platforms such as gephi (gephi.org) also provides new possibilities not only for engaging visualization of data but also developing data analysis, social network analysis, link analysis and measures of centrality in networks. These tools are providing increasingly powerful ways of understanding key questions in development research and can allow you to identify and analyze commonalities and differences arising in your analysis. For instance, if you were conducting research into the power of and connections between humanitarian organizations working on post-conflict reconstruction in a particular region, you could compare the links identified between donors and recipient organizations as detailed on websites or in financial/annual reports with those identified and promoted during interviews with organizational representatives. Such comparative analysis may uncover interesting dynamics of ideology, politics and funding that would otherwise remain hidden.

Quantitative data analysis

Depending on your level of knowledge and confidence, it is possible to utilize a wide array of methods to analyze quantitative data. Some of these are most easily carried out by hand, others using basic programs such as Excel, while more advanced methods require more advanced computer software packages such as SPSS (Statistical Package for the Social Sciences). We provide a basic introduction to a few common techniques here: if you want to develop these tools further you will find Field's (2013) *Discovering Statistics Using IBM SPSS Statistics* a useful companion text.

At the most basic level, knowing how many responses you have to questions – and the number of each type of response – is essential. You may opt to simply tally the number of responses or you could consider a slightly more detailed approach of using code numbers for each questionnaire in your tally sheet: this allows you to easily check back to the original sources to follow up on specific queries or to readily identify where missing data are.

Once you have basic statistical data such as the number and type of responses to a question, you can begin to perform some simple analysis. What tools you use for this analysis will depend on the variables and types of data you are working with. There are two key types of variable – categorical and continuous. Categorical variables are those with discrete, multiple categories and can be understood as nominal (more than two categories but no intrinsic ordering – for example, eye colour, month of birth), dichotomous (only two categories, e.g. questions that require a 'yes'/'no' response), or ordinal (more than two categories that can be ranked but not given an objective value, for instance providing predetermined responses of very poor, poor, ok, good, excellent to a question on access to clean drinking water). Continuous variables may be interval (have a numerical value and can be measured along a continuum) or ratio (these are interval variables with a point of zero that means there is nothing of the variable (e.g. age, distance, temperature in Kelvin) and that do not have an intrinsic ordering (e.g. eye colour, gender) and that you can calculate a ratio between measurements/values).

One of the most obvious tools here is to calculate percentages. Although useful in giving an accessible summary of findings, be cautious of using percentages with very small samples where a small change in the number of responses produces a large shift in percentages. In these situations, percentages can be misleading and it is often useful to provide both the percentage and actual number. Depending on how you conducted your sampling, you should be careful about using these descriptive statistics to generalize from your sample to the population.

Calculating percentage is a straightforward process and uses the formula:

(100/Sample size) × number of responses

So for a sample of 185 households with 68 respondents stating they had access to electricity, 79 had access to water and 167 used paraffin lanterns for lighting, we can calculate the percentages (rounded to the nearest whole number) for these:

Electricity: (100/185) × 68   =   37%

Water: (100/185) × 79   =   43%

Paraffin: (100/185) × 167 =   90%

Other useful basic descriptive statistics include averages and measures of spread. The mode average is the most common response in a dataset. The median average is the middle value of a dataset when all the data are arranged in order of size. The mean average is the calculation of the result of an equal distribution of values and is calculated by adding all of the values together and dividing by the number of cases. While knowing the mean average is useful, it can suffer from being skewed by outlying data points, which is why knowing the *spread* of the data is important. The simplest way of doing this is to subtract the lowest value from the highest value in your dataset: this gives you the range of your dataset.

A more useful measure of spread is the standard deviation. The standard deviation indicates how dispersed all of the data are around the mean. To calculate the standard deviation, you would need to work out the deviation of each data point from the mean, square them (in order to get rid of the negative numbers, otherwise the next step would result in a score of zero) and then add these up. This figure is then divided by number of data points minus one, to give the variance, and then you take the square root of this figure to produce the standard deviation. The smaller the standard deviation compared to the mean, the less dispersed the data points are from the mean and vice versa.

Going one step further, you can use some fairly straightforward tools to begin to explore whether correlations exist between variables and if so, how strong this relationship is. Given that much quantitative research is concerned with identifying and analyzing relationships between variables, this is a key tool to learn. There are multiple ways of conducting this kind of analysis depending on the kind of data being analyzed and the number of variables being analyzed. The easiest way to identify correlations is to use a cross-tabulation. The easiest way to complete a cross-tabulation is using software such as SPSS, particularly for large datasets. However, you can complete a cross-tabulation by hand: to do this you need to complete a contingency table with the two variables being compared and to tally the number of responses that meet the intersecting categories (i.e. in Table 14.1 tally up how many times a 'yes' response to 'Completed primary education' occurs for each of the income brackets).

Table 14.1 *Example cross-tabulation table*

Education – completed primary school	Income less than $2/day	Income $2/day to $4.99/day	Income $5/day to $9.99/day	Income above $10/day	Total
Yes	8 (6%)	12 (10%)	21 (17%)	16 (13%)	57 (46%)
No	39 (32%)	20 (16%)	5 (4%)	3 (2%)	67 (54%)
Total	47 (38%)	32 (26%)	26 (21%)	19 (15%)	124

To complete such a table by hand for each combination of variables in a survey would be incredibly time consuming so we would encourage you to use computer software to assist with this process. You will also then be able to conduct significance tests fairly easily on your correlation. Significant tests are vital when dealing with contingency tables and claims of correlation. The most common test used with contingency tables is a Pearson's chi-squared test which is used to test goodness of fit (whether an observed frequency distribution varies from a theoretical distribution) and independence (whether paired observations are independent of each other). Computer programs such as SPSS will run chi-squared and other tests for you and you can use these to check for significance of correlation. You should remember that chi-squared tests will not work if you have multiple cells in the contingency table with low frequencies (in such a situation, you may want to combine categories in order to ensure you have high enough frequencies across your contingency table).

Contingency tables only work for two variables (you can use them for three but it gets rather messy and complicated) but it is rare that you will be investigating a mono-causal effect. In other words, it is more common for you to be concerned with working out the relative strengths of influence of multiple factors on an outcome. In other words, if you wanted to analyze the drivers of household poverty, you may want to consider if levels of education, health, physical location, expenditure, schooling costs etc. were influential. In order to understand how powerful these drivers are compared to one another, a contingency table would be of no use. Instead, you would need to conduct multivariate regression analysis. These advanced statistical analysis tools, as well as methods such as logistical regression analysis, are beyond the scope of this book: if you want to explore these powerful analytical tools, we would suggest you consult Field (2013) *Discovering Statistics Using IBM SPSS Statistics* for an accessible introduction.

Whichever statistical tools you use to analyze your data, you will need to consider statistical inference. This is a way of finding out how confident you can be in your findings: in other words, when you generalize from your sample to the population how likely is it that your findings are going to be wrong? This tool allows you to talk about confidence levels or statistical significance levels – in social sciences the generally accepted cut-off point for 95% confidence level (or probability of being wrong $p = 0.05$) – generated as outputs from statistical software programs.

What to take from this chapter

This chapter has reflected on the potential for using and integrating multiple methods into your research design. While this approach requires an investment of time and energy, it can significantly strengthen your research findings – as long as the methods you use are appropriate for the questions being addressed and the data being sought. We then turned to thinking about analysis of qualitative and quantitative data and provided you with a brief overview of some key methods and concepts.

If you want to explore the ideas covered in this chapter in more detail, you may find these further readings useful:

Field, A. (2013) *Discovering Statistics Using IBM SPSS Statistics*. London: Sage.
Gilbert, N. (2008) *Researching Social Life*. London: Sage.
Montello, D. and Sutton, P. (2013) *An Introduction to Scientific Research Methods in Geography and Environmental Studies*. London: Sage.
Teddlie, C. and Tashakkori, A. (2009) *Foundations of Mixed Methods Research: Integrating Quantitative and Qualitative Approaches in the Social and Behavioral Sciences*. London: Sage.

Part III
PRESENTING AND WRITING UP RESEARCH

15

VISUALIZING DATA

Effectively and concisely communicating results is central to good research. Graphics, maps, and visualizations can be convincing tools in development research because they allow powerful points to be concisely communicated to both professionals and non-experts. This chapter begins with two stories that highlight some of the impacts that visualizations of research can have. It then moves to a discussion of core considerations to reflect on when creating visualizations. Your positionalitity as a researcher, the audiencing of the work, a reflection on bias and absences, as well as embedded assumptions and best practices are all discussed within the context of visualizing research results. The chapter then considers how inputs can be used, some of the most important tools that can be employed and the range of outputs that can be created. This chapter will be useful for anyone seeking an overview of how graphics might be useful in their work and seeking guidance on how to start making their own visualizations. Chapter 13 covered some of the important ways in which we map spatial data, so readers may also want to consult that chapter if they are working with geographic information.

The following two anecdotes about the discursive power of visualizations illustrate the power and importance of visualizing data in development research.

The first story is about a map of Wikipedia (see Figure 15.1). The map was created as an outcome from a project that Mark engaged in to understand the uneven geographies of Wikipedia. The research involved in generating the underlying data was relatively complex and resulted in millions of points of data, each representing a location on the earth's surface and the corresponding name of a Wikipedia article at that place.

There are numerous ways that these data could have been presented. For instance, the data could have been put in a spreadsheet or we could have identified some of the more interesting facets of the data and described them. However, we opted for a relatively simple approach: we counted the number of Wikipedia articles in every country and placed them on a map of the world.

This simple map told a simple story. Namely, poor countries are largely left out of 'peer-produced' networks of knowledge. However, the impact of the map far exceeded anything that we would have ever expected. Newspapers and media outlets from around the world decided to reproduce the image and write about the story that it told.

The old adage that 'a picture is worth a thousand words' perhaps comes close to describing some of the ways that our map allowed the message that it conveyed to quickly travel to a range of different media. It is unlikely that our work would have had the same

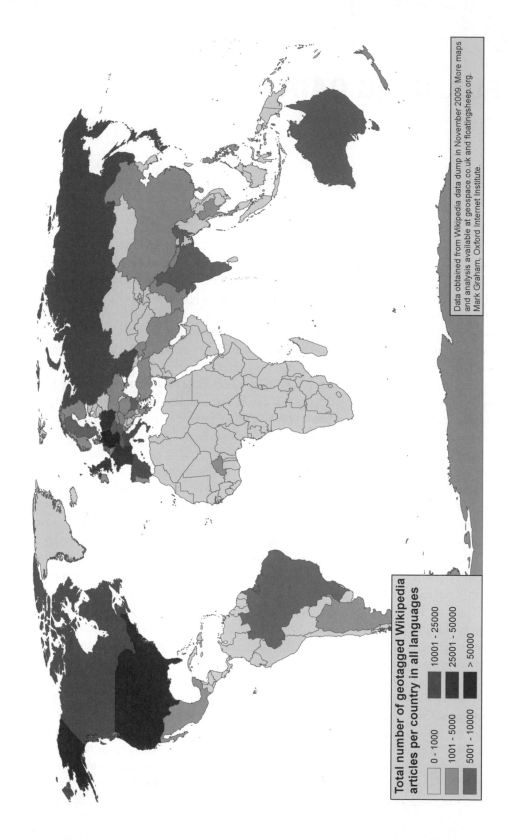

Figure 15.1 The spatiality of the construction of Wikipedia knowledge

Total number of geotagged Wikipedia articles per country in all languages

0 - 1000
1001 - 5000
5001 - 10000
10001 - 25000
25001 - 50000
> 50000

Data obtained from Wikipedia data dump in November 2009. More maps and analysis available at geospace.co.uk and floatingsheep.org. Mark Graham, Oxford Internet Institute.

impact or our message reached as many people if we had decided to simply describe our data without an accompanying visualization.

Our second story is about an even more influential visualization and comes from a story recounted by the *New York Times* journalist Nicholas Kristof. He describes how in the late 1990s he wrote a series of articles about disease in the global south. During a trip to Africa with Bill and Melinda Gates, Kristoph wrote:

> Bill and Melinda recently reread those pieces, and said that it was the second piece in the series, about bad water and diarrhoea killing millions of kids a year, that really got them thinking of public health. Great! I was really proud of this impact that my worldwide reporting and 3,500-word article had had. But then Bill confessed that actually it wasn't the article itself that had grabbed him so much – it was the graphic. It was just a two column, inside graphic, very simple, listing third world health problems and how many people they kill. But he remembered it after all those years and said that it was the single thing that got him redirected toward public health.
>
> No graphic in human history has saved so many lives in Africa and Asia. (http://www.nytimes.com/2008/02/25/business/media/25asktheeditors.html?_r=0 &pagewanted=all)

The purpose of these research stories is to illustrate some of the power that images, maps and data visualizations can have in the world. We can use visualizations to explore the multitude of meanings and ideas embedded in any particular dataset, show patterns and trends and to persuade and support arguments and recommendations.

Visualizing data

What does it mean to visualize data? Simply, it means to communicate information visually by abstracting text and numbers into static pictures, interactive graphics, animations and collaborative visualizations. The visualization of data allows us to view, analyze, change and communicate historical, spatial (for more detail on visualizing spatial data through mapping, see Chapter 13), qualitative and quantitative results and data. Visualizations turn datasets and bits of information into viewable landscapes. Visualizations can also be used to explore data (e.g. helping us when we do not even know what we are looking for in the data or what questions we should be asking), gain insights about data (e.g. showing us patterns or trends) and for communicating particular messages to policy makers, the media and the general public.

While much of this book has focused on how to obtain, extract or create data and information, it is also often the case that we are simply overwhelmed by it all. It generally is not a problem to look at a paragraph or page or even a book. But what if the data that we are dealing with are the size of a library? What if rather than selective or sampled data we are instead dealing with transactional data that means it is impossible to gain any sort of overview from just looking at them?

We can sometimes be so immersed in data that they become unmanageable. What visualizations do is allow us to step back or step away from the primary data in order to get an alternate or aggregate perspective. In doing so, visualizations allow for a form of knowledge compression: packing insights from otherwise incomprehensibly large datasets into a small visual space. Sometimes looking at original sources can be insightful, but the enormous size of some datasets can make doing so overwhelming or impossible. A visualization abstracts patterns from underlying sources and it can also be used to show gaps, absences, silences and outliers. If used effectively, visualizations can show us not just what is there or what the data tell us, but also what they do not.

At other times, we make visualizations not because datasets are too large but simply to effectively and visually convey messages and arguments. Visualizations resonate with many of us because of how they can quickly tell a story in way that pages of text or numbers often cannot. Most humans are surprisingly adept at seeing and making meaning from patterns, being able to recognize changes and even noticing subtle differences. As such, visualizations are also often created for a lay audience and communicate messages, trends, and patterns, illustrating information such as 'there is more X here than there', 'the interactions tend to be between these people and those people', 'the problem is there', etc.

Importantly, you do not need to be an expert in design or computer science or statistics in order to make useful and appealing visualizations. Expertise in relevant disciplines can certainly help, but there are also a range of tools that allow even the most inexperienced to transform data and information into visual narratives. However, the proliferation of easy to use visualization tools also comes with pitfalls. We need to carefully consider how we move from data to abstractions of data, as the design decisions that we make can have important impacts on the people who view and use our work.

The remainder of this chapter will guide you through *why*, *when* and *where* we might want to use visualizations and *how* we might go about making them. The tools that you use and the methods you adopt all fundamentally change and shape what you see and by extension what you know and what you do. As such, it is important to realize the power, pitfalls and potentials wrapped up in visualization tools.

Why use data visualizations?

Before discussing some of the ways in which you might visualize some of your own data, it is useful to expand on some of the various reasons you might have for using them. Your goals, purposes and audiences will all have significant impacts on how you make, use and interpret visualizations. In other words, think about the effects that you want your visualization to have in the world. What do you want it to do?

If we consider your audience, at the most straightforward level, you could be creating a visualization for yourself. This might be useful in cases when you want to explore and make meaning of your own data. Such visualizations can be used to highlight patterns, trends, anomalies and key ideas. Otherwise, it is likely that you will be creating the visualization for someone else. Instead of using the visualization for *exploration*, you will be using it for *explanation*.

It is at that point that your responsibilities as a researcher again become important. Because visualizations have the power to convince, their creator(s) necessarily have to heed a range of important considerations.

First, is consideration of the very purpose of the visualization itself and the effects and audiences that we want them to have. For instance, we could think about the difference between *scientific visualizations* (which are concerned with accuracy) and *information visualizations* (which are concerned with communicating in non-complex ways) as ways of organizing this discussion. In doing so, we are not attempting to assert that these are the only types of visualization. Neither are we claiming that all visualizations can be easily put into one of the two types: significant overlaps inevitably exist. Nonetheless, it is important to consider how you might want to position a visualization between those categories. How much detail and nuance do you want to sacrifice for the sake of clarity? The ways that you approach and address that question should be clear not just to yourself, but also to the viewers and users of your visualization.

Second, visualizations have to be clear about not just what they show (the 'known knowns'), but also should allow the user to understand what they do not show (the 'known unknowns'), as well as attempt to address some of their prospective biases and flaws (the 'unknown unknowns'). Particularly within the contexts of development studies, when research questions often centre on issues of inclusion, equity, fairness and participation, this attention to what is not shown should be used as not just an afterthought, but rather as a key element of the visualization. Think, for instance, of a pie chart that illustrates the proportion of people in a survey who responded 'yes' and 'no.' However, what if that question were particularly controversial and a significant number of people chose to abstain from answering? Or what if there was a distinct demographic bias in the proportion of people that choose to respond?

Figure 15.2, for instance, shows two very different ways of visualizing the same data in response to a survey question that asked 'should contraception be made freely available in your community?' Think about how the two different ways of visualizing the same data convey very different messages. (Note that these visualizations derive from the same survey that contains 31 'yes' responses, 69 'no' responses and 80 no responses.)

Third, and relatedly, visualizations need to be alert and attuned to biases that are inevitably embedded into, and reproduced by, them. Any data that we choose to use are necessarily created and interpreted under specific social conditions and therefore can never be a purely objective and unbiased representation of the external world. As such, it is your job to make some of those biases known and explicit. What do we know about how the data we are using were collected? What sorts of question were asked or omitted? What sorts of error might be unintentionally (or intentionally) present in the data? What level of accuracy and precision were your data collected under and how does this affect how you present your results? These are just a few of the questions that you need to consider when conveying the potential biases in your results to your audiences.

Fourth, it is important to be alert to the many assumptions that are often both built into visualizations and form the social contexts within which visualizations are interpreted. Some of these assumptions might seem obvious. For example, larger object size is often interpreted to mean larger amounts of the variables that it represents, or links between nodes in network diagrams are usually assumed to indicate that a relationship of some

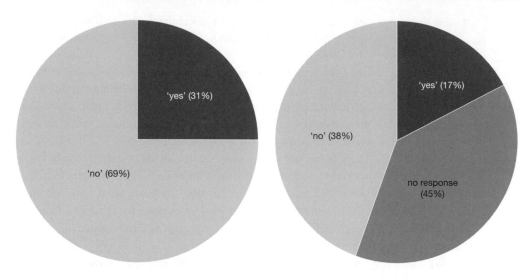

Figure 15.2 Pie charts including and excluding non-respondents

sort exists between the variables that those nodes represent. Different types and styles of visualization also come with their own assumptions about what they should be used for.

When visualizing compositions, relationships, distributions and comparisons you might employ very different visualization styles. For instance, to visualize compositions you might use stacked charts and pie charts, while scatter charts and bubble charts provide a more effective visualization of relationships. Comparisons are generally more effectively communicated through bar charts and line charts, while histograms and scatter charts provide clear communication of distributions. Figure 15.3 provides a more detailed overview of how you might want to match the data that you have, the messages that you intend to convey and the type of visualization that you employ.

Fifth, it is crucial to be reflexive about your subjectivity and biases as the visualization creator. It is relatively easy to change some of the original intent of underlying data, messages, and signals, and visualizations necessarily always selectively tell some stories and conceal others. Paradoxically, people tend to treat visual representations as truth documents; they are frequently presented as evidence and fact. A significant responsibility therefore rests in the hands of their creators. Visual representations are never pure reflections of *the* truth. This does not mean that a designer should invent stories or results that did not emerge from the underlying data.

Furthermore, designers have a responsibility to be explicit about some of the core decisions that were made about what narratives to present and which to omit. Being transparent about the processes involved in creating the visualization is also good practice. This can include, where possible, linking to underlying data sources to allow users to dig deeper and reach their own conclusions.

Finally, we always need to reflect on whether or not a visualization is appropriate for the data and for the contexts in which the visualization will be employed. Sometimes visualizations offer moments of insight, paths to new questions or hidden answers and they allow us to understand things that we might not have been able to see in primary

Chart suggestions – a thought starter

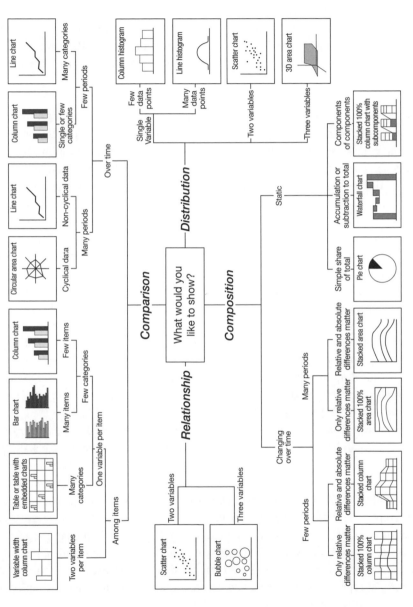

Figure 15.3 Chart suggestions

© A. Abela 2010, used with permission: www.ExtremePresentation.com

Box 15.1 Benjamin Hennig: Geographic visualization in social sciences – or *draw more maps!*

'It is important to change perspectives so that different methods are seen to be complementary, emphasizing the additive rather than divisive attributes of quantitative methods, qualitative methods and visualization (mainly GIS and cartography)' (ESRC 2013: 16). This quote, taken from a recent report on the state of human geography in the UK by the Economic and Social Research Council (ESRC), highlights the importance of visualization not only in (human) geography, but social sciences in general.

Geography and cartography are closely intertwined. Popular notions of geography often consist of images of a map, perhaps a mental map of a place or of the world, of cities, rivers, mountains and continents. This popular view of geography may be irritating to the academic who feels that the discipline consists of much more depth, breadth and maturity than the pictures that maps are sometimes disregarded as. But this somewhat childish view of geography must not be dismissed. Images of maps stick with us because they are such powerful and effective methods of communication and of conceiving of space. Map-reading skills are developed by most of us in our childhood and become our understanding of space. Using maps is therefore a chance to change people's understandings of our world.

It is interesting to consider how far the discipline of human geography appears to have distanced itself from maps over recent times, resulting almost in a form of cartophobia. Several papers over the last years showed a decline in map use and mapping practices in high-profile geographic journals. Cartographic skills as a natural expertise of a geographer seems to have vanished in many places, as have the theoretical and practical elements of geographic data visualization. Do many geographers 'prefer to write theory rather than employ critical visualizations', as Perkins (2004: 385) notes?

The views of our world have changed considerably since the 1970s, and perhaps also the availability of new perspectives over the internet has helped to challenge the perspectives that we have been exposed to in our childhood days or that we simply just became accustomed to. But, nevertheless, if geographers deploy cartographic visualization at all, many still rely very often on conventional cartography (and sometimes do so in rather bad practice that does not take theoretical notions of visualization into account). Especially in human geography, this is not always useful for understanding the complexity of the world and requires new and unconventional approaches to turn the diversity of research outcomes into meaningful representations. Here geovisualization turns from being a simple tool into a field of research in itself.

Social scientists must rediscover maps and other forms of geographic data visualization if they want to change perspectives. This is not only a challenge of finding suitable methods of visualizing results, but starts with the research question and with the way in which and what kind of data are gathered to answer the research questions. Part of visualization are new methods of data analysis and geospatial data visualization can as such become part of the process of investigating patterns and trends in complex data. Making maps is not as profane as it often sounds. Geographic data visualization is a way to communicate complex academic thinking, but also enables us to find new ways to understand the complex and diverse nature of geography and critically question the practices that we have become accustomed to.

Turner (2006: 2) states that 'Cartography enabled and recorded exploration and discovery for ages.' Modern cartographic methods go beyond drawing traditional maps, helping us to find patterns in ever growing amounts of data and developing new concepts of displaying findings: these are the characteristics of the new exploration of the world (Hennig 2013). The conclusion, therefore, can only be this: Draw more maps!

Visit www.viewsoftheworld.net for some practical examples of geographic data visualizations.

texts and databases. Certainly Benjamin Hennig (Box 15.1) of the WorldMapper group would advocate for greater use of maps as a means of representing developmental and other data in an accessible and compelling way, as is clear from his reflection on his involvement with the world mapper project (see Chapter 13 for extensive guidance on spatial data and producing maps). However, sometimes they do none of those things and we then need to ask whether the distortions, abstractions and the manipulation of data presented through a visualization are of sufficient use to justify those movements away from primary sources and data.

After all of these considerations, how do we actually make visualizations? In the following section, we will review *inputs*, *tools* and *outputs*.

Inputs

In many cases, you might want to create visualizations from primary data that you collect yourself (e.g. interviews that you code, locations that you visited or surveys that you conducted). In that case, it is important to remember that your data may not necessarily begin as nicely categorized numbers in spreadsheets. It is possible to design visualizations from both qualitative and quantitative data, as well as information collected at nominal, ordinal, interval and ratio scales of measurement.

Nonetheless, it is important to organize your data in ways that are both explicit and clear. For instance, it is easy to succumb to the temptation to rely on default file names and default variable names that leave you with a folder full of files that are labelled with names such as 'export_04.csv' and 'export_05.csv', and variable names like 'var1' and 'merge_08.' You may be intimately familiar with your data while creating those files, but it is not uncommon that you may want to return to your data and visualization one

Box 15.2 Online open data sources

The internet has facilitated mass storage of, and accessibility to, large datasets that are of interest to development researchers. Below we list a few online data sources that you might find useful, while recognizing that due to the rapidly changing nature of the internet this list may become rapidly outdated. However, we hope that it also has usefulness as an indicative list that illustrates the types of information that you may be able to relatively easily integrate into your own work:

• *AidData*   http://www.aiddata.org
  – a broad collection of data about development aid and the impacts of that aid.

• *Data.gov*   http://www.data.gov/opendatasites
  – a list of open data created by public agencies in 43 countries, and many more regions and cities.

• *Datasift*   http://datasift.com
  – provides access to a range of social media datasets unavailable on other platforms. However, unlike the other links presented here, they charge for access to most of their data.

• *Datacatalogs.org*   http://datacatalogs.org
  – performs a similar function to Data.gov, but is curated by a small group of open data experts and links to a significantly different set of data.

• *Google Public Data*   http://www.google.com/publicdata
  – contains a broad collection of international datasets that are presented through an interface that has useful search and visualization capabilities.

• *Guardian Data Store*   http://www.theguardian.com/data
  – an eclectic and disorganized collection of data, but can occasionally make available important development-related data not available on other large platforms.

• *OpenStreetMap*   http://www.openstreetmap.org
  – primarily a store of geographic data, but can be a fantastic resource if you need to export and use information about the locale(s) in which you are doing research.

• *UN Data*   http://data.un.org
  – access to many of the United Nations' store of statistical datasets.

• *World Bank Data*   http://data.worldbank.org
  – free access to a range of World Bank data.

• *World Health Organization*   http://www.who.int/research/en
  Data and Statistics
  – open data on global health.

day. In such a situation, you would want to know exactly what all of your data point to and represent. This is especially the case if you ever plan to share your work or upload your data into any sort of repository (a requirement that is increasingly becoming mandated by funding agencies and science councils around the world).

It is therefore good practice to always create *metadata*. Metadata literally means 'data about data'. That is, you would want to create an associated store of information that describes where your data came from, what they are, how they were created, when they were created and any other bits of information that may be useful for understanding exactly what your data represent. Metadata are especially useful in cases where you plan to share your data with partners and collaborators in order to avoid misunderstandings and faulty inferences.

Finally, it is often the case that you may want to visualize data from existing collections. Only a few years ago, it was extremely difficult to get freely available information that might be of interest to scholars and practitioners of development. But in the last few years, there has been a powerful and influential 'open data' movement that has transformed the types of data that can be accessed around the world (see Box 15.2 for suggestions of some sources of these data). Unfortunately, low-income countries are the very parts of the world that have traditionally been most characterized by poor availability of data.

But it remains possible that the rapid move to encourage open data in all aspects of development policy and practice might go some way to changing those informational imbalances. As such, we would encourage you to attempt a thorough internet search for data that you are interested in analyzing and visualizing before looking for alternatives or contemplating collecting the data yourself.

Tools

Just as the availability of data has exploded in the last few years, so too have the tools through which visualizations can be created. The list of proprietary, open-source, locally installed and online tools that are available is now far too extensive to list here. However, the remainder of this chapter will provide you with an important overview of key available capabilities as well as the core skills required to use them. (Chapter 13 contains a more detailed overview of tools to visualize and map spatial data.)

General purpose spreadsheet software that is available on a majority of computers (such as Microsoft's *Excel*, *Calc* in OpenOffice and Google's *Spreadsheets*) can offer surprisingly useful capabilities for creating straightforward charts, graphs, tables and histograms. These tools are designed to have a very gentle learning curve in order to maximize their user numbers. Unfortunately, the default settings in such software tend to produce hideous and confusing outputs. But with even a little effort to customize outputs, powerful, rigorous and effective visualizations can be made.

A range of more specialist software exists to cater for users that have familiarity with particular methods and data types. *SPSS* and *MATLAB* for statistics, *Gephi and Pajek* for networks, *NVivo* and *ATLAS.ti* for qualitative data and *ArcGIS* and *Quantum GIS* for geographic data. (Note that these are only illustrative examples. There are hundreds of other programs and tools that could have been mentioned.) These tools tend to have steep learning curves, but, as a result, allow a much greater degree of control over the visualizations that are created.

For those familiar with computer programming, languages such as *JavaScript*, *Processing* and *R* offer the potential of creating particularly powerful and bespoke solutions to visualization-related problems. Some JavaScript libraries that offer particularly useful solutions are *D3.js* (which offers a high degree of control over visual outputs), *Leaflet* (which is useful for creating interactive maps) and *Sigma* (for creating network graphs).

There are also a range of web-based tools that can be helpful in creating visualizations. Some tools offer only the ability to generate on-the-fly graphics and have limited or no ability to export those visualizations (for instance, the hugely popular Gapminder website). Other online visualization tools, in contrast, are starting to afford a much greater degree of customizability. A few online visualizations tools that you may want to explore include:

- *CartoDB* (www.cartdb.com) – a cloud computing platform that offers web mapping tools.
- *Geocommons* (http://geocommons.com) – a platform and data store focused on spatial data.
- *Google Fusion Tables* (http://tables.googlelabs.com) – tools that are especially useful for quickly visualizing large datasets. These tools are also designed to be useful in contexts that have multiple users.
- *Tableau* (http://www.tableausoftware.com) – multipurpose tools for visualizing a range of types of data.

Finally, it is often useful to apply a certain amount of post-processing to any graphic that you create and export from a visualization tool. This can be done either for aesthetic reasons or to add additional elements (such as descriptions, title or other design elements) that could not be created in the source tool. A few of the most commonly used graphics software tools include: *Adobe Illustrator* (for editing vector graphics), *Adobe Photoshop* (for editing raster graphics), GIMP (a free and open source raster graphics editor) and Inkscape (open source and free software for vector graphics editing).

Outputs

When creating visualizations you will need to make some crucial decisions early in the design process. First, is the decision about whether you are creating a *raster* or a *vector* graphic.

Raster graphics are comprised of thousands (or millions) of pixels (tiny squares), each one of which is assigned a specific colour. The benefits of using raster graphics in visualizations are that you can make very precise edits and capture very fine details (when using high resolutions). The key downsides of raster graphics are that they tend to create large file sizes (it is not unusual for raster files to take up hundreds of megabytes or more) and that they can become blurry when enlarged (this is because there are always only a finite number of pixels that make up your graphic). Common raster file formats include *.tiff*, *.jpg*, *.gif.* and *.png*.

Vector graphics, in contrast, work very differently. They employ points, lines, polygons and the mathematical relationships between those elements to create visual images. This allows them to become infinitely scalable. The computer does not have to guess what colours to use when you enlarge an image and can simply calculate exactly how to scale an image from the point, lines and polygons that it is made from. By not having to save a value for every pixel in an image, they also tend to have much smaller file sizes than raster graphics. Common vector file types include *.ai*, *.eps*, *.svg* and *.pdf.*

As a rule of thumb, if you want to create a visualization for use on the web, you would create it in one of the most common raster formats (although modern web browsers now also provide support for .svg images). But if creating a visualization for any sort of printed output (especially if it is to be printed in large format), it is likely that you would want to employ vector graphics. These are decisions that are useful to consider before starting work on your graphic, as they inevitably impact what you can do with it.

Another core decision to be made has arisen with the increasing sophistication of visualization tools: namely, whether to create static, dynamic, animated or interactive graphics. Many of the tools described earlier in this section allow the outputting of non-static graphics that you can easily embed into websites. A central benefit of using interactive or dynamic visualizations is that you can allow users to have a much greater degree of control over the information that you present to them. As a result, the number of messages and stories that your visualization can convey are multiplied and far less limited than employing a static graphic. However, this very interactivity can also have problematic effects. By ceding control over some dimensions of your visualization, the potentials for misinterpretations of your data increase. This is a trade-off that you might have to make, but one that might also be ameliorated by offering careful descriptions and metadata.

When thinking about how to design and structure the type of visualization that you decide on, there are some important considerations to bear in mind. We create visualization to guide viewers to important information or messages. By explicitly thinking about this process, we can better guide the process of choosing what to include and make visible and what to exclude in any particular visualization. How do we evaluate data visualizations? Are there good and bad ones? If so, what are some of the principles underpinning good examples? There are no universally accepted ways of effectively visually conveying the most important messages, but there are some general guidelines that are helpful in telling effective stories with visualizations:

Context: Be clear about whether the visualization should stand alone and convey meaning without any surrounding (con)text. Although some of the most powerful graphics speak for themselves, data visualizations do not always have to stand alone. They can also be used to effectively draw out or bring attention to particular points in a broader discussion.

Attention: Highlight the most important features. This can be done with size, colour, orientation, depth of field and other visual variables.

Simplicity of message: Concomitantly reduce clutter and superfluous detail. Most visualizations can only really tell one or two stories.

Simplicity of design: Visualizations are rarely effective if they contain more than a handful of fonts, colours, sizes and other elements.

Accuracy: Visualizations are never 'truth documents', but they can sometimes actively mislead. For instance, if using geographic area in a map to represent relative quantities of a variable, then ensure that the areas you use do actually correspond to differences in the underlying data.

Clarity: Clearly label all meaningful elements. Make sure that users of your visualization can understand what your axes are showing, what your nodes and links represent, etc.

Do something different: While this point may seem to contradict some of the above guidelines, you can really capture people's attention, and therefore enable your work or message to have more impact, by avoiding some common (and ugly) graphics like pie charts (we realize that we're not following our own advice by using Figure 15.2 as an example earlier in the chapter). In other words, instead of relying on software defaults, think about why you are using particular colours, shapes and layouts and recognize the infinite ways of graphically representing information instead of relying on software defaults.

Also useful is to consider how you might want to strike a balance between explicit and implicit guidance in your visualization:

Explicit guidance can include statements like 'look at the blue squares'. This can be effective because it leaves little room for ambiguity about a message. However, in some ways it defeats the purpose of a visualization. Your graphics are no longer able to speak for themselves.

Implicit guidance, in contrast, is built into the visualization itself. An example of this could be the way that bright red dots (high values) would stand out in a canvas of dark grey dots (low values). The implicit guidance here is to look at the red dots. This implicit guidance therefore only works if the differences between the target object (red dots) and other graphical elements are significant (people would be less drawn to bright red dots if all other dots were another shade of red). Also important is that the differences between other distracting elements are minimized.

What to take from this chapter

Visualizations are tools for telling stories and revealing messages and the key to making effective ones is to be clear about the stories that you want to tell. The creation of any visualization is necessarily an act of signification. Visualizations are a bridge between the data and an audience. This does not mean that your data can speak for themselves or offer anything other than selective representations of the people, places, patterns, processes that they represent. Your visualization can be used to both convey those selective messages in a concise way and to highlight all that has been left out.

This chapter began with two stories that demonstrated the discursive power of visualizations. By clearly and concisely illustrating important messages, they are able to both convey and convince. It then moved to a discussion of why we create visualizations, and some of the key considerations to heed when doing so. The chapter then proceeded to review how inputs can be used, some of the most important tools that can be employed and the range of outputs that can be created.

The chapter has also alerted you to some of the central problems that are entangled into any efforts to visualize data. Things that are not easily countable and measurable tend not be stored in datasets that can ultimately be visualized. Unfortunately, it is those uncountable data that are often needed for the most important and interesting research questions. This seems to be especially the case in development studies, where we work in sometimes data-poor environments. Where data do exist, there tend to be many constraints of both tools employed and the data themselves. Furthermore, because of the technical skill that can sometimes be required to move away from defaults and templates, we can see disconnects between visualizations and the messages, questions and interventions that they are supposed to be addressing. Creators of visualization are just as responsible for addressing the silences, gaps, and omissions in their work as they are for creating compelling and insightful narratives.

There are books, courses, and degrees devoted to visualization research and practice and this chapter could necessarily only offer a summary. But, by providing an overview and a summary of visualization practices and pitfalls, we hope that this chapter has offered a useful starting point for anyone contemplating using visualizations as a component of their research.

If you want to explore the ideas covered in this chapter in more detail, you may find these further readings useful:

Bostock, M. (2013) *D3 Gallery.* https://github.com/mbostock/d3/wiki/.

Jackson, D. and Simpson, R. (2012) *D_City: Digital Earth - Virtual Nations - Data Cities.* Canberra: National Library of Australia.

Tufte, E. (1990) *Envisioning Information.* Cheshire, CT: Graphics Press.

Tufte, E. (2001) *The Visual Display of Quantitative Information* (2nd ed.). Cheshire, CT: Graphics Press.

Ware, C. (2008) *Visual Thinking: For Design.* Burlington, MA: Morgan Kaufmann Publishers.

Yau, N. (2011) *Visualize This: The Flowing Data Guide to Design, Visualization, and Statistics.* Indianapolis, IN: John Wiley.

16

WRITING FOR DIFFERENT AUDIENCES AND KNOWLEDGE EXCHANGE

Communicating your research findings is an essential part of the research process: even if you have the most groundbreaking and insightful research findings in the world, they will be meaningless if no one ever engages with them. Thus, one key message underpins this final chapter: unwritten findings will be read by no one, poorly written results may be read by a few, but accessible and well-written findings can be read and understood by the widest audience.

For many student research projects, your audience may be incredibly limited (your supervisor/course leader, the staff marking your work and perhaps your parents), but even with this limited audience, it is still important to communicate your findings efficiently and effectively. At the most basic level, good communication of research findings in assessed work can improve your marks! For some pieces of student work, most commonly fieldclass projects and dissertations/theses, you may be able to identify and engage with a more diverse set of audiences. Development-based fieldclass findings and dissertations may be of interest to host communities or organizations with whom the research was conducted, local or international NGOs working in a related field and, in some cases, perhaps media outlets (including student media and departmental newsletters).

If you are going to reach these other audiences then you need to write up and present your findings in different ways in order to engage various groups of readers by making your findings accessible to them. The same set of practices apply for postgraduate students, postdoctoral researchers and academic staff: for your research to engage a variety of audiences, you need to present it in different ways to appeal to audiences with different levels of background knowledge and understanding and varying interests.

This chapter provides insights and guidance into communicating your research findings to a range of audiences. We begin by exploring the importance of engaging with different audiences before considering how to do this. To this end, we provide some pointers on common features of good writing and then address some specific strategies and techniques used in communicating with different audiences. We address components of writing for an academic audience, both in terms of assessed work and for publication, communicating with policy and practitioner audiences, engaging the press and media, as well as reflecting on some of the possibilities afforded by digital media. We then bring the chapter to a close by considering a broader set of questions relating to knowledge exchange and transferable skills linked to experiences and practices of research and fieldwork in development.

Why engage with different audiences?

Communicating research findings to diverse audiences is an important process and one that has received varying levels of attention over time. In South Africa, the role of academics in challenging oppression, promoting social justice and contributing to transitional nation building has entrenched engagement and dialogue with non-academic stakeholders in professional practice since the 1980s to the present (see, for instance, Parnell 1997; Oldfield et al. 2004; Connell 2007; Mather 2007). Meanwhile, British scholars have engaged in an uneven manner with debates around the role of academia in public life, the possibilities for activist academics and the nature and purpose of knowledge production, while in the US, ventures such as the Public Political Ecology Lab at the University of Arizona (ppel.arizona.edu) have contributed to efforts to convey academic research findings to the broader public, act as a contact hub for interested parties and to provide methods and media training to academics to enhance the accessibility and impact of their work.

At the time of writing, the 'impact agenda' is providing political and economic impetus to British academics' efforts to ensure their research is increasingly relevant to and used by policy, practitioner and community audiences. While the manifestation of this impetus is subject to extensive debate and critique of an emphasis on economic outcomes, a broader view of this agenda emphasizes the importance of ensuring a wider societal engagement with and impact of your research to influence policy, support social 'progress' and enhance community development and education (see Pain et al. 2011; Slater 2012). Gordon Dabinett's reflection (Box 16.1) on the role of the recently established Research Exchange for the Social Sciences at the University of Sheffield reiterates the growing importance of ensuring academic work has relevance and merit beyond the academy and indicates how the activities of bodies such as RESS can inspire your own actions.

Such debates are hardly new to development scholars, many of whom cite a sense of moral obligation or desire for their work to make a difference to the lives of those they engage with through their research as driving forces behind their work. The growing influence of post-colonial thinking on development research serves to further entrench these efforts (see Chapter 2). Against both this disciplinary backdrop and broader shifts in the academic landscape, it is likely that you will want and/or need to try to engage multiple audiences with your research. In order to do this, you will need to deploy various skills and techniques aimed at increasing the engagement of different audiences with your work.

To this end, you will need to think about the potential audience types for your research and the characteristics of these groups and the ways in which they might engage with your research. Once you have identified these different potential audiences, you need to think about why they might have an interest in your findings and then how you will frame and present these findings to them. Think for a moment about the different types of audience you may be trying to engage: academics, students, printed press, broadcast media, policy makers, politicians, opinion makers, government, non-governmental organizations and civil society, businesses and the general public. You cannot hope to engage all of these groups in the same way; you must tailor your outputs to the relevant audience.

Box 16.1 Gordon Dabinett: Reflections on the purpose and practice of research exchange

Purpose

The Research Exchange for the Social Sciences (RESS) was set up at the University of Sheffield in 2012 in an attempt to address two major questions that will resonate with you as a development researcher:

- How can academic research make a larger contribution to addressing major societal problems and issues?
- How can users of research have greater influence over what is researched and how that research can be made more relevant to their daily lives and decision making?

As one of five 'faculty gateways' funded by the Higher Education Innovation Fund (HEIF) to promote impact, innovation and knowledge exchange activities in the university, RESS has sought to widen, deepen and scale up the impact of social science research.

The recent increase in attention devoted to impact in UK higher education sets the wider context for this initiative. A major impetus has been generated by the inclusion of impact as a measure within the UK Research Excellence Framework (REF2014), one mechanism used to allocate state research funds to universities. More immediate and recurring demands are shaped by the expectations of the key research councils that require research proposals to outline not only how research will be disseminated, but also how funded research will lead to impacts. These pressures and expectations are not confined to faculty-level organizations but will affect you – shaping your professional practices and efforts to secure research funding.

Practice

In response to this changing landscape, RESS has set out to support and enable impact activities across the 13 departments and schools that constitute social science in the university. This results in a rich but challenging mix of inter-disciplinary research, but also a wide diversity of potential beneficiaries and partners in the public, private and voluntary sectors, at local, regional, national and international scales. Our ability, vision and willingness to draw together diverse stakeholders including national government departments and social enterprises can offer you inspiration in thinking about how to build your own networks and collaborations to give your research the strongest impact possible.

Similarly, the four key areas for RESS activities can be seen as mirrors for developing your own toolkit for impact. Our work focuses on: (i) reactive responses to internal and external enquiries; (ii) routine support of impact activities, such as public engagement events, raising awareness of funding calls, undertaking impact training for doctoral students and assisting the writing of collaborative research bids; (iii) proactive activities through the development of specific projects involving long-term relationships and outcomes, and building new ways of working with selected strategic research partners; and (iv) the promotion of innovation, through the testing of models and best practices, with particular emphasis on exploring methodologies for 'co-produced research'. Each of these areas of work is also a means through which you can seek to develop your own dissemination and impact practices.

The key lessons from our experience that can help you do this would be:

- Creating successful impact can involve high transaction costs (such as meetings with partners, presentations with collaborators and engagement with communities) that go beyond the usual costs associated with research design, implementation and reporting. These 'transactions' are essential in building mutual trust between partners over time but are not always fully recognized.
- Better research does not necessarily lead to better impacts. Securing greater impact is therefore a long-term goal and requires specific practices.

In approaching these concerns, consider if you are writing to people with no knowledge about the topic and/or context of the research, for students with a basic engagement with the work, for the community in which you conducted fieldwork who intimately know the context of the research or for a media outlet or for a government policy advisor. Each audience will have a different level of background knowledge, a different set of priorities and interests, differing amounts of time and energy to engage with your findings, be interested in different aspects of the findings and with a view to using them in very different ways.

When designing a research communication strategy, you must remember to honour the promises you have made – whether to funding bodies, gatekeepers or local hosts – regarding the sharing of research findings. Failure to do so reflects badly on your own professionalism as well as contributing to collective views towards academics and other researchers who may wish to work with those organizations or communities in the future. In short, if you have pledged to deliver a research report or policy brief to someone, then you must do this and it can be beneficial to plan ahead and think about the best way to present your findings (written, visual, verbal) and best timing for this (sooner rather than later, but perhaps not at a time of particular stress or importance for the community or organization). Underpinning all of these concerns, however, is a need to develop appropriate materials that are well written.

Common features of good writing

Good writing is time consuming and hard work, but investing time in developing good writing practices and techniques can make a significant difference to the accessibility and impact of your research. In this part of the chapter, we briefly consider a number of aspects of the writing process and specific practices that can help you get your message across in a more succinct, powerful and accessible manner. Before we turn to those techniques, it is useful to think briefly about writing as a process and the barriers that can develop to this.

Many students and academics suffer from writer's block in some form during the research process, perhaps due to a sense of being overwhelmed by the task at hand or a lack of confidence in their ability to produce a document of the expected quality. This can often manifest itself in an inability to begin or recommence writing, displacement of writing through procrastination and by finding other tasks to accomplish, or the continued and over-detailed revision of a small section of writing. While we cannot give you the flash of inspiration or crucial insight that will get you through this, we hope that the suggestions and pointers below will help your writing process.

Writing is very often an individual process and one we each approach in our own way. It is important, therefore, to find a space in which you feel comfortable writing and in which you can turn off your emails, the internet and other distractions and focus on your writing. It is often helpful to develop a routine that meets your productivity cycle – this might be to write little and often, to write first thing in the morning or in the late afternoon: whatever works for you. As well as developing a routine, it can be useful to break down the writing project into specific components (a book into chapters and then into subsections; an essay into sections and then into paragraphs) and tasks linked to each component (finding sources on X, consolidating your notes on Q) and develop a plan and timetable to tackle these discrete activities.

Once you start writing, there are a number of techniques that can increase the accessibility and readability of your writing. If you are wondering about the importance or relevance of this section to a book on development research, take a moment to think about the last ten academic journal papers you read. How many of them were easy to follow? How many included sentences or paragraphs that were confusing or failed to add to the argument being made? How many used overly complicated terms and language to make a simple point? In short, academic writing has a reputation for being complex, overly wordy, confusing and – at times – exclusionary and deliberately obtuse. Indeed, at times it seems that academics think that their work can only be meaningful and profound if no one else can understand it. Surely the inverse is true, that the hallmark of good academic writing is to make complex ideas and compelling arguments in a way that is accessible and engaging to as many people as possible.

Clarity in writing

So, how do you set about doing this? How can you make your writing elegant and accessible? A number of common techniques are advocated in many of the available style and writing guides. Try to be open yet assertive in your writing. Good writing will include a variety of sentence lengths and types to build a clear narrative and argument. However, do try to break long, complex sentences into multiple, shorter sentences with fewer subordinates.

Another way of helping the reader follow your argument is to remove unnecessary and redundant words (this also helps you meet word count limits on your essays and articles!). Removing commonly used redundant words (e.g. can, therefore, thus, often, not, frequent) and phrases (e.g. replacing 'The question as to whether' with 'Whether' or 'This is a policy which' with 'This policy') will make your writing far more reader friendly.

You should also be positive in your writing and avoid non-committal and hesitant terms (for example, rewriting the statement 'Basic income grant payments did not often arrive on time' as 'Basic income grant payments were usually late' reduces the number of words and changes the text from being *passive* to being *active*). Using an *active voice* in your writing produces a more assertive, direct essay. This approach to writing means that the subject of the sentence performs the action of the verb, the benefits of which can be seen in this example:

Passive = Spending on education was decreased as a result of the structural adjustment policy.
Active = The structural adjustment policy resulted in a decrease in education spending.

The active voice makes this statement clear and concise, improving readability and cutting the word count. Passive writing is generally longer winded and tends to confuse and obscure the meaning and message being presented. Using an active voice and being clear and definite in your writing (for instance, changing 'The policy had a number of unexpected negative outcomes' to 'The policy was flawed') will make your writing more concise, compelling and accessible.

An interesting challenge can be to take a section of your writing and aim to cut the word length in half without losing the core argument and meaning. In many cases, it is possible to do this by removing redundant words, emphasizing the active voice and ensuring an explicit focus on the core argument. While doing this, make sure you pay attention to the correct use of punctuation and ensure that your sentences are grammatically correct. There are many more tips and techniques than we can fit into this chapter but some useful further resources are noted in Box 16.2.

Planning and writing

Following the advice above will help make your writing accessible and engaging but these efforts amount to little if your composition and structure are poor. One of the most important parts of the writing process is planning your work. This includes identifying your core argument, the materials needed to make this argument and the structure of your argument. The first step is to consider what it is you want to say in a particular piece of writing and to ensure that the argument is substantive yet focused. Once you have a coherent idea of your overarching argument you should identify the 'red thread(s)' of your argument. These are the key messages that you make throughout the paper to support your argument. Identifying these will help you ensure that the narrative, from introduction to conclusion, is developed around your core idea(s).

Box 16.2 Accessible writing

There are many other tips and suggestions that could be made to assist in making your writing more accessible. We do not have the space for them here, but recommend that you look at some of the following sources for further advice and guidance:

The American government provides a useful hub of resources for promoting the use of plain language to enhance communication, as well as humorous examples underling why good writing is important: http://www.plainlanguage.gov/howto/index.cfm. Another useful hub for information, examples and guidance on plain language and accessible writing is provided by the Plain Language Network: http://www.plainlanguagenetwork.org/.

You may also find Helen Sword's website 'The Writer's Diet' an interesting and useful resource, in particular the 'Writer's Diet Test' tool which provides a basic analysis of your writing style and areas that might benefit from attention: writersdiet.com. Her book *Stylish Academic Writing* is also particularly useful (Sword, H. (2012) *Stylish Academic Writing*. Cambridge, MA: Harvard University Press).

Further resources of use include:

Gibaldi, J. (2003) *MLA Handbook for Writers of Research Papers* (6th ed.). New York: MLA.

Peck, J. and Coyle, M. (1999) *The Student's Guide to Writing: Grammar, Punctuation and Spelling*. Basingstoke: Macmillan.

Strunk, W. (2006) *The Elements of Style*. New York: Dover Publications.

Truss, L. (2003) *Eats, Shoots and Leaves: The Zero Tolerance Approach To Punctuation*. London: Profile Books.

You can then think about the flow of your piece, what (sub)sections you want in the document and how they act as building blocks to the construction of your argument while guiding the reader through the logic of your discussion. By this stage, you can begin to think about individual paragraphs, their content and how this contributes to the development of the overall argument. Once you have done this, you can then turn to the writing itself. It is important to remember that writing is an iterative process and you should recognize that the first version of the document will benefit from editing and redrafting. During this process, it can be helpful to get feedback and comments from colleagues in order to gain constructive advice on content, structure and flow at an early stage. This collegial 'peer reviewing' is useful both in terms of developing and strengthening your writing and as a gentle precursor to experiences of peer review when submitting work for academic publication.

Box 16.3 Daniel Brockington: Learning to co-produce knowledge

My academic career has been blessed by several long periods of fieldwork. Even better, three of those periods have involved long stays in rural Tanzania. The first, for my PhD, was almost two years long (1995–6). The second entailed 14 months in a remote corner of southern Tanzania, in the Rukwa valley in 1999–2000. The final Tanzanian stay was more recent, a sabbatical year in 2012–3, in Hanang District, where my wife was born. The fourth was less of a traditional fieldtrip, but it involved lots of time interviewing professionals from different sectors in the UK involved in constructing celebrity advocacy.

There is a sharp contrast between the timetable-driven and sometimes extractive work of my first trip, with the gentler pace, if not more arduous work, of the second. My PhD, in biological anthropology, entailed repeat-round surveys of households in different villages throughout the year. Then my sample sizes for the household surveys were just a fraction too large. I was anxious lest the schedule slip and that made me nervous and irritable. It also made me a rather poor researcher. I may well have collected a good many data but I did not really understand properly the much richer social context of the numbers.

It was interacting with Jim Igoe that changed all that – albeit slowly and gradually. Jim was a social anthropologist from Boston and had a different approach to timetables from mine. Writing together was an interesting experience from which we both learnt a great deal. But talking about fieldwork was more of a one-sided revelation.

It was as a result of that learning that my postdoctoral research in Rukwa was so different. There was much more participation observation, much more chance to try to understand lifeworlds around me. I was helped here by the relative isolation. Without mobile phones or email, keeping in touch meant letter writing and a six- to eight-hour walk up the escarpment every month or so to pick up mail, talk to government officials and phone home. It was not quite Papua New Guinea, but it allowed me to dwell in a place and for people to get used to me.

And this meant that I began to perceive how and why people might talk back and alter my research and its agenda. In Rukwa, when various youth representatives from opposition parties came to find out what I had concluded, tell me my thoughts on local institutional failure were obvious and ask me to conclude that they needed more farm implements, well, that in itself was interesting. Perhaps the best thing about fieldwork change in recent years has been the extent to which I have come to realize that research can be co-produced. It is, to some extent, negotiated and debated. This is what has made my fourth fieldwork episode, in the offices of development NGOs, media professionals and around the celebrity industries so interesting. For these settings are all about negotiating with stakeholders who can take a keen interest in your findings.

Collaborative writing

We often view writing as a lone experience, from undergraduate essays to sole-authored academic papers. However, writing with others can be a productive practice. It can also be frustrating and challenging. In thinking about writing collaboratively, you need to consider why you want to do this, what audience(s) you are writing for and whether the aims/agenda of those involved in the writing are the same. Within development research collaborative writing with non-academic partners can be viewed as an important way of ensuring findings reach non-academic audiences. This practice can also be seen as a mechanism for capacity development and 'giving back' to a host community or organization. Daniel Brockington's reflections (Box 16.3) on his changing writing styles and engagement illustrates this well, reminding us that co-production of research findings can be beneficial to multiple parties.

Whether you are writing with an academic colleague or non-academic partner, you will need to manage the dynamics of this relationship. There is no single blueprint on best practice for collaborative writing but a few tips may be useful. Ensure that all parties understand what it is they are agreeing to do (in terms of the audience being addressed, the type of document being written, the deadline to which this needs to be done, what their expected contribution will be) and that they have the required resources to do this. Work collectively to plan the document – its focus, flow and content – and identify which tasks and aspects of the paper will be the responsibility of different authors. From this it can be useful to develop a rough timetable for writing and to plan how you will share ideas and content at different stages of the process. This is useful to prevent authors heading into different areas or moving away from the agreed focus and narrative and also as a means of ensuring contributors feel that they are working collaboratively. From our experience, clarity in planning and communication throughout the writing process are vital to ensuring a productive and (as far as possible) enjoyable experience.

Writing for academic audiences

In one guise or another, it is likely that you will have to present your research findings to an academic audience. This might be in the form of a course essay, a dissertation or an academic journal article. While there are specific differences to each of these document types, there are a surprising number of commonalities in required content and expected structure.

Writing for an academic audience means you will need to include more theoretical and technical information than if you were writing for the general public. This requirement brings with it a set of stylistic expectations that, at their worst, produce dry, turgid prose full of complex terminology and obfuscated arguments. In other words, text that is difficult and dull to read (as evidenced, deliberately, in the previous sentence). There is no need for academic prose to be like this and we would echo Sword's (2012) call for you – as an author – to try and inject some style into your writing and to follow some of the tips outlined earlier and those in her book, *Stylish Academic Writing*.

From the outset, you will have some guidance and instructions that frame your writing. Whether you are writing a term paper or a journal article, you will have a word limit or guide. Knowing your word limit from the start is important – it helps you identify how much material you can cover in a paper and inform how you plan your work. As well as a word limit, many modules and most (if not all) journals provide you with a style guide (for some modules with no individual instructions, this guidance may be located in an overarching course handbook or assessment regulations). These guides may include details on formatting, use of (sub)titles, spelling, presentation of dates and numbers and how to present quotes or other data. It is likely that you will also be given guidance on the referencing style and it will save you a lot of time and energy to closely adhere to this guidance from the outset. A poorly formatted reference list is often seen as a sign of a rushed and poorly executed piece of writing, leaving the marker or reader with a negative sense of your work.

Telling a story: Generic structure and content

This next section provides a basic overview of a generic structure to a piece of writing. Not all components are needed in all pieces of academic writing, but the general principle remains intact: your structure and writing should provide a coherent narrative. Ideally, your aim is to tell a complex story in a simple and accessible manner. To do this, there are various structural building blocks you can use (outlined below) in conjunction with the tips presented above (and in the additional resources noted) to produce an accessible and engaging story.

Title

Your title should give some sense of what your writing is about. It may be that a particular quote or phrase is used as a hook to attract the reader's attention.

Abstract

A very brief section, usually only a few hundred words, that outlines the core argument and findings of your work. It may also include a brief contextual detail and comment on methodology. The main purpose is to give a potential reader a very brief overview of your work and, in doing that, to encourage them to read the whole paper.

Keywords and research highlights

Keywords and research highlights are used to help search engines and readers to find articles on particular topics and with particular findings. You should spend some time thinking about the best way of using these to attract the attention of those searching for papers by place, topic, theory, method and concept.

Acknowledgements

This section should acknowledge any financial support for the research, as well as those who have helped facilitate the research – including dissertation supervisors. If your work has been reviewed or commented on, you would also acknowledge this support here.

Introduction

Your introduction provides an outline to the reader of what your research addresses, the context for the work and indicates the key objectives of your research. It is also useful if your introduction signposts the structure of the document, providing the reader with a sense of the narrative that flows through the work and highlighting the core argument for them. There are various ways of crafting the introduction, you may want to use a short anecdote or example to set the scene or highlight the importance of the topic or you may want to locate the importance of the research in the context of some overarching debate or data.

Literature review

A literature review provides the basis for your work. While your case study or specific research questions or methodological approach may be new and innovative, there will be existing literature of relevance through which you need to position your study. A good literature review will not simply provide a shopping list or impartial summary of existing materials. Rather, it will synthesize this existing material and begin to develop the argument and narrative that is then carried through in your findings and discussion.

Methodology

The methodology outlines how you went about doing your research. It should provide enough detail to demonstrate the rigour and validity of your research and indicate to the reader how they might be able to repeat the work. This section is generally a factual outline of what you did but you may include some short reflexive comments about the experience of doing the research and how you mitigated any challenges or problems.

Findings and discussion

Depending on the style guide provided by your department or the journal to which you intend to submit your work, you may find that your results and discussion should be presented in separate sections. In this scenario, your results section reports the factual findings of your research and then the discussion section provides discussion and analysis of these and develops your key findings. In a situation where your findings and discussion are integrated into the same section, you can develop the discussion and analysis alongside your results, weaving these together as you draw out your key findings and main argument.

Conclusions

Your conclusions should bring together and highlight the key findings and overall argument of the paper. This is also a point at which you can point towards further questions or possibilities for additional research to develop the work further.

Reviews and feedback

It is common to get feedback on your academic writing. For course essays, this may be through formative or summative feedback after you have submitted the paper. For dissertations, this can be through comments on work in progress. For journal articles and book manuscripts, this is generally through peer review comments prior to publication. Regardless of the situation critical comments on a paper can be hard to take: all three authors of this book can attest to the range of emotions elicited by receiving feedback on written work.

When you have spent time crafting a piece of writing, you can become quite protective towards it. Therefore, when a reviewer is critical of your writing you may go through an initial sense of anger and hostility ('how can they say that, they clearly have not understood the paper', 'that is not fair') as well as annoyance and perhaps a sense of resignation that your work is not good enough. From experience, we would suggest that once you have had a first read of the reviews, it is good practice to leave the reviews and the submitted work to one side for a few days. Doing this lets the dust settle and your emotional reaction to fade. When you then return to the comments, it is usually with greater clarity and calmness, allowing you to engage with them more thoughtfully and constructively.

At this stage, it is useful to identify and differentiate the key and peripheral or basic critiques made of your work. You can then begin to formulate a response to them, determining which ones need addressing and which ones are less important or perhaps cannot be addressed at all. It is also useful to reflect on the types of comment and critique made of your different pieces of work over time. You may be able to identify common trends or themes across these that can be used to adapt and improve your written work in general.

Engaging policy makers and communities

A common audience for development researchers is policy makers and practitioners: readers who are unlikely to wade through a long academic paper. In order to engage these audiences, you need to produce short documents (often between one and four pages) that capture the key aspects, findings and arguments of your research in an accessible but robust manner. A policy brief must be tailored to the target audience, providing the non-specialist reader with a favoured policy option and the rationale for this in a document that can be read and understood within a few minutes. A useful way to conceptualize this imperative is to imagine that you are trying to get the attention and convince a politician of the veracity and effectiveness of your policy option in the time it takes them to walk between meetings: you have less than five minutes to get your message across and to be convincing and compelling in doing this.

In constructing a policy brief, it is useful to look at existing examples and templates (see Box 16.4 for links) but we can identify a number of common features to effective policy briefs. You need to include an executive summary (a short summary of the purpose and recommendations of the brief), a statement of the issue being addressed, brief background information (key need-to-know material for the decision maker), what the existing policies (if any) are, new/further policy options and the comparative merits and weaknesses of each and your recommendation. You should also indicate on behalf of which organization you are producing the brief and any other vested interests. You should make sure that your key recommendations and rationale for them are provided succinctly in a visible place on the first page of the policy brief.

In order to gain the reader's attention and to communicate ideas simply and quickly you may also want to include images, maps or diagrams. The adage that a picture tells a thousand words is pertinent here – some well-selected and appropriate photos can draw the reader's eye to the document. The use of diagrams, flow or organizational charts, data tables and charts and other visual representation of data or findings can be useful tools for communicating findings, relationships and recommendations succinctly. You may find it useful to look at a range of policy briefs produced by other researchers, academics and NGOs and to think critically about what you think works well in the different documents produced.

In addition to producing findings in a format designed for policy makers and civil society organizations, development researchers often want to provide a version of their findings back to their host community. In some instances, accessible policy briefs and reports may be useful for local government officials or chiefs, but may not be interesting or accessible to the general public. In these situations, you may consider a number of other mechanisms for sharing findings and even involving communities in the analysis of your findings.

Box 16.4 Policy briefs

The UK's *Economic and Social Research Council*: http://www.esrc.ac.uk/news-and-events/publications/evidence-briefings/index.aspx.

The *Overseas Development Institute* has an extensive online archive of their briefing documents across a range of development issues: http://www.odi.org.uk/publications/briefings.

The UK's *Department for International Development*'s Research for Development platform hosts an array of policy briefs and other materials designed to communicate research findings to policy makers and government officials: http://www.dfid.gov.uk/r4d/index.asp.

The *OECD Development Centre*'s *ilibrary* hosts an archive of development-related policy briefs in both English and French: http://www.oecd-ilibrary.org/development/oecd-development-centre-policy-briefs_20771681.

The *Institute for Development Studies*' archive of policy briefs can be found here: http://www.ids.ac.uk/publications/ids-series-titles/ids-policy-briefings.

Potential mechanisms for these activities could include setting up a temporary exhibition of photos or other images generated both by the researcher and by participants during the fieldwork and that is open to local community members. As noted in the autophotography section earlier, working with participants to involve them in the analysis phase of your research can both strengthen your research and facilitate the sharing of findings. Exhibitions of photos can also be useful ways of engaging audiences 'at home' with your research. It is possible that your department or university library may have some space you could use for such an exhibition. Such practices need not be confined to static images, but can include the production, screening and discussion of ethnographic films as a tool to communicate findings to specific non-academic audiences (for an example of this, see Franzen 2013). While there are certain ethical dilemmas to be resolved in relation to this practice, such events can ensure your work is engaged with by local communities and can be used to generate discussion and conversation over your findings, including the possibility to elicit feedback and further insights/commentary from host communities and research participants.

You may also arrange to hold workshops or village meetings with your participants to discuss findings and your analysis of them in a way that allows participants to comment on and contribute to your thinking.

Engaging the media: Press releases and broadcast media

Dealing with the media can involve reacting to media approaches as well as reaching out and making contact with media outlets yourself. These efforts are important as the exposure gained by getting your research into the media spotlight can help you influence policy, contribute to public debate, raise your reputation (and that of your department/ university) and assist you in gaining further/future funding.

While the advice below will provide you with some initial ideas and basic guidance, there are many more detailed resources freely available online and we encourage you to make use of these (a useful starting point is the ESRC's online toolkit of resources for engaging with the media: ESRC (no date) *Working with the Media: A Best Practice Guide*. Swindon: ESRC. http://www.esrc.ac.uk/_images/Working_with_the_Media_ tcm8–2674. pdf). In addition, it is likely that your university will have its own media team, which will have networks of contacts in the press, provide guidance, support and training for dealing with the media and have strategies for promoting news in print, broadcast and digital media. Your university's media team can be an invaluable resource and it is worth speaking with this team when you are developing any media engagement.

Press releases

Press releases are a common way of attempting to engage media interest in your research. They are a means for you to reach out to journalists to put your work on the public agenda. In preparing a press release, you need to think about why a journalist (and their audience – a section of the general public) would be interested in your work and what audience will your work appeal to? Will it appeal to student, local, national or international press?

Print, online or broadcast media? The general press or to specialist audiences (i.e. magazines or journals)? Having a realistic sense of who may be interested in your research will help you tailor your press release and to target appropriate media outlets. If you are based at a university, it is likely that there is a press office or media team that can assist with this. You may also want to begin to develop your own directory of press contacts at different levels who deal with more specialist coverage in your field.

Crucially, you need to think about how to sell your story. What is it about a news story that interests you – what makes you stop and read an article? Most often readers are looking for a human interest angle, either that or cute cats on YouTube videos. The serious point here is that to interest a journalist your story needs a hook – an answer to the 'so what?' question – and often this hook is a human interest angle. For development scholars, this human interest angle can often be related to how your research findings affect everyday life (and death) or linked to a current hot topic or major news item.

A priority for you in writing a press release is to develop this hook, the content that makes the story newsworthy so that it stands out from the hundreds of other press releases crossing the journalist's desk. Using examples and illustrative detail can make the findings seem 'real' to readers. You also need to ensure you use plain language (to use an acronym – KISS: keep it simple and straightforward) and highlight key facts, figures and quotes.

In writing a press release, you need to:

- Capture interest with the headline (this should be short and catchy, ideally short enough to fit into a tweet).
- Provide core details in the first paragraph (answering the five key questions: who, what, where, when, why).
- Offering more detail and explanation in later paragraphs (think about how you read the news – how often do you read to the end of an article compared to just the first paragraph or two?).
- Include your contact details in a press release, as this allows journalists and editors to follow up on the story in order to gain further details or clarity, to seek quotes or soundbites, or even to invite you to talk about your work 'on air'.
- Include other 'notes for editors' detailing any associated materials and publications that they may want/need to access, details of funding sources and methodology.

Even if you produce a well-written press release on a topic of popular interest and present groundbreaking research findings, there is no guarantee that this will get any coverage. In part, this depends on whether the journalist thinks your story will interest their readers and help their media outlet be profitable. In part, it also depends on the timing of the press release – there are slow news days and busy news days. On a day when there is little other news around, your story may have a stronger change of gaining coverage. On a day with lots of news, those days referred to by politicians and civil servants as 'days to bury bad news', your press release is less likely to get picked up and given media space.

Handling media enquiries

If you have distributed a press release, this may generate further enquiries from the media to follow up on a story relating to your research findings. You may also find yourself being contacted unexpectedly by the press to comment on a current affairs topic or news story that relates to your field of expertise. In either situation, you need to be able to handle the enquiry promptly, in a professional and well-informed manner. Dan's experience of contributing to broadcast media programmes is of receiving an email or phone call in the morning, often between 9 and 10am, with an initial request to participate in the recording of the programme later that day, giving him four hours not only to prepare to participate in a debate but also to travel to the television studio for filming. So, how best can you prepare for and handle media enquiries?

First, when a journalist contacts you, find out who they are, who they are working for and why they contacted you. Then find out as much as you can from them about the type of contribution they want from you – is it a quick telephone interview for a short comment on your research or a topical news story, is it a more in-depth interview, are you being asked to participate in a panel discussion or debate? What is the story about and what is the angle of the story? Who else is the journalist talking to and/or who else is being invited to participate in the discussion or debate? What kinds of sub-topics and questions are they likely to want you to speak about? When is the deadline for your contribution, how will they record it and are there any logistical considerations (such as travelling to a radio or television studio) that need to be considered for you to be able to meet the deadline?

Once you have this background information, you can decide whether you are willing and able to provide the input and commentary being requested. If you are uncomfortable with the angle of the story or that the topic is outside of your area of expertise, then you are under no obligation to participate and you can decline the invitation. If you do agree to participate then spend some time preparing yourself by thinking through your research and knowledge on the topic being addressed and how you might handle certain obvious or tricky questions. If possible, practise your responses out loud before the recording or interview. Remember to be open and honest in your answers – if you do not know the answer, say so. You will have seen and heard enough interviews in which politicians appear insincere, evade the question being asked or turn the debate to a different topic: do you want to come across in the same way? Other tips to remember – do not allow journalists to put words into your mouth or to reframe and alter the meaning of your comments. Keep your answers succinct and to the point; not only does this keep a discussion or interview moving but it also gives a stronger impression of control and conviction in your handling of the topic and the debate.

Other basic things to think about include: making sure you arrive in good time for an interview or recording; try to watch a previous edition of the programme before appearing, so that you have an idea of style and focus; watch/listen to your own appearance afterwards and reflect on how well you did; listen to instructions and follow them; do not have notes with you during the interview; do not assume the viewer knows much about the subject (but also do not treat them as if they know nothing, this comes across as arrogant and patronising); and turn your mobile phone off!

Engaging digital and social media

The rise of digital and social media provides further tools for you to broadcast your research findings and through which to engage different audiences. Websites, blogs, Twitter and Facebook all provide different platforms through which to promote your research and communicate quickly with different types of audience. Depending on how you set these up each platform provides the possibility for interaction and feedback, the generation of ideas and content and the sharing of related materials and resources.

An increasingly common practice is for researchers to build websites through which to demonstrate their research findings. These sites are of varying complexity and containing differing levels of detail and content. At the most basic, a webpage or website can provide a landing page for someone interested in your work to see who you are and what you are working on. Additional content may include more detailed summaries of research findings, databases, lists of and links to publications, biographical details, podcasts or video clips and so on. A well maintained webpage can provide a useful profile for your work and there are a range of free and low-cost web-hosting solutions you can use for this (a useful source of guidance on writing web content can be found here: http://knowhownonprofit.org/how-to/how-to-write-accessible-web-content).

Another common option is to develop a blog related to your research (again there are a range of free and low-cost hosting options). Blogs are usually relatively cheap, simple and straightforward to set up, requiring a lot less time and skill than a website. They provide a space through which you can reflect and commentate on, narrate, analyze and discuss your research and relevant materials. Be aware, however, that if you set up a blog it does need regular updating: they are by nature very dynamic and need regular upkeep in order to keep an audience (in this regard they are harder work than a website, which can be far more static in its content without looking quickly outdated). You might find the reflections posted by Salma Patel on the London School of Economics and Political Science's 'Impact blog' a useful beginners guide to many of these concerns: http://blogs. lse.ac.uk/impactofsocialsciences/2012/08/10/im-an-academic-and-desperately-need-an-online-presence-where-do-i-start/.

Microblogging (of which Twitter is perhaps the best known example) is another way of generating interest and communicating findings. Although you cannot provide detailed findings or results through Twitter (as your posts are limited to 140 characters), you can use this to promote your research through links to your publications, website or blog, reflecting on current debates or conference papers, as well as by participating in debates and (re)tweeting about relevant news items or discussions. You can also use this as a networking tool, to connect with other researchers working in the field and to follow their work though their tweets and links. As with blogs, a good Twitter presence requires regular posts and maintenance otherwise your profile will rapidly diminish. In order to boost your profile, you should employ techniques to produce good tweets (descriptive text plus link plus hashtag is a basic but good piece of advice) and to generate a following (retweeting, following other people, asking questions, responding to questions, tweeting links to interesting news items).

Transferability and exchange: Broader issues

In order to increase the impact of your research there are various other strategies, in addition to those outlined above, you can adopt. The effectiveness and efficiency of engagement can be dramatically improved by establishing and developing networks of users (be these journalists, policy makers, activists, etc.) and to involve these where possible throughout the research process (from concept to dissemination). Over time you will come to recognize specific barriers to engaging with different groups and you can develop strategies to overcome these.

At the same time, it is useful to consider how the skills and knowledge you develop through research and fieldwork in development are themselves transferable to other contexts. A common complaint from many employers of university graduates it that the graduates do not recognize and are unable to explain the skills that they have. In light of this, this final section reflects on some of the transferable skills you will develop through research and fieldwork in development.

Organizational skills

The preparation, planning and day-to-day management of field research demonstrates a range of organizational skills. In identifying a research topic, questions, site, methodology and then implementing the research design, you inherently demonstrate skills in *decision making*, *priority/goal setting*, *time management* and *project management*. If you reflect on how you carry out fieldwork, you should be able to identify evidence of these skills. Think about how you determined which tasks to complete as a priority, how you managed various resources (financial, time, transport) to get to research sites and conduct various research activities and how you worked without constant direction or supervision to complete fieldwork and then analyze and present your findings.

Numeracy and technology

Depending on the methods used in your research, you will be able to draw on different evidence of numeracy- and IT-related skills. If you have used quantitative methods then you can draw on these experiences to evidence *numeracy* and *statistical analysis* skills, as well as knowledge of specific *computer software*. Similarly, use of locational data and practices of visualizing data and findings provide evidence you can use to demonstrate your skills in *data analysis and presentation* as well as with *computer software*. Even without the use of these forms of data and presentation, your background research will have involved identifying, accessing and analyzing literature and existing data sources.

Communication skills

Communication skills are developed through research and fieldwork in development in a number of ways. The presentation of research findings potentially involves a variety of forms of communication. Essays, project or research reports, policy briefs and journal articles are all evidence of *written communication skills*. To provide more depth to your claims to skills development, you can reflect on how each of these forms of presentation requires an identification of audience and development of materials in a style appropriate to this audience. You may also be able to point to *visual communication skills* (the development of maps, multimedia communication of research findings, or other means of visualizing data) and/or to *oral presentations* that evidence additional communication skills.

During the conducting of fieldwork, you may also develop *inter-cultural communication skills* and *negotiation* and *conflict management skills*. These may be developed through working across cultural boundaries, interacting with different types of stakeholder and overcoming various power inequalities and dynamics, as well as through negotiating access to research sites.

Research skills

This skills set is perhaps the most obvious set of skills to have developed and you can reflect on your research to identify examples of *project design and implementation, ethics and risk assessment, research methods, analytical skills, reasoning skills* and *problem solving*. Implicit within these skills categories are a number of other attributes such as logic, decision making, time management, motivation and perseverance. These might be evidenced through discussion of the way in which you overcame a particular barrier or hurdle to a research method or question, how you recognized an emergent risk and conducted a dynamic risk assessment and shifted the location or approach of your research to mitigate this risk.

Professional skills

A range of slightly less easily defined skills are also developed through research and fieldwork in development. Through these experiences, you will have encountered a range of contexts and engagements that will have increased your awareness and sensitivity to different issues and challenges. Adopting a reflective approach to your research – as evidenced through a research diary, for instance – contributes to the development of *personal reflection* skills and increased *self-awareness*. It is also likely that you will have developed *networking skills* by working with academics and other researchers/students as well as practitioners, policy makers, community leaders and other stakeholders.

What to take from this chapter

This chapter should have indicated to you the potential range of audiences for your research and the importance of trying to make your research findings accessible and engaging for audiences outside academia. Underpinning our discussion of how to engage with different audiences, there lie a number of tips and techniques to help with the writing process and style that we hope you have also picked up. Finally, the closing section of the chapter should have indicated to you the range of transferable skills you will have developed through conducting development research. Overall, this chapter has aimed to help you to both develop a set of transferable skills *and* to recognize how many of the other skills you have developed in your research and fieldwork experiences are also important transferable skills that you can use to further your career in a range of contexts.

If you want to explore the ideas covered in this chapter in more detail, you may find these further readings useful:

Becker, H. (1986) *Writing for Social Scientists: How To Start and Finish Your Thesis, Book or Article*. Chicago: University of Chicago Press.

Bowden, J. (2011) *Writing a Report: How to Prepare, Write and Present Really Effective Reports*. Oxford: How To Books.

Craswell, G. and Poore, M. (2012) *Writing for Academic Success*. London: Sage.

Hall, G. (2013) *How To Write A Paper*. Oxford: Wiley-Blackwell.

Murray, R. (2005) *Writing for Academic Journals*. Maidenhead: Open University Press.

Silvia, P. (2007) *How To Write A Lot: A Practical Guide to Productive Academic Writing*. Washington, DC: APA.

FINAL THOUGHTS:
YOUR RESEARCH JOURNEY

This book will, we hope, have given you both a sense of the realities of doing development research and a sense of enthusiasm for the research endeavour. In planning our own fieldwork – whether on individual or group projects or leading student fieldtrips – we regularly work through many of the questions, concerns and experiences outlined in this book. We know that development research can be complex, uncertain and challenging. We also know that, because of these very same factors, development research is often enjoyable and rewarding. We, too, keep learning through our fieldwork: none of us is the mythical 'perfect researcher'. Our reflexive engagement with what has worked well or poorly, of how we have encountered and dealt with new or unexpected challenges, is one of the most underrated pleasures and inspirations of working in this field. We hope you too will be inspired by fieldwork.

Throughout this book, we have focused primarily on your role and experiences as a researcher. It is vital, however, to remember that your research is not only about you – it involves and affects many other people, including your family, friends and supervisor, and – most importantly – those with whom you are conducting your research. At various points, we have encouraged you to reflect on the rationale for your work, the ways in which you might manage and mitigate the burden you place on host communities or organizations, how you can try to ensure your research is relevant to the host community, the possibilities of co-producing knowledge, and various techniques for sharing and exploring your findings with non-academic audiences. In concluding this book, this is an opportune moment to reiterate that your research practices and engagement with 'the field' requires sensitive engagement with and consideration for research participants and their communities.

There are many ways in which you can recognize and integrate the concerns and needs of the individuals, communities and organizations that are involved with your research. These possibilities extend from involving these stakeholders in the conception and design of your research, through the use of participatory methods, to the provision of methodological training and the development and dissemination of research findings to multiple audiences in appropriate forms. Our reflections here have been brief and we would encourage you to explore these questions in greater depth: you may find the reflections in Desai and Potter (2006) and Scheyvens (2014) accessible and useful.

If you want to challenge yourself further, you might consider exploring some more nuanced debates. Earlier in the book we highlighted the influence of post-colonial thinking on development research, both in terms of how idea(l)s of development are generated and by whom and in ensuring development research is accountable, relevant and accessible to those involved in this process. These debates are developing quickly and are ones of which you should be aware. Layered on to these concerns are increasing pressures to ensure academic research has 'impact' beyond the academy: regardless of whether you are an undergraduate or postgraduate student or a postdoctoral fellow or lecturer you can also engage with and contribute to such debates and practices.

Alternatively, you might consider how questions of responsibility are implicated in development research, policy and practice. How responsibility is understood and related to research is a complex and contested arena. The post-colonial turn has emphasized researchers' responsibilities to distant others (Smith 2002) and a number of academics have argued for research to be more sensitive to social, political, economic and cultural contexts, outcomes and progressive change (Massey 2004; Hobson 2006). Although such debates can often feel rather esoteric and abstract, the underlying question of morals and obligations are often at the heart of development concerns and research: from development aid as a moral obligation in the light of historical exploitation of colonial and imperial ventures, to arguments over the (in)equality of contemporary world systems and economic and political structures.

In short, what we are trying to say in closing this book is that development research engages with and opens up a vast array of questions and debates, across many scales and connected to multiple audiences and stakeholders. It is this vista of possibilities that excites and inspires many development researchers: we hope you, too, will be inspired by your own experience of development research.

REFERENCES

Abbott, D. (2006) 'Disrupting the 'whiteness' of fieldwork in geography', *Singapore Journal of Tropical Geography*, 27: 326–41.

Adedze, A. (2010) 'Portraits in the hands of strangers: colonial and postcolonial postcards as vignettes to African women's history', *Afrika Zamani*, 18–19: 1–16.

Adler, P. (1975) 'The transitional experience: an alternative view of culture shock', *Journal of humanistic psychology*, 15: 13–23.

Akbar, D., van Horen, B., Minnery, J. and Smith, P. (2007) 'Assessing the performance of urban water supply systems in providing potable water for the urban poor', *International Development Planning Review*, 29: 299–318.

Akresh, R., Lucchetti, L. and Thirumurthy, H. (2012) 'Wars and child health: evidence from the Eritrean-Ethiopian conflict', *Journal of Development Economics*, 99: 330–40.

Ansell, N. (2001) 'Producing knowledge about 'Third World women': the politics of fieldwork in a Zimbabwean secondary school', *Ethics, Place and Environment*, 4: 101–16.

Ansell, N., Hajdu, F., van Blerk, L. and Robson, E. (2012) 'Learning from young people about their lives: using participatory methods to research the impacts of AIDS in southern Africa', *Children's Geographies*, 10: 169–86.

Archives of Exile (2010) Exhibition website. http://www.hrionline.ac.uk/archiveofexile/.

Ardilly, P. and Tillé, Y. (2006) *Sampling Methods: Exercises and Solutions*. New York: Springer.

Ashmore, P., Craggs, R. and Neate, N. (2012) 'Working-with: talking and sorting personal archives', *Journal of Historical Geography*, 38: 81–9.

Atkinson, P., Coffey, A., Delamont, S., Lofland, J. and Lofland, L. (eds.) (2001) *Handbook of Ethnography*. London: Sage.

Avenell, D., Noble, M. and Wright, G (2009) 'South African datazones: a technical report about the development of a new statistical geography for the analysis of deprivation in South Africa at small area level'. Working Paper 8. University of Oxford: Centre for the Analysis of South African Social Policy.

Bachmann, V. (2011) 'Participating and observing: positionality and fieldwork relations during Kenya's post-election crisis', *Area*, 43: 362–8.

Barker, D. (2006) 'Field surveys and Inventories', in Desai, V. and Potter, R. (eds.) *Doing Development Research*. London: Sage.

Barnard, A. (1992) *Hunters and Herders of Southern Africa: A Comparative Ethnography of the Khoisan peoples*. Cambridge: Cambridge University Press.

Barro, R. and Lee, J. (2013) 'A new data set of educational attainment in the world, 1950–2010', *Journal of Development Economics*, 104: 184–98.

Barry, M. and Rüther, H. (2005) 'Data collection techniques for informal settlement upgrades in Cape Town, South Africa', *URISA Journal,* 17(1): 43–52.

Baruah, B. (2010) 'Women and globalization: challenges and opportunities facing construction workers in contemporary India', *Development in Practice*, 20: 31–44.

Bassett, K. (1996) 'Postmodernism and the crisis of the intellectual: reflections on reflexivity, universities, and the scientific field', *Environment and Planning D: Society and Space*, 14: 507–27.

Batista, C., Lacuesta, A. and Vincente, P. (2012) 'Testing the "brain gain" hypothesis: micro-evidence from Cape Verde', *Journal of Development Economics*, 97: 32–45.

Becker, H. (1986) *Writing for Social Scientists: How To Start and Finish Your Thesis, Book or Article*. Chicago: University of Chicago Press.

Beegle, K., Carletto, C. and Himelein, K. (2012) 'Reliability of recall in agricultural data', *Journal of Development Economics*, 98: 34–41.

Beinart, W. and McKeown, K. (2009) 'Wildlife media and representations of Africa, 1950s to the 1970s', *Environmental History*, 14: 429–52.

Bell, K. (2013) 'Doing qualitative fieldwork in Cuba: social research in politically sensitive locations', *International Journal of Social Research Methodology*, 16: 109–24.

Belousov, K., Horlick-Jones, T., Bloor, T., Gilinskiy, Y., Golbert, V., Kostikovsky, Y. et al. (2007) 'Any port in a storm: fieldwork difficulties in dangerous and crisis-ridden settings', *Qualitative Research*, 7: 155–75.

Belzile, J. and Öberg, G. (2012) 'Where to begin? Grappling with how to use participant interaction in focus group design', *Qualitative Research*, 12: 459–72.

Bénit-Gbaffou, C. (2010) 'The researcher as the ball in a political game', *Carnets de Géographes*, 10: 1–10.

Biesele, M., Guenther, M., Hitchcock, R., Lee, R. and Macgregor, J. (1989) 'Hunters, clients and squatters: the contemporary socioeconomic status of Botswana Basarwa', *African Studies Monographs*, 9(3): 109–51.

Binns, T. (2006) 'Doing fieldwork in developing countries: planning and logistics', in Desai, V. and Potter, R. (eds.) *Doing Development Research*. London: Sage.

Black, M. (2007) *The No-Nonsense Guide to International Development*. Oxford: New Internationalist.

Blaikie, N. (2000) *Designing Social Research*. Cambridge: Polity Press.

Block, E. and Erskine, L. (2012) 'Interviewing by telephone: specific considerations, opportunities and challenges', *International Journal of Qualitative Methods*, 11: 428–45.

Blomley, N. (1994) 'Activism and the academy', *Environment and Planning D: Society and Space*, 12: 383–5.

Boesten, J. (2008) 'A relationship gone wrong? Research ethics, participation, and fieldwork realities'. NGPA Working Paper Series No. 20. London School of Economics.

Bollig, M. (1998) 'Moral exchange and self-interest: kinship, friendship and exchange among the Pokot (North West Kenya)', in Schweizer, T. and White, D. (eds.) *Kinship, Networks and Exchange*. Cambridge: Cambridge University Press.

Booysen, S. (2007) 'With the ballot and the brick: the politics of attaining service delivery', *Progress in Development Studies*, 7: 21–32.

Bornstein, E. (2007) 'Harmonic dissonance: reflections on dwelling in the field', *Ethnos*, 72: 483–508.

Bourgois, P. (1990) 'Confronting anthropological ethics: ethnographic lessons from Central America', *Journal of Peace Research*, 27: 43–54.

Bowden, J. (2011) *Writing a Report: How to Prepare, Write and Present Really Effective Reports*. Oxford: How To Books.

Brooks, S. (2005) 'Images of "wild Africa": nature tourism and the (re)creation of the Hluhluwe game reserve, 1930–1945', *Journal of Historical Geography*, 31: 220–40.

Brydon, L. (2006) 'Ethical practices in doing development research', in Desai, V. and Potter, R. (eds.) *Doing Development Research*. London: Sage.

Bulmer, M. and Warwick, D. (eds.) (1993) *Social Research in Developing Countries*. London: UCL Press.

Burgess, R. (1981) 'Keeping a research diary', *Cambridge Journal of Education*, 11: 75–83.

Burgess, R. (2002) *In the Field: An Introduction to Field Research*. London: Routledge.

Campbell, D. (2007) 'Geopolitics and visuality: sighting the Darfur conflict', *Political Geography*, 26: 357–82.

Carapico, S. (2006) 'No easy answers: the ethics of field research in the Arab world', *PS: Political Science and Politics*, 39: 429–31.

Caviglia-Harris, J., Hall, S., Mullna, K., MacIntyre, C., Bauch, S., Harris, D. et al. (2012) 'Improving household surveys through computer-assisted data collection: use of touch-screen laptops in challenging environments', *Field Methods*, 24: 74–94.

Chagnon, N. (1992) *Y_nomamö*. Fort Worth, TX: Harcourt Brace College Publishers.

Chakravarty, A. (2012) '"Partially trusting" field relationships: opportunities and constraints of fieldwork in Rwanda's postconflict setting', *Field Methods*, 24: 251–71.

Chambers, R. (1983) *Rural Development: Putting the Last First*. London: Longman.

Chambers, R. (1991) 'In search of professionalism, bureaucracy and sustainable livelihoods for the 21st century', *IDS Bulletin*, 22(4): 5–11.

Chambers, R. (1994) 'The origins and practice of participatory rural appraisal', *World Development*, 22: 953–69.

Chambers, R. (1997) *Whose Reality Counts? Putting the Last First*. London: Intermediate Technology.

Chambers, R. (2002) *Participatory Workshops: A Sourcebook of 21 Sets of Ideas and Activities*. London: Earthscan.

Chatterton, P. (2006) '"Give up activism" and change the world in unknown ways: or, learning to walk with others on uncommon ground', *Antipode*, 38: 259–81.

Chilcote, R. (1974) 'Dependency: a critical synthesis of the literature', *Latin American Perspectives*, 1: 4–29.

Child, J. (2005) 'The politics and semiotics of the smallest icons of popular culture: Latin American postage stamps', *Latin American Research Review*, 40: 108–37.

Chouinard, V. (1994) 'Reinventing radical geography: is all that's left right?', *Environment and Planning D: Society and Space*, 12: 2–6.

Chowdhury, M., Ghosh, D. and Wright, R. (2005) 'The impact of micro-credit on poverty: evidence from Bangladesh', *Progress in Human Geography*, 5: 298–309.

Clark, G. (2006) 'Field research methods in the Middle East', *PS: Political Science and Politics*, 39: 417–24.

Clark, T. (2011) 'Gaining and maintaining access: exploring the mechanisms that support and challenge the relationship between gatekeepers and researchers', *Qualitative Social Research*, 10: 485–502.

Clifford, J. and Marcus, G. (eds.) (1986) *Writing Culture: The Poetics and Politics of Ethnography*. Berkeley: University of California Press.

Cloke, P. (2002) 'Deliver us from evil? Prospects for living ethically and acting politically in human geography', *Progress in Human Geography*, 26: 587–604.

Connell, R. (2007) 'The heart of the problem: South African intellectual workers, globalization and social change', *Sociology*, 41: 11–28.

Cook, I. and Crang, M. (1995) 'Doing ethnographies'. CATMOG.

Cook, K. and Nunkoosing, K. (2008) 'Maintaining dignity and managing stigma in the interview encounter: the challenge of paid-for participation', *Qualitative Health Research*, 18: 418–27.

Cook, N. (2009) 'It's good to talk: performing and recording the telephone interview', *Area*, 41: 176–85.

Cooke, B. and Kothari, U. (eds.) (2001) *Participation: The New Tranny?* London: Zed Books.

Cope, M. (2010) 'Coding transcripts and diaries', in Clifford, N., French, S. and Valentine, G. (eds.) *Key Methods in Geography*. London: Sage.

Corbin, J. and Morse, J. (2003) 'The unstructured interview: issues of reciprocity and risks when dealing with sensitive topics', *Qualitative Inquiry*, 9: 335–54.

Corbin Dwyer, S. and Buckle, J. (2009) 'The space between: on being an insider-outsider in qualitative research, *International Journal of Qualitative Methods*, 8: 54–63.

Corbridge, S. (1993) 'Marxisms, modernities and moralities: development praxis and the claims of distant strangers', *Environment and Planning D*, 11: 449–72.

Cornelissen, S. (2005) 'Producing and imagining "place" and "people": the political economy of South African international tourist representation', *Review of International Political Economy*, 12: 674–99.

Craggs, R. (2008) 'Situating the imperial archive: the Royal Empire Society Library, 1868–1945', *Journal of Historical Geography*, 34: 48–67.

Craggs, R. (2014) 'Development in a global-historical context', in Desai, V. and Potter, D. (eds.) *The Companion To Development Studies*. London: Routledge.

Craib, R. (2010) 'The archive in the field: document, discourse, and space in Mexico's agrarian reform', *Journal of Historical Geography*, 36: 411–20.

Cramer, C., Hammond, L. and Pottier, J. (eds.) (2011) *Researching Violence in Africa: Ethical and Methodological Challenges*, Leiden: Brill.

Crane, L., Lombard, M. and Tenz, E. (2009) 'More than just translation: challenges and opportunities in translingual research', *Social Geography*, 4: 39–46.

Crang, M. and Cook, I. (2007) *Doing Ethnographies*. London: Sage.

Craswell, G. and Poore, M. (2012) *Writing for Academic Success*. London: Sage.

Crush, J. (1995) *Power of Development*. London: Routledge.

Cupples, J. (2002) 'The field as a landscape of desire: sex and sexuality in geographical fieldwork', *Area*, 34: 382–90.

Cupples, J. and Kindon, S. (2003) 'Far from being "home alone": the dynamics of accompanied fieldwork', *Singapore Journal of Tropical Geography*, 24: 211–28.

Curran, S. (2006) 'Ethical considerations for research in cross-cultural settings', in Perecman, E. and Curran, S. (eds.) *A Handbook for Social Science Field Research: Essays and Bibliographic Sources on Research Design and Methods*. London: Sage.

Czaja, R. and Blair, J. (2005) *Designing Surveys: A Guide to Decisions and Procedures*. London: Pine Forge.

Davis, D. (2003) 'What did you do today? Notes from a politically engaged anthropologist', *Urban Anthropology and Studies of Cultural Systems and World Economic Development*, 32: 147–73.

De Leeuw, S. (2012) 'Alive through the looking glass: emotion, personal commotion and reading colonial archives long the grain', *Journal of Historical Geography*, 38: 273–81.

De Vaus, D. (2002) *Surveys in Social Research*. London: Routledge.

Denzin, N. and Lincoln, Y. (eds.) (2005) *The SAGE Handbook of Qualitative Research*. London: Sage.

Denzin, N., Lincoln, Y. and Smith, L. (eds.) (2008) *Handbook of Critical and Indigenous Methodologies*. London: Sage.

Desai, V. and Potter, R. (eds.) (2006) *Doing Development Research*. London: Sage.

Desai, V. and Potter, R. (2014) *The Companion to Development Studies* (3rd ed.). Abingdon: Routledge.

Dhanju, R. and O'Reilly, K. (2013) 'Human subjects research and the ethics of intervention: life, death and radical geography in practice', *Antipode*, 45: 513–16.

Dittmer, J. and Gray, N. (2010) 'Popular geopolitics 2.0: towards new methodologies of the everyday', *Geography Compass*, 4: 1664–77.

Dodman, D. (2003) 'Shooting in the city: an autophotographic exploration of the urban environment in Kingston, Jamaica', *Area*, 35: 293–304.

Dodson, B. (2010) 'Locating xenophobia: debate, discourse and everyday experience in Cape Town, South Africa', *Africa Today*, 56: 2–22.

Dritsas, L. and Haig, J. (2013) 'An archive of identity: the Central African archives and Southern Rhodesian history', *Archival Science*, 14(1): 35–54.

Duncan, J., Hoffman, T., Rohde, R., Powell, E. and Hendricks, H. (2006) 'Long-term population changes in the Giant Quiver Tree, *Aloe pillansii* in the Richtersveld, South Africa', *Plant Ecology*, 185: 73–84.

Dwyer, C. and Davies, G. (2010) 'Qualitative methods III: animating archives, artful interventions and online environments', *Progress in Human Geography*, 34: 88–97.

Dyer, S. and Demeritt, D. (2009) 'Un-ethical review? Why it is wrong to apply the medical model of research governance to human geography', *Progress in Human Geography*, 33: 46–64.

Dyrness, A. (2008) 'Research for change versus research as change: lessons from a *mujerista* participatory research team', *Anthropology and Education Quarterly*, 39: 23–44.

Economic and Social Research Council (ESRC) (2013) *International Benchmarking Review of UK Human Geography*. Swindon: ESRC.

Edensor, T. (1998) *Tourists at the Taj: Performance and Meaning at a Symbolic Site*. London: Routledge.

Edwards, R. (1998) 'A critical examination of the use of interpreters in the qualitative research process', *Journal of Ethnic and Migration Studies*, 24: 197–208.

Elmhirst, R. (2012) 'Methodological dilemmas in migration research in Asia: research design, omissions and strategic erasures', *Area*, 44: 274–81.

Elwood, S. and Martin, D. (2000) '"Placing" interviews: location and scales of power in qualitative research', *Professional Geographer*, 52: 649–57.

Endfield, G. (2010) 'Through marsh and mountain: tropical acclimatization, health and disease and the CMS mission to Uganda, 1875–1920', *Journal of East African Studies*, 4: 61–90.

Engin, M. (2011) 'Research diary: a tool for scaffolding', *International Journal of Qualitative Methods*, 10: 296–306.

England, K. (1994) 'Getting personal: reflexivity, positionality and feminist research', *Professional Geographer*, 46: 80–89.

Ensign, J. (2003) 'Ethical issues in qualitative health research with homeless youths', *Journal of Advanced Nursing*, 43: 43–50.

Ergun, A. and Erdemir, A. (2010) 'Negotiating insider and outsider identities in the field: "insider" in a foreign land; "outsider" in one's own land', *Field Methods*, 22: 16–38.

Escobar, A. (1992) 'Reflections on "development": grassroots approaches and alternative politics in the third world', *Futures*, 24: 411–36.

Escobar, A (1995) *Encountering Development: The Making and Unmaking of the Third World*. Princeton, NJ: Princeton University Press.

Escobar, A. (2008) 'The problematization of poverty', in Chari, S. and Corbridge, S. (eds.) *The Development Reader*. Oxford: Routledge.

ESRC (no date) *Working with the Media: A Best Practice Guide*. Swindon: ESRC. http://www.esrc.ac.uk/_images/Working_with_the_Media_tcm8–2674.pdf.

Evans-Pritchard, E. (1940) *The Nuer: A Description of the Modes of Livelihood and Political Institutions of a Nilotic People*. Oxford: Clarendon Press.

Eyben, R., Harris, C. and Pettit, J. (2006) 'Introduction: exploring power for change', *Institute for Development Studies Bulletin*, 37: 1–10.

Ezzy, D. (2010) 'Qualitative interviewing as an embodied emotional performance', *Qualitative Inquiry*, 16: 163–70.

Faria, C. and Good, R. (2012) 'The importance of everyday encounters: young scholars reflect on fieldwork in Africa', *African Geographical Review*, 31: 63–6

Farnsworth, J. and Boon, B. (2010) 'Analysing group dynamics within the focus group', *Qualitative Research*, 10: 605–24.

Fernandez, R. and Ocampo, J. (1974) 'The Latin American revolution: a theory of imperialism, not dependence', *Latin American Perspectives*, 1: 30–61.

Ferreira, S. (1999) 'Crime: a threat to tourism in South Africa', *Tourism Geographies*, 1: 313–24.

Field, A. (2013) *Discovering Statistics Using IBM SPSS Statistics*. London: Sage.

Findlay, A. (2006) 'The importance of census and other secondary data in development studies', in Desai, V. and Potter, R. (eds.) *Doing Development Research*. London: Sage.

Fink, A. (1995) *How to Sample in Surveys*. London: Sage.

Flick, U. (2009) *An Introduction to Qualitative Research* (4th ed.). London: Sage.

Flyvberg, B. (2001) *Making Social Science Matter: Why Social Inquiry Fails and How It Might Succeed Again*. Cambridge: Cambridge University Press.

Foddy, W. (1993) *Constructing Questions for Interviews and Questionnaires: Theory and Practice in Research*. Cambridge: Cambridge University Press.

Fowler, F. (2002) *Survey Research Methods*. London: Sage.

Frank, A. (1973) 'The development of underdevelopment', *Monthly Review*, 18: 4–17.

Franzen, S. (2013) 'Engaging a specific, not general, public: the use of ethnographic film in public scholarship', *Qualitative Research*, 13: 414–27.

Fraser, A. (2012) 'The 'throwntogetherness' of research: reflections on conducting fieldwork in South Africa', *Singapore Journal of Tropical Geography*, 33: 291–300.

Frayne, B. (2010) 'Pathways of food: mobility and food transfer in Southern African cities', *International Development Planning Review*, 32: 291–310.

Fujii, L. (2012) 'Research ethics 101: dilemmas and responsibilities', *PS: Political Science and Politics*, 45: 717–23.

Fuller, H. (2008) '*Civitatis Ghaniensis Conditor*: Kwame Nkrumah, symbolic nationalism and the iconography of Ghanaian money 1957 – the Golden Jubilee', *Nations and Nationalism*, 14: 520–41.

Fuller, I., Edmondson, S., France, D., Higgit, D. and Ratinen, I. (2006) 'International perspectives on the effectiveness of geography fieldwork for learning', *Journal of Geography in Higher Education*, 30: 89–101.

Fürsich, E. and Robins, M. (2004) 'Visiting Africa: constructions of nation and identity on travel websites', *Journal of Asian and African Studies*, 39: 133–52.

Galasinski, D. and Kozlowska, O. (2010) 'Questionnaires and lived experience: strategies of coping with the quantitative frame', *Qualitative Inquiry*, 16: 271–84.

Garton, S. and Copland, F. (2010) '"I like this interview; I get cakes and cats!": the effect of prior relationships on interview talk', *Qualitative Research*, 10: 533–51.

Gasper, D. (2000) 'Development as freedom: taking economics beyond commodities – the cautious boldness of Amartya Sen', *Journal of International Development*, 12: 989–1001.

Gaventa, J. (2006) 'Finding the spaces for change: a power analysis', *Institute for Development Studies Bulletin*, 37: 23–33.

Geertz, C. (1972) 'Deep play: notes on the Balinese cockfight', *Daedalus*, 101: 1–37.

Geertz, C. (1973) *The Interpretation of Cultures*. New York: Fontana Press.

Geertz, C. (ed.) (1983) *Local Knowledge: Further Essays in Interpretative Anthropology*. New York: Basic Books.

Gentile, M. (2013) 'Meeting the "organs": the tacit dilemma of field research in authoritarian states', *Area*, 45: 426–32.

Gibaldi, J. (2003) *MLA Handbook for Writers of Research Papers* (6th ed.). New York: MLA.

Gibson, D. (2000) 'Seizing the moment: the problem of conversational agency', *Sociological Theory*, 18: 368–82.

Gilbert, N. (2008) *Researching Social Life*. London: Sage.

Glewwe, P. and Kremer, M. (2006) 'Schools, teachers, and education outcomes in developing countries', in Hanushek, E. and Welch, F. (eds.) *Handbook of the Economics of Education, Volume 2*. Amsterdam: North Holland.

Gobo, G. (2011) 'Glocalizing methodology? The encounter between local methodologies', *International Journal of Social Research Methodology*, 16: 417–37.

Goebel, A. (1998) 'Process, perception and power: notes from "participatory" research in a Zimbabwean resettlement area', *Development and Change*, 29: 277–305.

Gold, R. (1997) 'The ethnographic method in sociology', *Qualitative Inquiry*, 3: 388–402.

Gottlieb, A. (2006) 'Ethnography: theory and methods', in Perecman, E. and Curran, S. (eds.) *A Handbook for Social Science Field Research: Essays and Bibliographic Sources on Research Design and Methods*. London: Sage.

Gough, K.V., Langevang, T. and Namatovu, R. (2014) Researching entrepreneurship in low-income settlements: the strengths and challenges of participatory methods, *Environment and Urbanisation*, 26: 297–311.

Graham, M. and Shelton, T. (2013) 'Geography and the Future of Big Data; Big Data and the Future of Geography', *Dialogues in Human Geography*, 3: 255–261.

Gratton, M-F. and O'Donnell, S. (2011) 'Communication technologies for focus groups with remote communities: a case study of research with First Nations in Canada', *Qualitative Research*, 11: 159–75.

Grauerholz, L., Barringer, M., Colyer, T., Guittar, N., Hecht, J., Rayburn, R. et al. (2013) 'Attraction in the field: what we need to acknowledge and implications for research and teaching', *Qualitative Inquiry*, 19: 167–78.

Gregory, D., Johnston, R., Pratt, G., Watts, M. and Whatmore, S. (2009) *Dictionary of Human Geography*, Chichester: Wiley-Blackwell.

Griffith, D. (2008) 'Ethical considerations in geographic research: what especially graduate students need to know', *Ethics, Place and Environment*, 11: 237–52.

Grix, J. (2004) *The Foundations of Research*. Basingstoke: Palgrave Macmillan.

Groes-Green, C. (2012) 'Ambivalent participation: sex, power and the anthropologist in Mozambique', *Medical Anthropology*, 31: 44–60.

Gruley, J. and Duvall, C. (2012) 'The evolving narrative of the Darfur conflict as represented in *the New York Times* and *The Washington Post*, 2003–2009', *GeoJournal*, 77: 29–46.

Gunaratnam, Y. (2003) *Researching 'Race' and Ethnicity: Methods, Knowledge and Power*. London: Sage.

Haer, R. and Becher, I. (2011) 'A methodological note on quantitative field research in conflict zones: get your hands dirty', *International Journal of Social Research Methodology*, 15: 1–13.

Hall, C. (2011) 'In cyberspace can anybody hear you scream? Issues in the conduct of online fieldwork', in Hall, C. (ed.) *Fieldwork in Tourism: Methods, Issues and Reflections*. Abingdon: Routledge.

Hall, G. (2013) *How To Write A Paper*. Oxford: Wiley-Blackwell.

Hall, S. (2009) '"Private life" and "work life": difficulties and dilemmas when making and maintaining friendships with ethnographic participants', *Area*, 41: 263–72.

Halperin, S. (2013) *Re-Envisioning Global Development: A Horizontal Perspective*. London: Routledge.

Hammersley, M. (2000) *Taking Sides in Social Research: Essays on Partisanship and Bias*. London: Routledge.

Hammersley, M. (2009) 'Against the ethicists: on the evils of ethical regulation', *International Journal of Social Research Methodology*, 12: 211–25.

Hammersley, M. and Atkinson, P. (1995) *Ethnography: Principles in Practice* (2nd ed.). London: Routledge.

Hammett, D. (2008) 'The challenge of a perception of "un-entitlement" to citizenship in post-apartheid South Africa', *Political Geography*, 27. 652–68.

Hammett, D. (2010) 'Zapiro and Zuma: a symptom of an emerging constitutional crisis in South Africa?', *Political Geography*, 29: 88–96.

Hammett, D. (2011a) 'Resistance, power and geopolitics in Zimbabwe', *Area*, 43: 202–10.

Hammett, D. (2011b) 'British media representations of South Africa and the 2010 FIFA World Cup', *South African Geographical Journal*, 93: 63–74.

Hammett D. (2012) 'Envisaging the nation: the philatelic iconography of transforming South African national narratives', *Geopolitics*, 17: 526–52.

Hammett, D. (2013) 'A crisis of representation? British media framing of South Africa's destination image', *Tourism Geographies*, 16(2): 221–36.

Hammett, D. (2014) 'Expressing nationhood under conditions of constrained sovereignty: postage stamp iconography of the Bantustans', *Environment and Planning A*.

Hammett, D. and Sporton, D. (2012) 'Paying for interviews? Negotiating ethics, power and expectation', *Area*, 44: 496–502.

Hanna, P. (2012) 'Using internet technologies (such as Skype) as a research medium: a research note', *Qualitative Research*, 12: 239–42.

Harrison, J., MacGibbon, L. and Morton, M. (2001) 'Regimes of trustworthiness in qualitative research: the rigors of reciprocity', *Qualitative Inquiry*, 7: 323–45.

Hartman, W., Silvester, S. and Hayes, P. (1998) *The Colonising Camera: Photographs in the Making of Namibian History*. Cape Town, University of Cape Town Press.

Harvey, D. (1995) 'Militant particularism and global ambition', *Social Text*, 42: 69–98.

Harvey, W. (2011) 'Strategies for conducting elite interviews', *Qualitative Research*, 11: 431–441.

Hay, I. (2001) 'Critical geography and activism in higher education', *Journal of Geography in Higher Education*, 25: 141–46.

Hazareesingh, S. (2012) 'Cotton, climate and colonialism in Dharwar, western India, 1840–1880', *Journal of Historical Geography*, 38: 1–17.

Heald, S. (2008) 'Embracing marginality: place-making versus development in Gardenton, Manitoba', *Development in Practice*, 18: 17–29.

Hecht, T. (1998) *At Home in the Street: Street Children of Northeast Brazil*. Cambridge: Cambridge University Press.

Hemmings, A. (2006) 'Great ethical divides: bridging the gap between institutional review boards and researchers', *Educational Researcher*, 35: 12–18.

Hennig, B. (2013) *Rediscovering the World: Map Transformations of Human and Physical Space*. Berlin: Springer.

Hennink, M. (2007) *International Focus Group Research: A Handbook for the Health and Social Sciences*. Cambridge: Cambridge University Press.

Herrick, C. (2010) 'Lost in the field: ensuring student learning in the "threatened" geography fieldtrip', *Area*, 41: 108–16.

Hickey, S. and Mohan, G. (eds.) (2004) *Participation: From Tyranny to Transformation?* London: Zed Books.

Hitchcock, R. (2002) '"We are the first people": Land, natural resources and identity in the Central Kalahari, Botswana', *Journal of Southern African Studies*, 28: 797–824.

Hitchcock, R., Sapignoli, M. and Babchuk, W. (2011) 'What about our rights? Settlements, subsistence and livelihood security among Central Kalahari San and Bakgalagadi', *International Journal of Human Rights*, 15: 62–88.

Hobson, K. (2006) 'Environmental responsibility and the possibilities of pragmatist-orientated research', *Social and Cultural Geography*, 7: 283–98.

Hodge, J. (2007) *Triumph of the Expert: Agrarian Doctrines of Development and the legacies of British Colonialism*. Columbia, OH: Ohio University Press.

Hoffman, M., Rohde, R., Duncan, J. and Kaleme, P. (2010) 'Repeat photography, climate change and the long-term population dynamics of tree aloes in Southern Africa', in Webb, R., Boyer, D. and Turner, R. (eds.) *Repeat Photography: Methods and Applications in the Geological and Ecological Sciences*. Washington, DC: Island Press.

Hoffman, M.T. and Rohde, R.F. (2011) 'Long-term changes in the vegetation of southern Africa as revealed by repeat photography', in Zeitsman, L. (ed.) *Observations of Large-scale and Long-term Trends in Southern Africa*. South African Environmental Observation Network (SAEON).

Holt, A. (2010) 'Using the telephone for narrative interviewing: a research note', *Qualitative Research*, 10: 113–21.

Hong, S., Strlic, M., Ridely, I., Ntanos, K., Bell, B. and Casser, M. (2012) 'Climate change mitigation strategies for mechanically controlled repositories: the case of the National Archives, Kew', *Atmospheric Environment*, 49: 163–70.

Hongslo, E., Rohde, R. and Hoffman, T. (2009) 'Landscape change and ecological processes in relation to land-use in Namaqualand, South Africa, 1939 to 2005', *South African Geographical Journal*, 91: 63–74.

Hoogendoorn, G. and Visser, G. (2012) 'Stumbling over researcher positionality and political-temporal contingency in South African second-home tourism research', *Critical Arts: South-North Cultural and Media Studies*, 26: 254–71.

Hopkins, P. (2007) 'Thinking critically and creatively about focus groups', *Area*, 39: 528–35.

Hoyo, H. (2012) 'Fresh views on the old past: the postage stamps of the Mexican bicentennial', *Studies in Ethnicity and Nationalism*, 12: 19–44.

Iphofen, R. (2009) *Ethical Decision-Making in Social Research*. Basingstoke: Palgrave Macmillan.

Irvine, A. (2011) 'Duration, dominance and depth in telephone and face-to-face interviews: a comparative exploration', *International Journal of Qualitative Methods*, 10: 202–20.

Irvine, A., Drew, P. and Sainsbury, R. (2013) '"Am I not answering your questions properly?" Clarification, adequacy and responsiveness in semi-structured telephone and face-to-face interviews', *Qualitative Research*, 13: 87–106.

Irwin, K. (2006) 'Into the dark heart of ethnography: the lived ethics and inequality of intimate field relationships', *Qualitative Sociology*, 29: 155–75.

Irwin, R. (2007) 'Culture shock: negotiating feelings in the field', *Anthropology Matters*, 9: 1–11.

Israel, M. and Hay, I. (2006) *Research Ethics for Social Scientists*. London: Sage.

Jacobs, N. (2002) 'The colonial ecological revolution in Southern Africa: the case of Kuruman', in Dovers, S., Edgecombe, R. and Guest, B. (eds.) *South Africa's Environmental History: Cases and Comparisons*. Columbia, OH: Ohio University Press.

Jakobsen, H. (2012) 'Focus groups and methodological rigour outside the minority world: making the method work and its strengths in Tanzania', *Qualitative Research*, 12: 111–30.

Jennings, M. (2006) 'Using archives', in Desai, V. and Potter, R. (eds.) *Doing Development Research*. London: Sage.

Jensen, S. (2008) *Gangs, Politics and Dignity in Cape Town*. Oxford: James Currey.

Johnsen, S., May, J. and Cloke, P. (2008) 'Imag(in)ing "homeless places": using auto-photography to (re)examine the geographies of homelessness', *Area*, 40: 194–207.

Jolly, R. (2005) 'The UN and development thinking and practice', *Forum for Development Studies*, 1: 49–73.

Kaufmann, J. (2002) 'The informant as resolute overseer', *History in Africa*, 29: 231–55.

Kawulich, B. (2011) 'Gatekeeping: an ongoing adventure in research', *Field Methods*, 23: 57–76.

Kay, R. and Oldfield, J. (2011) 'Emotional engagements with the field: a view from area studies', *Europe-Asia Studies*, 63: 1275–93.

Kent, M., Gilberston, D. and Hunt, C. (1997) 'Fieldwork in geography teaching: a critical review of the literature and approaches', *Journal of Geography in Higher Education*, 21: 313–32.

Kenyon, E. and Hawker, S. (1999) '"Once would be enough": some reflections on the issue of safety for lone researchers', *International Journal of Social Research Methodology*, 2: 313–27.

Kevane, M. (2008) 'Official representations of the nation: comparing the postage stamps of Sudan and Burkina Faso', *African Studies Quarterly*, 10(1): 71–94. http://africa.ufl.edu/asq/v10/v10i1a3.htm.

Kiragu, S. and Warrington, M. (2012) 'How we used moral imagination to address ethical and methodological complexities while conducting research with girls in school against the odds in Kenya', *Qualitative Research*, 13: 173–89.

Kitchin, R., and Tate, N. (2000) *Conducting Research in Human Geography*. London: Pearson.

Koch, N. (2013) 'Introduction – field methods in "closed contexts"': undertaking research in authoritarian states and places', *Area*, 45(4): 390–95.

Kovats-Bernat, J.C. (2002) 'Negotiating dangerous fields: pragmatic strategies for fieldwork amid violence and terror', *American Anthropologist*, 104: 208–22.

Kreike, E. (2003) 'Hidden fruits: a social ecology of fruit trees in Namibia and Angola 1880s–1990s', in Beinart, W. and McGregor, J. (eds.) *Social History and African Environments*. Oxford: James Currey.

Kulick, D. and Wilson, M. (eds.) (1995) *Taboo: Sex, Identity and Erotic Subjectivity in Anthropological Fieldwork*. London: Routledge.

Kuntz, A. and Presnall, M. (2012) 'Wandering the tactical: from interview to intraview', *Qualitative Inquiry*, 18: 732–44.

Lambert, J. (2013) *Digital Storytelling: Capturing Lives, Creating Community*. London: Routledge.

Langevang, T. (2007) 'Movements in time and space: using multiple methods in research with young people in Accra, Ghana', *Children's Geographies*, 5: 267–82.

Laurier, E. (2003) 'Participant observation', in Clifford, N. and Valentine, G. (eds.) *Research Methods in Human and Physical Geography*. London: Sage.

Laws, S., Harper, C., Jones, N. and Marcus, R. (2013) *Research for Development: A Practical Guide*. London: Sage.

Leach, M. and Fairhead, J. (2000) 'Fashioned forest pasts, occluded histories? International environmental analysis in West African locales', *Development and Change*, 31: 35–59.

Leadbeater, C. (2000) *Living on Thin Air: The New Economy*. London: Penguin.

Lee-Treweek, G. and Linkogle, S. (eds.) (2000) *Danger in the Field: Risk and Ethics in Social Research*. London: Routledge.

Lees, L. (1999) 'Critical geography and the opening up of the academy: lessons from "real life" attempts', *Area*, 31: 377–83.

Lefevere, A. and Bassnett, S. (1990) 'Introduction: Proust's grandmother and the thousand and one nights: the "cultural turn" in translation studies', in Bassnett, S. and Lefevere, A. (eds.) *Translation, History and Culture*. London: Pinter.

Lekgoathi, S. (2009) 'You are listening to Radio Lebowa of the South African Broadcasting Corporation: vernacular radio, Bantustan identity and listenership, 1960–1994', *Journal of Southern African Studies*, 35: 575–94.

Letouzé, E., Meier, P. and Vinck, P. (2013) *New Oil & Old Fires: Reflections on Big Data for Conflict Prevention*. New York: International Peace Institute (IPI).

Lewin, E. and Leap, W. (eds.) (1996) *Out in the Field: Reflections of Lesbian and Gay Anthropologists*. Urbana, IL: University of Illinois Press.

Lincoln, Y. (2009) 'Ethical practices in qualitative research', in Mertens, D. and Ginsberg, P. (eds.) *Handbook of Social Research Ethics*. London: Sage.

Linke, A. (2013) 'The aftermath of an election crisis: Kenyan attitudes and the influence of individual-level and locality violence', *Political Geography*, 37: 5–17.

Lloyd-Evans, S. (2006) 'Focus groups', in Desai, V. and Potter, R. (eds.) *Doing Development Research*. London: Sage.

Lombard, M. (2012) 'Using autophotography to understand place: reflections from research in urban informal settlements in Mexico', *Area*, 45: 23–32.

Long, N. (2001) *Development Sociology. Actor Perspectives*. London and New York: Routledge.

Longhurst, R. (2010) 'Semi-structured interviews and focus groups', in Clifford, N., French, S. and Valentine, G. (eds.) *Key Methods in Geography*. London: Sage.

Lopez, G., Figueroa, M., Connor, S. and Maliski, S. (2008) 'Translation barriers in conducting qualitative research with Spanish speakers', *Qualitative Health Research*, 18: 1729–37.

Lorimer, H. (2003) 'Telling small stories: spaces of knowledge and the practice of geography', *Transactions of the Institute of British Geographers*, 28: 197–217.

Lunn, J. (ed.) (2014) *Fieldwork in the Global South: Ethical Challenges and Dilemmas*. London: Routledge.

McAreavey, R. and Das, C. (2013) 'A delicate balancing act: negotiating with gatekeepers for ethical research when researching minority communities', *International Journal of Qualitative Methods*, 12: 113–31.

McAreavey, R. and Muir, J. (2011) 'Research ethics committees: values and power in higher education', *International Journal of Social Research Methodology*, 14: 391–405.

McClintock, A. (1995) *Imperial Leather: Race, Gender and Sexuality in the Colonial Contest*. New York: Routledge.

Maclean, K. (2007) 'Translation in cross-cultural research: an example from Bolivia', *Development in Practice*, 17: 784–90.

McDowell, L. (1998) 'Elites in the city of London: some methodological considerations', *Environment and Planning A*, 30: 2133–46.

McDowell, L. (2001) 'Working with young men', *Geographical Review*, 91: 201–14.

McEwan, C. and Bek, D. (2006) '(Re)politicizing empowerment: lessons from the South African wine industry', *Geoforum*, 37: 1021–34.

McKeagney, N. (2001) 'To pay or not to pay: respondents' motivation for participating in research', *Addiction*, 96: 1237–38.

McLafferty, S. (2010) 'Conducting questionnaire surveys', in Clifford, N., French, S. and Valentein, G. (eds.) *Key Methods in Geography*. London: Sage.

Madge, C. and O'Connor, H. (2004) 'Online methods in geography educational research', *Journal of Geography in Higher Education*, 28: 143–52.

Madriz, E. (1998) 'Using focus groups with lower socioeconomic status Latina women', *Qualitative Inquiry*, 4: 114–28.

Mahajan, S. (2011) 'Beyond the archives: doing oral history in contemporary India', *Studies in History*, 27: 281–98.

Malam, L. (2005) 'Embodiment and sexuality in cross-cultural research', *Australian Geographer*, 35: 177–83.

Malinowksi, B. (1922) *Argonauts of the Western Pacific: An Account of Native Enterprise and Adventure in the Archipelagos of Melanesian New Guinea*. London: Routledge.

Mandiyanike, D. (2009) 'The dilemmas of conducting research back in your own country as a returning student – reflections of research fieldwork in Zimbabwe', *Area*, 41: 64–71.

Markowitz, F. and Ashkenazi, M. (eds.) (1999) *Sex, Sexuality and the Anthropologist*.Urbana, IL: University of Illinois Press.

Marshall, C. and Rossman, G. (2006) *Doing Qualitative Research*. London: Sage.

Massey, D. (2004) 'Geographies of responsibility', *Geografiska Annaler*, 86: 5–18.

Mather, C. (2007) 'Between the "local" and the "global": South African geography after apartheid', *Journal of Geography in Higher Education*, 31: 143–59.

Mauthner, N. and Doucet, A. (2003) 'Reflexive accounts and accounts of reflexivity in qualitative data analysis', *Sociology*, 37: 413–31.

Mawdsley, E. (2008) 'Fu Manchu versus Dr Livingstone in the dark continent? Representing China, Africa and the West in British broadsheet newspapers', *Political Geography*, 27: 509–29.

Mayoux, L. (2001) 'Participatory methods', Application Guidance Note. EDIAIS. http://library.uniteddiversity.coop/Measuring_Progress_and_Eco_Footprinting/Participatory Methods.pdf.

Mayoux, L. and Chambers, R. (2005) 'Reversing the paradigm: quantification, participatory methods and pro-poor impact assessment', *Journal of International Development*, 17: 271–98.

Mercer, C., Mohan, G. and Power, M. (2003) 'Towards a critical political geography of African development', *Geoforum*, 34: 419–36.

Mero-Jaffe, I. (2011) '"Is that what I said?" Interview transcript approval by participants: an aspect of ethics in qualitative research', *International Journal of Qualitative Methods*, 10: 231–47.

Meth, P. (2003a) 'Entries and omissions: using solicited diaries in geographical research', *Area*, 35: 195–205.

Meth, P. (with Malaza, K.) (2003b) 'Violent research: the ethics and emotions of doing research with women in South Africa', *Ethics, Place and Environment*, 6: 143–59.

Meth, P. (2004) 'Using diaries to understand women's responses to crime and violence', *Environment and Urbanization*, 16: 153–64.

Mikecz, R. (2012) 'Interviewing elites: addressing methodological issues', *Qualitative Inquiry*, 18: 482–93.

Mikkelsen, B. (2005) *Methods for Development Work and Research: A New Guide for Practitioners*. London: Sage.

Mills, S. (2013) 'Cultural-historical geographies of the archive: fragments, objects and Ghoists', *Geography Compass*, (7/10): 701–13.

Mistry, J. and Berardi, A. (2012) 'The challenges and opportunities of participatory video in geographical research: exploring collaboration with indigenous communities in the North Rupununi, Guyana', *Area*, 44: 110–16.

Molony T. and Hammett, D. (2007) 'The friendly financier: talking money with the silenced assistant', *Human Organization*, 66: 292–300.

Montello, D. and Sutton, P. (2013) *An Introduction to Scientific Research Methods in Geography and Environmental Studies*. London: Sage.

Morrell, R., Epstein, D. and Moletsane, R. (2012) 'Doubt, dilemmas and decisions: towards ethical research on gender and schooling in South Africa', *Qualitative Research*, 12: 613–29.

Morrison, M. (2007) 'Using diaries in research', in Briggs, A. and Coleman, M. (eds.) *Research Methods in Educational Leadership and Management*. London: Sage.

Morrow, V., Boddy, J. and Lamb, R. (2014) 'The ethics of secondary data analysis: learning from the experience of sharing qualitative data from young people and their families in an international study of childhood poverty'. Novella Working Paper. http://eprints.ncrm.ac.uk/3301/.

Moseley, W. (2007) 'Collaborating in the field, working for change: reflecting on partnerships between academics, development organizations and rural communities in Africa', *Singapore Journal of Tropical Geography*, 28: 334–47.

Mosse, D. (1994) 'Authority, gender and knowledge: theoretical reflections on the practice of PRA', *Development and Change*, 25: 497–526.

Moyo, D. (2009) *Dead Aid: Why Aid Is Not Working and How There is Another Way for Africa*. London: Penguin.

Mukeredzi, T. (2011) 'Qualitative data gathering challenges in a politically unstable rural environment: a Zimbabwean experience', *International Journal of Qualitative Methods*, 11: 1–11.

Murphy, E. and Dingwall, R. (2007) 'Informed consent, anticipatory regulation and ethnographic practice', *Social Science and Medicine*, 65: 2223–34.

Murray, L., Pushor, D. and Renihan, P. (2012) 'Reflections on the ethics-approval process', *Qualitative Inquiry*, 18: 43–54.

Murray, R. (2005) *Writing for Academic Journals*. Maidenhead: Open University Press.

Mwangi, W. (2002) 'The lion, the native and the coffee plant: political imagery and the ambiguous art of currency design in colonial Kenya', *Geopolitics*, 7: 31–62.

Mwesige, P. (2009) 'The democratic functions and dysfunctions of political talk radio: the case of Uganda', *Journal of African Media Studies*, 1: 221–45.

Narayanasamy, N. (2009) *Participatory Rural Appraisal: Principles, Methods and Application*. London: Sage.

Nash, D. and Endfield, G. (2002) 'A 19th century climate chronology for the Kalahari region of central southern Africa derived from missionary correspondence', *International Journal of Climatology*, 22: 821–41.

Ndimande, B. (2012) 'Decolonizing research in postapartheid South Africa: the politics of methodology', *Qualitative Inquiry*, 18: 215–26.

Neimark, B. (2012) 'Finding that "eureka" moment: the importance of keeping detailed field notes', *African Geographical Review*, 31: 76–9.

Nelson, I. (2013) 'The allure and privileging of danger over everyday practice in field research', *Area*, 45(4): 419–25.

Nelson, V. (2005) 'Representations and images of people, place and nature in Grenada's tourism', *Geografiska Annaler B*, 87: 131–43.

Newhouse, L. (2012) 'Footing it, or why I walk', *African Geographical Review*, 31: 67–71.

Nichols, P. (1991) *Social Survey Methods: A Fieldguide for Development Workers*. Oxford: Oxfam.

Noble, M., Barnes, H., Wright, G., McLennan, D., Avenall, D., Whitworth, A. et al. (2009) *The South African Index of Multiple Deprivation 2001 at Datazone Level*. Pretoria: National Department for Social Development.

Nordstrom, C. and Robben, A. (eds.) (1995) *Fieldwork Under Fire: Contemporary Studies of Violence and Survival*. Berkeley, CA: University of California Press.

Noxolo, P. (2012) 'One world, big society: a discursive analysis of the Conservative green paper for international development', *Geographical Journal*, 178: 31–41.

Ntseane, P. (2009) 'The ethics of the researcher-subject relationship: experiences from the field', in Mertens, D. and Ginsberg, P. (eds.) *Handbook of Social Research Ethics*. London: Sage.

Oldfield, S., Parnell, S. and Mabin, A. (2004) 'Engagement and reconstruction in critical research: negotiating urban practice, policy and theory in South Africa', *Social and Cultural Geography*, 5: 285–99.

Olsen, W. (1993) 'Random sampling and repeat surveys in south India', in Devereux, S. and Hoddinot, J. (eds.) (1993) *Fieldwork in Developing Countries*. Boulder, CO: Lynne Reinner.

Pain, R., Kesby, M. and Askins, K. (2011) 'Geographies of impact: power, participation and potential', *Area*, 43: 183–88.

Pain, R., Kesby, M. and Askins, K. (2012) 'The politics of social justice in neoliberal times: a reply to Slater', *Area*, 44: 120–23.

Paleker, G. (2010) 'The B-Scheme subsidy and the "black film industry" in apartheid South Africa, 1972–1990', *Journal of African Cultural Studies*, 22: 91–104.

Panelli, R. and Welch, R. (2005) 'Teaching research through field studies: a cumulative opportunity for teaching methodology to human geography undergraduates', *Journal of Geography in Higher Education*, 29: 255–77.

Park, A. (2006) 'Using survey data in social science research in developing countries', in Perecman, E. and Curran, S. (eds.) *A Handbook for Social Science Field Research: Essays and Bibliographic Sources on Research Design and Methods*. London: Sage.

Parker, M. (2007) 'Ethnography/ethics', *Social Science and Medicine*, 65: 2248–59.

Parnell, S. (1997) 'South African cities: perspectives from the ivory tower of urban studies', *Urban Studies*, 34: 891–906.

Pearson, A. and Paige, S. (2012) 'Experiences and ethics of mindful reciprocity while conducting research in sub-Saharan Africa', *African Geographical Review*, 31: 72–5.

Peck, J. and Coyle, M. (1999) *Student's Guide to Writing: Grammar, Punctuation and Spelling*. Basingstoke: Macmillan.

Penrose, J. (2011) 'Designing the nation: banknotes, banal nationalism and alternative conceptions of the state', *Political Geography*, 30: 429–40.

Pentland, S. (2012) 'Reinventing society in the wake of big data. Interview with The Edge'. www.edge.org/conversation/reinventing-society-in-the-wake-of-big-data

Perecman, E. and Curran, S. (eds.) (2006) *A Handbook for Social Science Field Research: Essays and Bibliographic Sources on Research Design and Methods*. London: Sage.

Perkins, C. (2004) 'Cartography – cultures of mapping: power in practice', *Progress in Human Geography*, 28: 381–91.

Perks, R. and Thomson, A. (eds.) (1998) *The Oral History Reader*. London: Routledge.

Perry, K. (2011) 'Ethics, vulnerability, and speakers of other languages: how university IRBs (do not) speak to research involving refugee participants', *Qualitative Inquiry*, 17: 899–912.

Petray, T. (2012) 'A walk in the park: political emotions and ethnographic vacillation in activist research', *Qualitative Research*, 12: 554–64.

Pettigrew, J., Shneiderman, S. and Harper, I. (2004) 'Relationships, complicity and representation: conducting research in Nepal during the Maoist insurgency', *Anthropology Today*, 20: 20–25.

Phillips, R. and Johns, J. (2012) *Fieldwork for Human Geography*. London: Sage.

Pienaar, J. and Visser, G. (2009) 'The thorny issue of identifying second homes in South Africa', *Urban Forum*, 20: 455–69.

Pieterse, J. (2010) *Development Theory* (2nd ed.). London: Sage.

Poland, B. (1995) 'Transcription quality as an aspect of rigor in qualitative research', *Qualitative Inquiry*, 1: 290–310.

Portelli, A. (1991) *The Death of Luigi Trastulli and Other Stories: Form and Meaning in Oral History*. New York: State University of New York Press.

Porter, G., Hampshire, K., Boudillon, M., Robson, E., Munthali, A., Abane, A. et al. (2010) 'Children as research collaborators: issues and reflections from a mobility study in sub-Saharan Africa', *American Journal of Community Psychology*, 46: 215–17.

Postlewait, H., Cain, K. and Thomson, A. (2006) *Emergency Sex (and Other Desperate Measures): True Stories from a War Zone*. London: Ebury Press.

Potter, R., Binns, T., Elliott, J. and Smith, D. (2008) *Geographies of Development: An Introduction to Development Studies*. Harlow: Pearson.

Potter, R., Conway, D., Evans, R. and Lloyd-Evans, S. (2012) *Key Concepts in Development Geography*. London: Sage.

Power, M. (2003) *Rethinking Development Geographies*. London: Routledge.

Pratt, M. (1992) *Imperial Eyes: Travel Writing and Transculturation*. Abingdon: Routledge.

Pretty, J. (1995) 'Participatory learning for sustainable agriculture', *World Development*, 23: 1247–63.

Price, D. (2002) 'Interlopers and invited guests: on anthropology's witting and unwitting links to intelligence agencies', *Anthropology Today*, 18: 16–21.

Price, P. (2012) 'Introduction: protecting human subjects across the geographic research process', *Professional Geographer*, 64: 1–6.

Punch, M. (1998) 'Politics and ethics in qualitative research', in Denzin, N. and Lincoln, Y. (eds.) *The Landscape of Qualitative Research: Theories and Issues*, Thousand Oaks, CA: Sage.

Razon, N. and Ross, K. (2012) 'Negotiating fluid identities: alliance-building in qualitative interviews', *Qualitative Inquiry*, 18: 494–503.

Reeves, C. (2010) 'A difficult negotiation: fieldwork relations with gatekeepers', *Qualitative Research*, 10: 315–31.

Rice, G. (2010) 'Reflections on interviewing elites', *Area*, 42: 70–75.

Rigg, J. (2006) 'Data from international agencies', in Desai, V. and Potter, R. (eds.) *Doing Development Research*. London: Sage.

Robles, M. and Keefe, M. (2011) 'The effects of changing food prices on welfare and poverty in Guatemala', *Development in Practice*, 21: 578–89.

Robson, E. (1997) 'From teacher to taxi driver: reflections on research roles in developing areas', in Robson, E. and Willis, K. (eds.) *Postgraduate Fieldwork in Developing Areas: A Rough Guide*. London: DARG, RGS.

Rogers, L. and Swadener, B. (1999) 'Re-framing the "field"', *Anthropology and Education*, 30: 436–40.

Rohde, R.F. (2010) 'Written on the surface of the soil: west Highland crofting landscapes during the 20th century', in Webb, R., Boyer, D. and Turner, R. (eds.) *Repeat Photography: Methods and Applications in the Geological and Ecological Sciences*. Washington, DC: Island Press.

Rostow, W. (1956) 'The take-off into self-sustained growth', *Economic Journal*, 66: 25–48.

Rostow, W. (1959) 'The stages of economic growth', *Economic History Review*, 12: 1–16.

Roulston, K. (2010) 'Considering quality in qualitative interviewing', *Qualitative Research*, 10: 199–228.

Roulston, K., de Marrais, K. and Lewis, J. (2003) 'Learning to interview in the social sciences', *Qualitative Inquiry*, 9: 643–68.

Routledge, P. (1996) 'Third space as critical engagement', *Antipode*, 28: 399–419.

Routledge, P. (2002) 'Travelling East as Walter Kurtz: identity, performance and collaboration in Goa, India', *Environment and Planning D: Society and Space*, 20: 477–98.

Rubenstein, S. (2004) 'Fieldwork and the erotic economy on the colonial frontier', *Signs*, 29: 1041–71.

Rubin, M. (2012) '"Insiders" versus "outsiders": what difference does it really make?', *Singapore Journal of Tropical Geography*, 33: 303–307.

Ruddick, S. (2004) 'Activist geographies: building possible worlds', in Cloke, P., Crang, P. and Goodwin, M. (eds.) *Envisioning Human Geographies*. London: Arnold.

Russell, M (1976) 'Slaves or workers? Relations between bushmen, Tswana and Boers in the Kalahari', *Journal of Southern African Studies*, 2(2): 178–97.

Ryan, L., Kofman, E. and Aaron, P. (2010) 'Insiders and outsiders: working with peer researchers in researching Muslim communities', *International Journal of Social Research Methodology*, 14: 49–60.

Ryen, A. (2009) 'Ethnography: constitutive practice and research ethics', in Mertens, D. and Ginsberg, P. (eds.) *Handbook of Social Research Ethics*. London: Sage.

Ryen, A. and Gobo, G. (2011) 'Managing the decline of globalized methodology', *International Journal of Social Research Methodology*, 16: 411–15.

Said, E. (1978) *Orientalism: Western Conceptions of the Orient*. London: Penguin.

Santos, T. (1970) 'The structure of dependence', *American Economic Review*, 60: 231–6.

Sawada, Y. and Loskhin, M. (2009) 'Obstacles to school progression in rural Pakistan: an analysis of gender and sibling rivalry using field survey data', *Journal of Development Economics*, 88: 335–47.

Scheyvens, H. and Nowak, B. (2003) 'Personal issues' in Scheyvens, R. and Storey, D. (eds.) *Development Fieldwork: A Practical Guide*. London: Sage.

Scheyvens, R. (2014) *Development Fieldwork: A Practical Guide* (2nd ed.). London: Sage.

Scheyvens, R. and Leslie, H. (2000) 'Gender, ethics and empowerment: dilemmas of development fieldwork', *Women's Studies International Forum*, 23(1): 119–30.

Scheyvens, R., Scheyvens, H. and Murray, W. (2003) 'Working with marginalised, vulnerable or privileged groups', in Scheyvens, R. and Storey, D. (eds.) *Development Fieldwork: A Practical Guide*. London: Sage.

Schmidt, L. (2011) *Using Archives: A Guide to Effective Research*. Society of American Archivists. http://www2.archivists.org/usingarchives.

Schuermans, N. and Newton, C. (2012) 'Being a young and foreign researcher in South Africa: towards a post-colonial dialogue', *Singapore Journal of Tropical Geography*, 33: 295–300.

Schumaker, L. (2001) *Africanizing Anthropology: Fieldwork, Networks, and the Making of Cultural Knowledge in Central Africa*. Durham, NC: Duke University Press.

Scott, S., Hinton-Smith, T., Härmä, V. and Broome, K. (2012) 'The reluctant researcher: shyness in the field', *Qualitative Research*, 12: 715–34.

Scott, S., Miller, F. and Lloyd, K. (2006) 'Doing fieldwork in development geography: research culture and research spaces in Vietnam', *Geographical Research*, 44: 28–40.

Sen, A. (1988) 'The concept of development', in Chenery, H., Srinivasan, T. and Streeten, P. (eds.) *Handbook of Development Economics*. London: Elsevier.

Sen, A. (1999) *Development as Freedom*. Oxford: Oxford University Press.

Sen, A. (2013) 'The ends and means of sustainability', *Journal of Human Development and Capabilities*, 14: 6–20.

Sharp, G. and Kremer, E. (2006) 'The safety dance: confronting harassment, intimidation and violence in the field', *Sociological Methodology*, 36: 317–27.

Sherif, B. (2001) 'The ambiguity of boundaries in the fieldwork experience: establishing rapport and negotiating insider/outside status', *Qualitative Inquiry*, 7: 436–47.

Shortt, N. and Hammett, D. (2013) 'Informal settlement upgrading in South Africa – outcomes for health and community well-being', *Journal of Housing and the Built Environment*, 28(4): 615–27.

Silvia, P. (2007) *How To Write A Lot: A Practical Guide to Productive Academic Writing*. Washington, DC: APA.

Simon, C., Mosavel, M. and van Stade, D. (2007) 'Ethical challenges in the design and conduct of locally relevant international health research', *Social Science and Medicine*, 64: 1960–69.

Simon, D. (2006) 'Your questions answered? Conducting questionnaire surveys', in Desai, V. and Potter, R. (eds.) *Doing Development Research*. London: Sage.

Sin, C. (2003) 'Interviewing in "place": the socio-spatial construction of interview data', *Area*, 35: 305–12.

Slater, T. (2012) 'Impacted geographers: a response to Pain, Kesby and Askins', *Area*, 44: 117–19.

Smiley, S. (2012) 'Population censuses and changes in housing quality in Dar es Salaam, Tanzania', *African Geographical Review*, 31: 1–16.

Smith, D. (2002) 'Responsibility to distant others', Desai, V. and Potter, R. (eds.) *The Companion to Development Studies*. London: Arnold.

Smith, F. (1996) 'Problematising language: limitations and possibilities in "foreign language" research', *Area*, 28(2): 160–66.

Smith, H. (2005) 'The relationship between settlement density and informal settlement fires: case study of Imizamo Yethu, Hout Bay and Joe Slovo, Cape Town Metropolis', in Oosterom, P., Zlatanova, S. and Fendel, E. (eds.) *Geo-information for Disaster Management*. Berlin: Springer.

Smith, L. (1999) *Decolonizing Methodologies: Research and Indigenous Peoples*. London: Zed Books.

Smith, M. (2006) *Beyond the 'African Tragedy': Discourses on Development and the Global Economy*. Farnham: Ashgate.

Speed, S. (2006) 'At the crossroads of human rights and anthropology: toward a critically engaged activist research', *American Anthropologist*, 108: 66–76.

Stoler, A. (2002) *Carnal Knowledge and Imperial Power: Race and the Intimate in Colonial Rule*. Berkeley, CA: University of California Press.

Strunk, W. (2006) *The Elements of Style*. New York: Dover Publications.

Sturge, K. (1997) 'Translation strategies in ethnography', *Translator*. 3(1), 21–38.

Sturges, J. and Hanrahan, K. (2004) 'Comparing telephone and face-to-face qualitative interviewing: a research note', *Qualitative Research*, 4: 107–18.

Sullivan, S. and Rohde, R. (2002) 'On non-equilibrium in arid and semi-arid grazing systems', *Journal of Biogeography*, 29: 1595–18.

Sutton, B. (2011) 'Playful cards, serious talk: a qualitative research technique to elicit women's embodied experiences', *Qualitative Research*, 11: 177–96.

Sword, H. (2012) *Stylish Academic Writing*. Cambridge, MA: Harvard University Press.

Takeda, A. (2012) 'Reflexivity: unmarried Japanese male interviewing married Japanese married women about international marriage', *Qualitative Research*, 13: 285–98.

Temple, B. (2002) 'Crossed wires: interpreters, translators, and bilingual workers in cross-language research', *Qualitative Health Research*, 12: 844–54.

Temple, B. (2006) 'Being bilingual: issues for cross-language research', *Journal of Peace Research*, 2: 1–15.

Temple, B. and Edwards, R. (2002) 'Interpreters/translators and cross-language research: reflexivity and border crossings', *International Journal of Qualitative Methods*, 1: 1–22.

Temple, B. and Young, A. (2004) 'Qualitative research and translation dilemmas', *Qualitative Research*, 4: 161–78.

Theis, J. and Grady, H. (1991) *Participatory Rapid Appraisal for Community Development: A Training Manual Based on Experiences in the Middle East and North Africa*. London: IIED and Save the Children.

Thomas, F. (2006) 'Stigma, fatigue and social breakdown: exploring the impacts of HIV/AIDS on patient and carer well-being in the Caprivi Region, Namibia', *Social Science and Medicine*, 63: 3174–87.

Thomson, S., Ansoms, A. and Murison, J. (eds.) (2012) *Emotional and Ethical Challenges for Field Research in Africa: The Story behind the Findings*. London: Palgrave Macmillan.

Tickell, A. (1995) 'Reflections on "activism and the academy"', *Environment and Planning D: Society and Space*, 13: 235–7.

Tieken, M. (2013) 'The distance to delight: A graduate student enters the field', *Qualitative Inquiry*, 19: 320–26.

Tolich, M. and Fitzgerald, M. (2006) 'If ethics committees were designed for ethnography', *Journal of Empirical Research on Human Research Ethics*, 1: 71–8.

Truman, H 1949 Inauguration speech. http://www.trumanlibrary.org/whistlestop/50yr_archive/inagural20jan1949.htm.

Truss, L. (2003) *Eats, Shoots and Leaves: The Zero Tolerance Approach To Punctuation*. London: Profile Books.

Turner, R. (2006) *Introduction to Neogeography*. Sebastopol, CA: O'Reilly Media.

Turner, S. (2013) 'Red stamps and green tea: fieldwork negotiations and dilemmas in the Sino-Vietnamese borderlands', *Area*, 45(4): 396–402.

Turner, V. (1967) *The Forest of Symbols: Aspects of Ndembu Ritual*. Ithaca, NY: Cornell University Press.

Twyman, C., Dougill, A., Sporton, D. and Thomas, D. (2001) 'Community fencing in open rangelands: self-empowerment in Eastern Namibia', *Review of African Political Economy*, 28: 9–26.

Twyman, C., Morrison, J. and Sporton, S. (1999) 'The final fifth: auto-biography, reflexivity and interpretation in cross-cultural research', *Area*, 31: 313–25.

Uddin, N. (2011) 'Decolonising ethnography in the field: an anthropological account', *International Journal of Social Research Methodology*, 14: 455–67.

USHA/UCEA (2011) *Guidance on Health and Safety in Fieldwork (including offsite visits and travel in the UK and overseas)*. Eastbourne and London: USHA/UCEA.

Van Blerk, L. (2006) 'Working with children in development', in Desai, V. and Potter, R. (eds.) *Doing Development Research*. London: Sage.

Van Donge, J. (2006) 'Ethnography and participant observation', in Desai, V. and Potter, R. (eds.) *Doing Development Research*. London: Sage.

Vanderbeck, R. (2005) 'Masculinities and fieldwork: widening the discussion', *Gender, Place and Culture*, 12: 387–402.

VeneKlasen, L. and Miller, V. (2002) *A New Wave of People, Power and Politics: the Action Guide for Advocacy and Citizen Participation*. Oklahoma City, OK: World Neighbours.

Vokes, R. (2010) 'Reflections on a complex (and cosmopolitan) archive: postcards and photography in early colonial Uganda, *c.*1904–1928', *History and Anthropology*, 21: 375–409.

Wallach, Y. (2011) 'Creating a country through currency and stamps: state symbols and nation-building in British-ruled Palestine', *Nations and Nationalism*, 17: 129–47.

Wallin, A.-M. and Ahlström, G. (2006) 'Cross-cultural interview studies using interpreters: systematic literature review', *Journal of Advanced Nursing*, 55: 723–35.

Ward, C., Bochner, S. and Furnham, A. (2001) *The Psychology of Culture Shock*. Hove: Routledge.

Warr, D. (2005) '"It was fun . . . but we don't usually talk about these things": analyzing sociable interaction in focus groups', *Qualitative Inquiry*, 11: 200–25.

Watson, E. (2004) '"What a dolt one is": language, learning and fieldwork in geography', *Area*, 36: 59–68.

Watts, M. (2006) 'In search of the Holy Grail: projects, proposals and research design, but mostly about why writing a dissertation proposal is so difficult', in Perecman, E. and Curran, S. (eds.) *A Handbook for Social Science Field Research: Essays and Bibliographic Sources on Research Design and Methods*. London: Sage.

Weiner, A. (1988) *The Trobrianders of Papua New Guinea*. Fort Worth, TX: Harcourt Brace College Publishers.

White, P. (2009) *Developing Research Questions: A Guide for Social Scientists*. Basingstoke: Palgrave Macmillan.

White, P. (2010) 'Making use of secondary data', in Clifford, N., French, S. and Valentine, G. (eds.) *Key Methods in Geography*. London: Sage.

Whiteford, L. and Trotter, T. (2008) *Ethics for Anthropological Research and Practice*. Long Grove, IL: Waveland Press.

Wieder, A. (2003) *Voices from Cape Town Classrooms: Oral Histories of Teachers Who Fought Apartheid*. New York: Peter Lang

Wieder, A. (2004) 'Testimony as oral history: lessons from South Africa', *Educational Researcher*, 33: 23–8.

Willems, W. (2008) 'Mocking the state: comic strips in the Zimbabwean Press', in Abbink, J. and van Dokkum, A. (eds.) *Dilemmas of Development: Conflicts of Interest and their Resolution in Modernizing Africa*. Leiden: African Studies Centre.

Williams, G. (2012) 'The disciplining effects of impact evaluation practices: negotiating the pressures of impact within an ESRC-DFID project', *Transactions of the Institute of British Geographers*, 37: 489–95.

Williams, G. (2013) 'Researching with impact in the global south? Impact-evaluation practices and the reproduction of "development knowledge"', *Contemporary Social Science*, 10(21): 58–75.

Williams, G. and Meth, P. (2010) 'Researching the global south in an age of impact', *Social Text Online*, 26 August.

Williams, G., Meth, P. and Willis, K. (2014) *Geographies of Developing Areas: The Global South in a Changing World*. Abingdon: Routledge.

Williamson, D., Choi, J., Charchuk, M., Rempel, G., Pitre, N., Breitkreuz, R. et al. (2011) 'Interpreter-facilitated cross-language interviews: a research note', *Qualitative Research*, 11: 381–94.

Wilmsen, E.N. (1989) *Land Filled with Flies: A Political Economy of the Kalahari*. Chicago: University of Chicago Press.

Wilson, K. (1993) 'Thinking about the ethics of fieldwork', in Devereux, S. and Hoddinot, J. (eds.) *Fieldwork in Developing Countries*. Boulder, CO: Lynne Reinner.

Wood, P. (1997) 'Ethnography and ethnology', in Barfield, T. (ed.) *Dictionary of Anthropology*. Oxford: Blackwell.

Yon, D. (2003) 'Highlights and overview of the history of educational ethnography'. *Annual Review of Anthropology*, 32: 411–29.

Young, L. and Ansell, N. (2003) 'Fluid households, complex families: the impacts of children's migration as a response to HIV/AIDS in southern Africa', *Professional Geographer*, 55: 464–79.

Young, L. and Barrett, H. (2001) 'Adapting visual methods: action research with Kampala street children', *Area*, 33: 141–52.

INDEX

Note: page numbers in *italic* type refer to Figures; those in **bold** type refer to Tables.

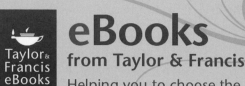